T0323529

Bio-optical Modeling and Remote Sensing of Inland Waters

Bio-optical Modeling and Remote Sensing of Inland Waters

Edited by

Deepak R. Mishra
University of Georgia (UGA), United States

Igor Ogashawara
Indiana University—Purdue University at Indianapolis (IUPUI),
United States

Anatoly A. Gitelson
Israel Institute of Technology, Israel; University of Nebraska-Lincoln
(UNL), United States

ELSEVIER

Elsevier
Radarweg 29, PO Box 211, 1000 AE Amsterdam, Netherlands
The Boulevard, Langford Lane, Kidlington, Oxford OX5 1GB, United Kingdom
50 Hampshire Street, 5th Floor, Cambridge, MA 02139, United States

Notices
Knowledge and best practice in this field are constantly changing. As new research and experience broaden
our understanding, changes in research methods, professional practices, or medical treatment may become
necessary.

Practitioners and researchers must always rely on their own experience and knowledge in evaluating and
using any information, methods, compounds, or experiments described herein. In using such information or
methods they should be mindful of their own safety and the safety of others, including parties for whom they
have a professional responsibility.

To the fullest extent of the law, neither the Publisher nor the authors, contributors, or editors, assume any
liability for any injury and/or damage to persons or property as a matter of products liability, negligence or
otherwise, or from any use or operation of any methods, products, instructions, or ideas contained in the
material herein.

British Library Cataloguing-in-Publication Data
A catalogue record for this book is available from the British Library

Library of Congress Cataloging-in-Publication Data
A catalog record for this book is available from the Library of Congress

ISBN: 978-0-12-804644-9

For Information on all Elsevier publications
visit our website at https://www.elsevier.com/books-and-journals

Working together
to grow libraries in
developing countries

www.elsevier.com • www.bookaid.org

Publisher: Candice Janco
Acquisition Editor: Louisa Hutchins
Editorial Project Manager: Hilary Carr
Production Project Manager: Punithavathy Govindaradjane
Cover Designer: Matthew Limbert

Typeset by MPS Limited, Chennai, India

Contents

List of Contributors

Federica Braga National Research Council of Italy, Institute of Marine Sciences (CNR-ISMAR), Venice, Italy

Vittorio E. Brando National Research Council of Italy, Institute of Atmospheric Sciences and Climate (CNR-ISAC), Rome, Italy

Mariano Bresciani National Research Council of Italy, Institute for Electromagnetic Sensing of the Environment (CNR-IREA), Milan, Italy

Ilaria Cazzaniga National Research Council of Italy, Institute for Electromagnetic Sensing of the Environment (CNR-IREA), Milan, Italy

Liesbeth De Keukelaere Flemish Institute for Technological Research (VITO), Mol, Belgium

Tonio Fincke Brockmann Consult, Geesthacht, Germany

Peter Gege Deutsches Zentrum für Luft- und Raumfahrt (DLR), Remote Sensing Technology Institute, Weßling, Germany

Claudia Giardino National Research Council of Italy, Institute for Electromagnetic Sensing of the Environment (CNR-IREA), Milan, Italy

Alexander A. Gilerson City University of New York, New York, NY, United States

Anatoly A. Gitelson Israel Institute of Technology, Haifa, Israel; University of Nebraska-Lincoln (UNL), Lincoln, NE, United States

Yannick Huot Université de Sherbrooke, Québec, QC, Canada

Kari Y. Kallio Finnish Environment Institute, Helsinki, Finland

Els Knaeps Flemish Institute for Technological Research (VITO), Mol, Belgium

Sampsa Koponen Finnish Environment Institute, Helsinki, Finland

Tiit Kutser University of Tartu, Tallinn, Estonia

Linhai Li University of California San Diego, La Jolla, CA, United States

Tim J. Malthus CSIRO Oceans and Atmosphere, Brisbane, QLD, Australia

Mark W. Matthews CyanoLakes (Pty) Ltd., Cape Town, South Africa; University of Cape Town, Cape Town, South Africa

Deepak R. Mishra University of Georgia (UGA), Athens, GA, United States

Marcos J. Montes U.S. Naval Research Laboratory, Washington, DC, United States

Wesley J. Moses U.S. Naval Research Laboratory, Washington, DC, United States

Igor Ogashawara Indiana University—Purdue University at Indianapolis (IUPUI), Indianapolis, IN, United States

Birgot Paavel University of Tartu, Tallinn, Estonia

Kaishan Song Chinese Academy of Sciences, Changchun, China

Sindy Sterckx Flemish Institute for Technological Research (VITO), Mol, Belgium

Abbreviations and Notations

ABBREVIATIONS

Acronym	Description
6S	Second Simulation of the Satellite Signal in the Solar Spectrum
ACOLITE	Atmospheric correction for OLI 'lite'
AERONET	Aerosol Robotic Network
AISA	Airborne Imaging Spectrometer for Applications
AOD	Aerosol Optical Depth
AOP	Apparent Optical Property
ARVI	Atmospherically Resistant Vegetation Index
ATREM	ATmospheric REMoval
AVIRIS	Airborne Visible/Infrared Imaging Spectrometer
BEAM	Basic ERS & ENVISAT (A)ATSR and MERIS
C2R	Case-2 Regional processor
CALIPSO	Cloud-Aerosol Lidar and Infrared Pathfinder Satellite Observations
CASI	Compact Airborne Spectrographic Imager
CDM	Colored Detrital Matter (CDOM + NAP)
CDOM	Colored Dissolved Organic Matter
chl-a	Chlorophyll-a
CZCS	Coastal Zone Color Scanner
DDV	Dense Dark Vegetation
DOC	Dissolved Organic Carbon
DOS	Dark Object Subtraction
EnMAP	Environmental Mapping and Analysis Program
ENVI	ENvironment for Visualizing Images
EU-WFD	European Water Framework Directive
EVI	Enhanced Vegetation Index
FAI	Floating Algal Index
FLAASH	Fast Line-of-sight Atmospheric Analysis of Spectral Hypercubes
FSS	Fixed Suspended Solids
FUB	Free University Berlin processor
ICOL	Improved Contrast between Ocean and Land
IGOS	Integrated Global Observing Strategy
HyspIRI	Hyperspectral Infrared Imager
IOP	Inherent Optical Property
IPCC	Intergovernmental Panel for Climate Change
ISS	Inorganic Suspended Solids
LAI	Leaf Area Index
LUT	Look-Up-Table

(Continued)

(Continued)

Acronym	Description
MEIS	Multispectral Electro-Optical Image Sensor
MERIS	Medium Resolution Imaging Spectrometer
MIP	Modular Inversion and Processing
MIVIS	Multispectral Infrared Visible Imaging Spectrometer
MSI	MultiSpectral Instrument
MSL	Mean Sea Level
MODIS	MODerate resolution Imaging Spectroradiometer
MODTRAN	MODerate resolution TRANsmission
MOMO	Matrix Operator Method
MUMM	Management Unit of the North Sea Mathematical Models
NAP	Nonalgal Particles
NASA	National Aeronautics and Space Administration
NDAVI	Normalized Difference Aquatic Vegetation Index
NDWI	Normalized Difference Water Index
NDVI	Normalized Difference Vegetation Index
NGAI	NIR-Green Angle Index
NIR	Near-infrared
OBIA	Object Based Image Analysis
OLCI	Ocean Land Colour Instrument
OLI	Operational Land Imager
OPERA	OPERational Atmospheric correction algorithm
PACE	Pre-Aerosol, Clouds and ocean Ecosystem
PAR	Photosynthetic Available Radiation
PC	Phycocyanin
PCA	Principal Components Analysis
POLYMER	POLYnomial based algorithm applied to MERIS
PRISMA	PRecursore IperSpettrale della Missione Applicativa
PSF	Point Spread Function
QAA	Quasi-Analytical Algorithm
QUAC	QUick Atmospheric Correction
RT	Radiative Transfer
RTE	Radiative Transfer Equation
SAR	Synthetic Aperture Radar
SAIL	Scattering by Arbitrary Inclined Leaves (model)
SAV	Submerged Aquatic Vegetation
SCAPE-M	Self-Contained Atmospheric Parameters Estimation for MERIS data
SeaDAS	SeaWiFS Data Analysis System
SeaWiFS	Sea-viewing Wide Field-of-view Sensor
SIMEC	SIMilarity Environment Correction
SIOCS	Sensor-Independent Ocean Colour Processor
SIOP	Specific Inherent Optical Property
SNAP	Sentinel Application Platform
SNR	Signal-to-Noise Ratio
SPM	Suspended Particulate Matter
SPOT	Satellite Pour l'Observation de la Terre
SUVA	Specific Ultraviolet Absorption
SWIR	Short Wave Infrared

(Continued)

(Continued)

Acronym	Description
TSM	Total Suspended Matter
UAV	Unmanned Aerial Vehicle
US-CWA	US Clean Water Act
UV	Ultraviolet
VI	Vegetation Index
VIIRS	Visible Infrared Imaging Radiometer Suite
VNIR	Visible-near-infrared
VSS	Volatile Suspended Solids

NOTATIONS

Symbol	Parameter	Units
λ	Wavelength	nm
ψ	Scattering angle	sr, deg
θ	Zenith angle	sr, deg
θ_{sun}	Sun Zenith angle in air	sr, deg
θ'_{sun}	Sun Zenith angle in water	sr, deg
φ	Azimuth angle	sr, deg
φ_{sun}	Sun Azimuth angle	sr, deg
z	Depth	m
z_B	Bottom depth	m
V	Volume	m^3
PAR	Photosynthetic available radiation	$W\ m^{-2}$, µmol photons $s^{-1}\ m^{-2}$
Q	Light distribution factor	sr
f	Geometrical light factor	—
Φ	Radiant flux	W
$d\Omega$	Solid angle element	sr
R_{rs}	Remote sensing reflectance	sr^{-1}
R_{rs}^w	At-surface remote sensing reflectance of water	sr^{-1}
R_{rs}^-	Remote sensing reflectance below water surface	sr^{-1}
R_{rs}^b	Remote sensing reflectance of bottom	sr^{-1}
R	Irradiance reflectance	—
R^b	Bottom albedo	—
L	Radiance	$W\ m^{-2}\ sr^{-1}\ nm^{-1}$
$L_{at\text{-}sens}$	Radiance received by a remote sensor	$W\ m^{-2}\ sr^{-1}\ nm^{-1}$
L_{atm}	Atmospheric path radiance	$W\ m^{-2}\ sr^{-1}\ nm^{-1}$
L_{sky}	Sky radiance	$W\ m^{-2}\ sr^{-1}\ nm^{-1}$
L_{spec}	Specular radiance from the water surface	$W\ m^{-2}\ sr^{-1}\ nm^{-1}$
L_{ref}	Reference panel radiance	$W\ m^{-2}\ sr^{-1}\ nm^{-1}$
L_u	Upwelling radiance	$W\ m^{-2}\ sr^{-1}\ nm^{-1}$
L_w	Water-leaving radiance	$W\ m^{-2}\ sr^{-1}\ nm^{-1}$
E	Irradiance	$W\ m^{-2}\ sr^{-1}\ nm^{-1}$
E_o	Scalar irradiance	$W\ m^{-2}\ sr^{-1}\ nm^{-1}$
E_d	Downwelling irradiance	$W\ m^{-2}\ sr^{-1}\ nm^{-1}$

(Continued)

(Continued)

Symbol	Parameter	Units
E_u	Upwelling irradiance	W m^{-2} sr^{-1} nm^{-1}
H_0	Extraterrestrial solar irradiance	W m^{-2} sr^{-1}
$k(\theta, \varphi)$	Direct attenuation coefficient of radiance	m^{-1}
K_d	Diffuse attenuation coefficient of downwelling irradiance	m^{-1}
K_u	Diffuse attenuation coefficient of upwelling vector irradiance	m^{-1}
K_{Lu}	Diffuse attenuation coefficient of upwelling vector radiance	m^{-1}
K_{PAR}	Diffuse attenuation coefficient of PAR	m^{-1}
a	Absorption coefficient	m^{-1}
a^*	Specific absorption coefficient	m^2 mg^{-1}
a_p	Absorption coefficient of particulate matter	m^{-1}
a_{phy}	Absorption coefficient of phytoplankton	m^{-1}
a_{PC}	Absorption coefficient of phycocyanin	m^{-1}
a_{NAP}	Absorption coefficient of nonalgal particles	m^{-1}
a_m	Absorption coefficient of minerogenic particles	m^{-1}
a_d	Absorption coefficient of organic detritus	m^{-1}
a_{CDM}	Absorption coefficient of colored detritus matter	m^{-1}
a_{CDOM}	Absorption coefficient of colored dissolved organic matter	m^{-1}
a_w	Absorption coefficient of pure water	m^{-1}
β	Volume scattering function	m^{-1} sr^{-1}
$\tilde{\beta}$	Scattering phase function	sr^{-1}
b	Scattering coefficient	m^{-1}
b_b	Backscattering coefficient	m^{-1}
b_f	Forward scattering coefficient	m^{-1}
$b_{b,w}$	Backscattering coefficient of pure water	m^{-1}
$b_{b,p}$	Backscattering coefficient of particulate matter	m^{-1}
$b_{b,phy}$	Backscattering coefficient of phytoplankton	m^{-1}
$b_{b,NAP}$	Backscattering coefficient of non-algal particles	m^{-1}
c	Beam attenuation coefficient	m^{-1}
a^*_{phy}	Specific absorption coefficient of phytoplankton	m^2 mg^{-1}
$b_b^*{}_{TSM}$	Specific backscattering coefficient of TSM	m^2 g^{-1}
$\varepsilon(\lambda_1, \lambda_2)$	Ratio of aerosol reflectances at two wavelengths	—
g	Asymmetry parameter	—
m	Complex refractive index	—
μ_0	Average cosine of refracted photons	—
n_w	Refractive index of water	—
n	Angström exponent of particle scattering	—
ρ_a	Dimensionless reflectance of atmospheric aerosols	—
ρ_w	Dimensionless reflectance of water	—
s	Spherical albedo of the Earth's atmosphere	—
S	Spectral slope of CDOM absorption	nm^{-1}
S_d	Spectral slope of detritus absorption	nm^{-1}
t_{spec}	Transmittance of specular radiance through the atmosphere	—
t_{sun}	Transmittance of solar radiation	—

(Continued)

(Continued)

Symbol	Parameter	Units
t_w	Transmittance of water-leaving radiance	—
η_F	Fluorescence quantum yield	—
ϖ_o	Single-scattering Albedo	—
[chl-a]	Chlorophyll-a concentration	$\mu g\ L^{-1}$
[TSM]	Total suspended matter concentration	$mg\ L^{-1}$
[NAP]	Nonalgal particles concentration	$mg\ L^{-1}$
[DOC]	Dissolved organic carbon concentration	$mg\ L^{-1}$

Chapter 1

Remote Sensing of Inland Waters: Background and Current State-of-the-Art

Igor Ogashawara[1], Deepak R. Mishra[2], and Anatoly A. Gitelson[3,4]
[1]Indiana University—Purdue University at Indianapolis (IUPUI), Indianapolis, IN, United States, [2]University of Georgia (UGA), Athens, GA, United States, [3]Israel Institute of Technology, Haifa, Israel, [4]University of Nebraska-Lincoln (UNL), Lincoln, NE, United States

1.1 INLAND WATERS

This book is designed to highlight the theories, past developments, and current state-of-the-art knowledge in bio-optical modeling of inland waters. This area of remote sensing research has been intensively developed in the last 30 years primarily using the concepts and theories from optical oceanography. The focus of this book is squarely placed on inland waters because of the lack of a coherent and comprehensive synthesis of multidecadal bio-optical research on this important environment. Inland waters are aquatic environments typically confined within the land boundaries and provide exceptionally important ecological, environmental, hydrological, and socioeconomic services to mankind and the environment. The Millennium Ecosystem Assessment, a taskforce initiated by United Nations Secretary-General in 2001, generated a list of services provided by or derived from inland waters which were divided into four categories: provisioning (i.e., food supply, water supply, and biodiversity); regulating (i.e., climate regulation, hydrological flows, and pollution control); cultural (i.e., recreational, aesthetic, and educational); and supporting (i.e., soil formation, nutrient cycling, and pollination) (Millennium Ecosystem Assessment MEA, 2005).

The term "inland waters" is used wherever possible unless a specific ecosystem type is mentioned, such as lakes, reservoirs, rivers, ponds, swamps, wetlands, and even coastal areas. Inland waters represent an extremely diverse environment including a broad array of shapes and sizes, and physical, chemical, and optical properties. For example, inland waters

Bio-optical Modeling and Remote Sensing of Inland Waters.
DOI: http://dx.doi.org/10.1016/B978-0-12-804644-9.00001-X

could be fresh (Caspian Sea, western Asia-eastern Europe; Great Lakes, USA), saline (Dead Sea, Middle East), or brackish (Lake Pontchartrain, USA; Lake Chilika, India). The extent and distribution of inland waters is poorly and unevenly known at the global scale, since their size varies from small (i.e., ponds) to very large (i.e., Great Lakes) often creating inconsistencies in detection and inventory over broad geographic scale. Verpoorter et al. (2014) used GeoCover product developed using imageries from Enhanced Thematic Mapper Plus (ETM$^+$) onboard Landsat-7 satellite to map all inland water bodies greater than $0.002\ km^2$. Their findings contained geographic and morphometric information for approximately 117 million inland aquatic systems with a combined surface area of approximately $5 \times 10^6\ km^2$. Although inland waters only comprise a small percentage of Earth's total land surface, they play an essential role in biogeochemical cycle and are very important in the history of mankind (Bastviken et al., 2011).

Inland water bodies serve as sentinels to changing environment, such as climate change, developmental pressure, and land use land cover change. Rapid and uncontrolled environmental change, such as deforestation and reduction of vegetation cover, nutrient pollution, drought, urbanization, and engineered modifications to watershed most often result in negative impact including accelerated eutrophication, proliferation of toxic blue-green algae, extreme turbidity and deterioration of water clarity, loss of aquatic benthos, and harmful effect on human and animal health. Considering the vital uses of inland waters, water quality management needs to be a top priority for environmental regulatory agencies around the world. The National Research Council (NRC) published a comprehensive report entitled "The Drama of the Commons" highlighting seven key challenges in environmental resource management (National Research Council NRC, 2002). Included among those were "low-cost enforcement of rules" and "monitoring the resource and users' compliance with rules," which highlighted the need for a low-cost environmental monitoring program using remote sensing technologies. Rapid monitoring surveys using remote sensing should be conducted frequently along with less frequent field-based methods as an effective water quality management strategy (Ostrom et al., 2003). As opposed to the traditional field-based methods to monitor water quality, which are usually costly and labor intensive, remote sensing offers a low-cost, high frequency, broad coverage, and practical alternative for water quality assessment and monitoring (Duan et al., 2010; Hadjimitsis and Clayton, 2009). The main advantage of remote sensing is its capability to perform frequent large-scale synoptic monitoring of water resources and, therefore, the development of new techniques or fine-tuning of existing bio-optical models (a.k.a. bio-optical algorithms), and methods for using remote sensor derived products are essential for accurate monitoring inland water resources and isolating the natural and anthropogenic stressors.

1.2   REMOTE SENSING OF INLAND WATERS

The utilization of remote sensing techniques to monitor aquatic environments was initiated during 1960s by analyzing ocean color under the assumption that chlorophyll-a (chl-a), a proxy for phytoplankton biomass, and surface temperature could be estimated remotely (Morel and Gordon, 1980; Gordon et al., 1988). Based on these assumptions, oceanographers started to remotely monitor the optical properties of water constituents, such as phytoplankton, colored dissolved organic matter (CDOM) and total suspended solids (TSS) (Jerlov, 1968; Preisendorfer, 1976; Morel, 2001). As a result, the concepts of hydrologic optics and radiative transfer theory were developed and formed the basis of what is currently known as bio-optical modeling. However, the application of these theories and concepts developed by oceanographers were applied to inland waters only during the last three decades concomitant to the extensive use of bio-optical models to monitor optically active water constituents, such as chl-a, TSS, and CDOM (Jerlov, 1968, 1976; Gordon and Morel, 1983; Morel, 2001).

Nevertheless, application of remote sensing techniques to inland waters can be quite different from open ocean waters mainly because of the variable composition of water constituents. Morel and Prieur (1977) proposed a two-tier classification of water bodies: Case 1 and Case 2. This classification was based on the reflectance ratio at 443 and 550 nm. The ratio should be greater than 1.0 for Case 1 and less than 1.0 for Case 2 waters. Gordon and Morel (1983) proposed new definitions for Case 1 and Case 2 waters. They classified Case 1 waters as waters whose optical properties are determined mainly by phytoplankton and the other covarying compounds, such as CDOM and detritus. Case 2 waters are waters whose optical properties are significantly influenced by other constituents, such as mineral particles and CDOM, and their concentrations do not covary with phytoplankton. However, this classification showed several problems, such as the misinterpretation that inland waters mostly belong to Case 2 category, when it is possible that these water bodies can be dominated by phytoplankton or its derivatives, which would make them Case 1 (Mobley et al., 2004). Despite of the criticisms about the classification, it was used widely by researchers to categories water types for a quick overview.

Because of the presence of multiple constituents at different composition [i.e., phytoplankton, nonalgal particles (NAP), CDOM, and detritus], the use of remote sensing for monitoring inland water quality has been far less successful compared to open oceans. The complex interaction among the water constituents, which is often intensified by anthropogenic actions, creates uncertainty in remote sensing models designed for inland waters. Within the same inland aquatic system, it is possible to have different regions dominated by different constituents. For example, Gurlin et al. (2011) showed that for different sampling locations in Fremont Lakes, USA, the absorption coefficients of water constituents, phytoplankton [$a_{phy}(\lambda)$], nonalgal particles [$a_{NAP}(\lambda)$], and CDOM [$a_{CDOM}(\lambda)$], varied considerably along with constituents concentrations (Fig. 1.1). In the blue region, the contribution of absorption coefficient of water [$a_w(\lambda)$] is minimal while the major contributors are $a_{CDOM}(\lambda)$ and $a_{NAP}(\lambda)$. $a_{phy}(\lambda)$ contribution in the blue region is mostly minor and varied based on the chl-a concentration. In the green region, the absorption coefficients of $a_{CDOM}(\lambda)$, $a_{NAP}(\lambda)$, and $a_{phy}(\lambda)$ are smaller than in the blue and $a_w(\lambda)$ is higher. Ocean color algorithms (e.g., OC4v4) that use blue and green spectral channels often provide a relatively accurate estimate of chl-a in ocean waters (Case 1) where the total nonwater absorption is dominated by phytoplankton. However, inland waters, which are optically complex, spectral channels in the blue-green region, are heavily affected by an intricate interaction between a variety of water constituents, such as phytoplankton, CDOM, detritus, and tripton, and cannot be used to resolve any constituent. Only in the red spectral region, beyond 650 nm, absorption is

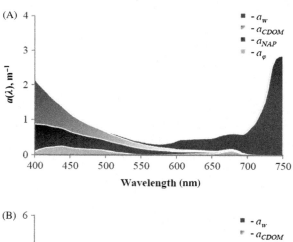

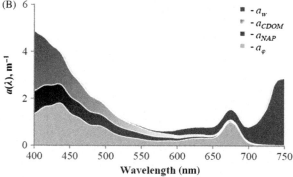

FIGURE 1.1 Absorption coefficients of phytoplankton, nonorganic particles, dissolved organic matter, and pure water, $a_{phy}(\lambda)$, $a_{NAP}(\lambda)$, $a_{CDOM}(\lambda)$, and $a_w(\lambda)$, for two sampling stations with chl-a concentrations of 4.6 mg m^{-3} (A) and 58.1 mg m^{-3} (B) ($a_{phy}(\lambda)$ is the same as a_φ in the figure). *Source: Figure 4.1 from Gurlin (2012).*

mainly governed by phytoplankton and pure water; therefore, red-near infrared (NIR) wavelengths have been found to be more appropriate for the development of bio-optical models for inland waters, particularly in waters where chl-a concentration is above $10\,\mu g/L$ (Gitelson, 1992; Mishra and Mishra, 2012).

Because of these complexities due to mixed constituents, commonly found in inland waters, this book is aimed at synthesizing the current state-of-the-art in terms of both knowledge and applications to monitor these aquatic environments.

1.3 FUNDAMENTAL BIO-OPTICAL PROPERTIES

Remote sensing of inland waters is based on the optical properties of water constituents (e.g., Morel, 2001). These properties can be divided into two categories: (1) properties that depend on the medium and the directional structure of the ambient light field, known as apparent optical properties (*AOPs*), and (2) those which depend only on the medium and are independent of the ambient light field, known as inherent optical properties (*IOPs*) (Preisendorfer, 1976). Optical properties, such as irradiance reflectance (R), above and below water remote sensing reflectance (R_{rs} or r_{rs} respectively), and various diffuse attenuation functions (K) are *AOPs*, since they vary according to the composition of the medium and the light field. Properties, such as absorption (a) and scattering (b) coefficients, which vary based on just the composition of the medium or constituents, are *IOPs*. A more detailed description about AOPs and IOPs is provided in Chapter 2 which covers physical principles and radiative transfer theory applied to inland waters.

For remote sensing of inland waters, the most commonly used *AOP* is R_{rs}. R_{rs} is defined as the ratio of water-leaving radiance (L_w) to downwelling irradiance (E_d), expressed in terms of *per steradian* (sr^{-1}). In case of IOPs, absorption properties of water constituents are the most commonly used parameters since instruments for measuring scattering in inland waters are not widely deployed and more research is needed to evaluate their performance in these optically complex waters.

To understand the optical properties of a medium (e.g., water column), one needs to recognize the Kubelka–Munk remission function. The function explains the infinite reflectance (R_∞) of an ideal layer of turbid media in which a further increase in thickness results in no noticeable difference in reflectance (Kubelka and Munk, 1931). R_∞ is proportional to the ratio of the absorption coefficient to the scattering coefficient of the media (Kubelka and Munk, 1931; Kortum, 1969). Studies have shown that the Kubelka–Munk remission function relates very closely (determination coefficient above 0.99) to the inverse reflectance of leaves (Gitelson et al., 2003) and water (Dall'Olmo et al, 2003). Thus, inverse reflectance may be used as proxy of absorption by phytoplankton pigments, NAP, CDOM, and water. Fig. 1.2

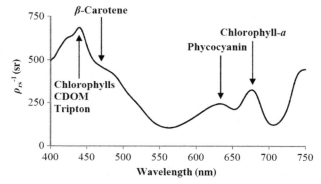

FIGURE 1.2 Typical inland waters inverse R_{rs} spectrum. The positions of the absorption peaks by constituents, such as chl-a, PC, NAP, and CDOM are labeled. *Source: Figure 3.16 from Gurlin (2012).*

shows distinctive spectral features of an inverse R_{rs} spectrum acquired from an inland water body, which has a high absorption in the blue spectral region due to the combined absorption by phytoplankton pigments, β-Carotene, NAP, and CDOM. The absorption peak around 625 nm is related to the presence of phycocyanin (PC), a proxy for cyanobacteria, and another absorption peak around 675 nm is related to the absorption by chl-a. These complex interactions between the constituents and their corresponding spectral features form the foundation for the development of bio-optical models for inland waters.

A list of research studies reporting absorption coefficients, such as a_{CDOM}, a_{NAP}, and a_{phy} from different inland waters is provided in Table 1.1. It is possible to further partition some of the primary absorption coefficients using decomposition algorithms. For example, a_{phy} can be decomposed into absorption coefficients of chl-a (a_{chl-a}) and PC (a_{PC}) (Simis et al., 2005, Mishra et al., 2013, 2014); and a_{NAP} can be partitioned into absorption coefficients of minerogenic particles (a_m) and organic detritus (a_d) (Peng and Effler, 2013).

IOPs vary not only across geographic regions but also within the same site (Table 1.1). For example, Gons et al. (2008) reported a_{CDOM} ranges from Finger Lakes, New York which varied from 0.06 to 67 m^{-1}. In addition, studies such as Matthews and Bernard (2013), which analyzed the three absorption coefficients in multiple lakes, demonstrated the complexities of interaction between constituents and their optical properties in inland waters. The complexity is mainly due to the spatial-temporal variability of the water constituents at the same site. In other words, the dominant constituent in the water column at a study site may not only change spatially across short distances but also across seasons and even daily (Yacobi et al., 1995; Huang et al., 2015). Multiple dominant constituents over relatively short period of

TABLE 1.1 List of Studies Reporting Absorption Coefficients of Optically Active Constituents Measured at Various Inland Water Bodies

Reference	Site	Location	Wavelength (nm)	Absorption Coefficient (m^{-1})
		a_{CDOM}		
Dall'Olmo and Gitelson (2005)	Eastern Nebraska water bodies	Nebraska, USA	440	0.5−4.4
Dekker (1993)	Vecht Lakes	The Netherlands	440	1.19−3.74[a]
Gitelson et al. (2008)	Nebraska Lakes (2005)	Nebraska, USA	440	0.3−1.77
Gons et al. (2008)	Finger Lakes	New York, USA	440	0.06−67
Gurlin et al. (2011)	Fremont Lakes (2008)	Nebraska, USA	440	0.46−1.46
Gurlin et al. (2011)	Fremont Lakes (2009)	Nebraska, USA	440	0.35−1.35
Kutser et al. (2005a)	Finnish Lakes	Southern Finland	420	1.28−7.74
Kutser et al. (2015)	Lake Mälaren	Stockholm, Sweden	443	2.7−4.0
Le et al. (2009)	Lake Tai	China	440	0.02−1.42
Matthews and Bernard (2013)	Lake Hartbeespoort	South Africa	442	0.63−4.13[b]
Matthews and Bernard (2013)	Lake Loskop	South Africa	442	0.75−1.87[b]
Matthews and Bernard (2013)	Lake Theewaterskloof	South Africa	442	1.22−2.49[b]
Mishra et al. (2014)	Catfish Ponds	Mississippi, USA	443	3.2−4.33[c]
Song et al. (2014)	Central Indiana Reservoirs	Indiana, USA	440	0.58−3.23
		a_{NAP}		
Dall'Olmo and Gitelson (2005)	Eastern Nebraska water bodies	Nebraska, USA	440	0.4−6.7[d]
Gitelson et al. (2008)	Nebraska Lakes (2005)	Nebraska, USA	440	0.13−4.34[d]
Gurlin et al. (2011)	Fremont Lakes (2008)	Nebraska, USA	675	0.003−0.294

(Continued)

TABLE 1.1 (Continued)

Reference	Site	Location	Wavelength (nm)	Absorption Coefficient (m^{-1})
Gurlin et al. (2011)	Fremont Lakes (2009)	Nebraska, USA	675	0.007–0.113
Le et al. (2009)	Lake Tai	China	440	0.57–7.16
Matthews and Bernard (2013)	Lake Hartbeespoort	South Africa	442	0.07–1.74[d]
Matthews and Bernard (2013)	Lake Loskop	South Africa	442	0.1–1.57[d]
Matthews and Bernard (2013)	Lake Theewaterskloof	South Africa	442	0.51–2.26[d]
		a_{phy}		
Dall'Olmo and Gitelson (2005)	Eastern Nebraska water bodies	Nebraska, USA	678	1.0–6.0
Gurlin et al. (2011)	Fremont Lakes (2008)	Nebraska, USA	675	0.04–4.45
Gurlin et al. (2011)	Fremont Lakes (2009)	Nebraska, USA	675	0.04–2.72
Le et al. (2009)	Lake Tai	China	675	0.06–3.99
Li et al. (2013)	Central Indiana Reservoirs	Indiana, USA	665	0.439–3.87[e]
Li et al. (2013)	Central Indiana Reservoirs	Indiana, USA	665	0.037–2.51[e]
Matthews and Bernard (2013)	Lake Hartbeespoort	South Africa	442	1.73–455.32
Matthews and Bernard (2013)	Lake Loskop	South Africa	442	0.05–11.12
Matthews and Bernard (2013)	Lake Theewaterskloof	South Africa	442	0.41–2.43
Mishra et al. (2014)	Catfish Ponds 1	Mississippi, USA	443	10.27–15.55
Mishra et al. (2014)	Catfish Ponds 1	Mississippi, USA	665	3.96–6.23

[a]Reported as aquatic humus absorption coefficient.
[b]Reported as gelbstoff absorption coefficient.
[c]Reported as colored detrital matter absorption coefficient.
[d]Reported as detrital absorption coefficient.
[e]Reported as total less water absorption coefficient (a_{t-w}).

time seen in inland water environments is not common for open ocean environments which makes the remote sensing of inland waters a challenging task. It can also be seen from Table 1.1, studies quantifying absorption coefficients of constituents in inland waters are fairly recent indicating that the development of robust models for the estimations of IOPs in inland waters is a relatively new field of research.

The variability in the composition of constituents and the associated IOPs within an aquatic system can affect the magnitude and shape of R_{rs}. To exemplify this, data from three water bodies with different trophic status are used: two tropical hydroelectric reservoirs (Itumbiara and Funil Reservoirs) in Brazil and catfish ponds in Mississippi, USA. The Itumbiara Hydroelectric Reservoir located in Central Brazil is an oligotrophic reservoir. Funil Hydroelectric Reservoir is a mesotrophic reservoir located in Southeast Brazil. The catfish ponds located at Delta Research Extension Center near Stoneville, MS, USA are hypereutrophic in nature.

Fig. 1.3 shows the average R_{rs} spectra normalized at 575 nm for the three study sites. The normalization to a specific wavelength is usually performed to allow the intercomparison of R_{rs} spectra from different water bodies. The R_{rs} spectrum for the oligotrophic Itumbiara reservoir showed a very distinct

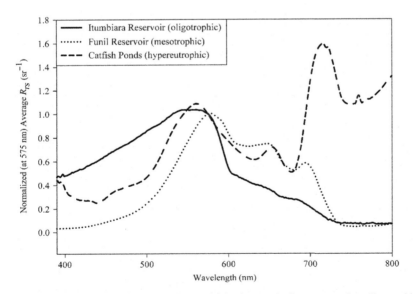

FIGURE 1.3 Average normalized R_{rs} for three inland water bodies representing oligotrophic waters (Itumbiara Reservoir, chl-*a*: 0.25−10.20 µg/L; TSS: 0.25−1.81 mg/L; DOC: 0.53−2.59 mg/L); mesotrophic waters (Funil Reservoir, chl-*a*: 4.92−52.78 µg/L; PC: 9.16−35.95 µg/L; TSS: 4.50−9.50 mg/L; DOC: 0.91−6.30 mg/L); and hypereutrophic waters (Catfish Ponds, chl-*a*: 59.40−1376.60 µg/L; PC: 68.13−857.08 µg/L).

spectral shape without a pronounced phytoplankton absorption feature typically observed between 665 and 685 nm. Spectral features such as low reflectance in blue and scattering in green are representative of waters with low chl-a concentration. The R_{rs} spectrum for the mesotrophic Funil Reservoir showed the lowest values in the blue wavelengths mainly due to a combined absorption by CDOM and chl-a. The trough around 625 nm is primarily associated with PC absorption (Schalles and Yacobi, 2000; Mishra et al., 2009; Mishra and Mishra, 2014). A pronounced trough at 675 nm caused by a strong chl-a absorption and when combined with the strong absorption by pure water in NIR, generates a peak near 700 nm. R_{rs} spectrum for the hypereutrophic Catfish Ponds with extremely high chl-a concentration showed the largest R_{rs} peak near 700 nm, mainly due to scattering by phytoplankton cells as well as two prominent troughs at 665 nm and around 620 nm caused by chl-a and PC absorption, respectively.

The variations in R_{rs} are related to the variations in IOPs, such as a_{phy}, a_{NAP}, and a_{CDOM} at each study site. The compositional variability of the IOPs can be visualized using ternary plots representing the absorption coefficients at three characteristic wavelengths, 440, 560, and 665 nm (Fig. 1.4).

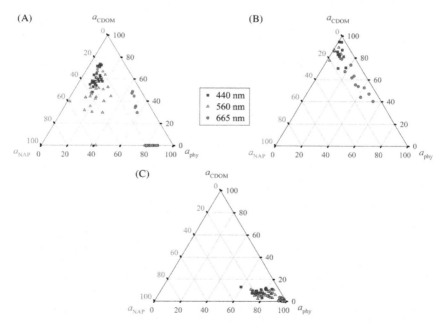

FIGURE 1.4 Ternary plots of absorption coefficients of CDOM, phytoplankton and NAP at three different wavelengths: 440 (square), 560 (triangle), and 665 nm (round). (A) oligotrophic Itumbiara Reservoir, Brazil; (B) eutrophic Funil Reservoir, Brazil; (C) hypereutrophic Catfish Ponds, USA.

Absorption coefficient at 440 nm is influenced by a combined absorption of CDOM, chl-*a* and tripton (Fig. 1.2). Absorption coefficient at 560 nm is usually governed by NAP absorption, and absorption coefficient at 665 nm is related to the absorption of chl-*a*. The ternary plot for Itumbiara Reservoir (Fig. 1.4A) showed a mix of CDOM and NAP dominance which can also be observed in its R_{rs} spectrum with only one prominent peak at around 550 nm (Fig. 1.3). The ternary plot for Funil Reservoir (Fig. 1.4B) showed that a_{CDOM} is the dominant IOP, which supported the very low value in the blue wavelength range observed in R_{rs} spectrum. The complete dominance of a_{phy} is clearly visible in the ternary plot for the hypereutrophic catfish ponds (Fig. 1.4C). These ternary plots exemplify the optical complexities of inland waters and how reflectance at certain spectral region (e.g., blue) is being influenced by multiple constituents. Because of that bio-optical models for inland waters are increasingly focused on red-NIR bands (e.g., Gurlin et al., 2011).

1.4 BIO-OPTICAL MODELS

The expression "bio-optical" was first used to describe the "state of ocean waters" (Smith and Baker, 1977). The term "state" is defined by Smith and Baker (1977) as the optical state of water, which is essentially controlled by the optical properties of the biological materials in the water column, mainly phytoplankton and its derivatives. However, the current use of the term "bio-optical" is often combined with models in remote sensing research. Bio-optical models are based on radiometric quantities or IOPs and AOPs, such as downwelling spectral solar and sky radiation and the absorption and scattering properties of constituents in the water column. Studies by Cox and Munk (1954), Petzold (1972), Jerlov (1968, 1976), Preisendorfer (1976), among numerous others, established the main theory of hydrologic optics (currently known as ocean optics) before or around the launch of one of the first Earth Observing Satellites, the Earth Resources Technology Satellites 1 (ERTS-1), in 1972. Subsequently, transformative research was conducted by Gordon et al. (1975) who developed the first bio-optical model. They used Monte Carlo simulation of the radiative transfer equations to develop relationship between AOPs and IOPs for oceanic waters. Since then, numerous other bio-optical models have been published for oceanic, coastal, and inland waters. Applications of these bio-optical models to inland waters have recently been highlighted by several studies. Although there is a rapid increase of studies on bio-optical modeling, there is a lack of consistency in terminology and classification of types of bio-optical models (Ogashawara, 2015).

Bio-optical models can be defined in two ways. The first definition refers to the various ways of describing the "bio-optical state" of the aquatic system (Morel, 2001). It means that the optical properties are a function of the biological and geomorphological activities in the water body. Therefore, bio-optical models are often aimed at deriving information about the biological and physical processes in the water body by establishing relationships between radiometric measurements and optically active constituents. According to Morel (2001), these models are usually classified as empirical and descriptive based on statistical relationships between radiometric quantities. The second definition for bio-optical models refers to the use of the radiative transfer theory to derive optical properties of constituents in the water column (Mobley, 2001). These bio-optical models are based on the quantification of IOPs, which can be used to derive the composition of constituents through a ratio of their absorption coefficients (a) and specific absorption coefficients (a^*), and backscattering coefficients (b_b) and specific backscattering coefficients (b_b^*). These bio-optical models are usually referred to as analytical models.

Bio-optical models have been used to monitor the inland water quality, mainly estimating concentrations of chl-a, or TSS as indicators of the trophic or turbidity status of the water body, respectively. Bio-optical models have also been used to quantify biogeochemical fluxes and the underlying environmental forcing (Kutser et al., 2005b, 2009, Giardino et al., 2010). These models combined with multiplatform observations (in situ, airborne, and space-borne) can be extremely useful in understanding the feedback between the water body and surrounding landscape.

1.4.1 Classification of Bio-optical Models

Different terms have been used to classify bio-optical models according to their formulation and goals (Odermatt et al., 2012) due to a lack of consistency in generally agreed terminology. Based on the formulation and final goals of the bio-optical models found in the literature, models may be classified into five broad categories as empirical, semi-empirical, semi-analytical, quasi-analytical, and analytical (Ogashawara, 2015).

Empirical and semi-empirical models are usually based on statistical relationships between in situ measurements of water constituents and radiometric data from satellite sensor or proximal remote sensing devices. The difference between empirical and semi-empirical models relies on the assumptions used during their development; empirical models focus only on the statistical estimators. The formulae used in empirical models are based on a combination of R_{rs} at different wavelengths, which will provide the best correlation between reflectance data and the concentration of optically active water constituents. These types of models typically use statistical techniques, such as neural networks, least squares, and stepwise regressions to extract the best relationship between R_{rs} and constituent concentrations. The selection of spectral bands does not follow any physical or optical principles of the IOPs or AOPs.

Semi-empirical models, on the contrary, are based on specific spectral features of absorption and scattering of the constituents governing reflectance (e.g., Morel and Gordon, 1980). An example of a physical assumption used in semi-empirical models is the R_{rs} peak near 700 nm commonly used to estimate chl-*a* concentration in inland and coastal waters (Gitelson, 1992, 2005; Gurlin et al., 2011; Mishra and Mishra, 2012). It forms at that spectral region because of minimal combined absorption by chlorophyll and water and, thus, scattering is mostly pronounced in the reflectance spectra. The magnitude of the peak increases and position shifts toward longer wavelength with increase of chl-*a* concentration (Gitelson, 1992). The physical principle behind the characteristics of the 700 nm peak forms the basis for band selection in semi-empirical models unlike pure statistical relationship in case of empirical models. The outputs of semi-empirical models are usually related to water constituents via statistical estimators.

Semi-analytical and quasi-analytical models rely on the inversion of the radiative transfer equations to establish relationships among AOPs and IOPs, which is computed through several analytical and empirical steps. Inverse models use AOPs, such as reflectance above or below water surface (R_{rs} or r_{rs}, respectively) to derive IOPs (Gordon et al., 1975, 1988):

$$r_{rs}(\lambda) = \frac{L_u(0^-, \lambda)}{E_d(0^-, \lambda)} = g_1 \left(\frac{b_b(\lambda)}{a(\lambda) + b_b(\lambda)} \right) + g_2 \left(\frac{b_b(\lambda)}{a(\lambda) + b_b(\lambda)} \right)^2 \quad (1.1)$$

where: $r_{rs}(\lambda)$ is the remote sensing reflectance just below water surface, $a(\lambda)$ is the total spectral absorption coefficient, $b_b(\lambda)$ is the total spectral backscattering coefficient, $L_u(0^-, \lambda)$ and $E_d(0^-, \lambda)$ are upwelling radiance and downwelling irradiance, respectively, g_1 and g_2 are geometrical factors.

The main difference between semi-analytical and quasi-analytical models is the process used to estimate $a(\lambda)$ and $b_b(\lambda)$. In semi-analytical models, the estimation of $a(\lambda)$ is computed by the sum of a_{phy}, a_{NAP} and a_{CDOM}. To derive $a(\lambda)$ in quasi-analytical models, knowledge about other absorption coefficients is not necessary since it estimates $a(\lambda)$ directly from R_{rs} and the other absorptions coefficients are computed from the spectral decomposition of the estimated $a(\lambda)$ (Lee et al., 2002). For the $b_b(\lambda)$ estimation, semi-analytical models usually compute it as the sum of the backscattering coefficients for each water constituent except CDOM. In case of quasi-analytical models, $b_b(\lambda)$ is typically computed based on the widely used expression (Gordon and Morel, 1983):

$$b_b(\lambda) = b_{b,w}(\lambda) + b_{b,p}(\lambda_0)\left(\frac{\lambda_0}{\lambda}\right)^{\eta} \quad (1.2)$$

where λ_0 is the target wavelength, $b_{b,p}$ is backscattering coefficients of suspended particles and η is known as Angström exponent and related to the particle size distribution (see Chapter 2 for detail). The outputs from semi-analytical and quasi-analytical models estimated IOPs are validated using IOPs derived from water samples via analytical methods. A purely analytical model is based only on the physical properties of constituents, for example the one proposed by Bricaud et al. (1995) for the estimation of chl-a concentration by using the ratio of a_{phy} and the specific absorption coefficient of phytoplankton (a_{phy}^*) (Table 1.2).

Table 1.2 presents examples of the five types of bio-optical models used to estimate chl-a in inland waters from remote sensing data. Chl-a is shown as an example since it is one of the most widely studied water constituent in remote sensing. The first example is an empirical model proposed by Allan et al. (2015) who used an average reflectance in green and red spectral bands of Landsat5-TM to establish relationship between band ratio and chl-a concentration. The two-band (Gitelson, 1992) and three-band (Dall'Olmo and Gitelson 2005; Gitelson et al., 2003) semi-empirical models (not shown in Table 1.2) used reflectance in red and NIR spectral regions and have been widely applied to predict chl-a in inland waters. Similarly, Mishra and Mishra (2012) proposed a semi-empirical model, normalized difference chlorophyll index (NDCI), based on the difference between reflectance at 665 nm (MEdium Resolution Imaging Spectrometer, MERIS band 7) and 709 nm (MERIS band 9). Semi-analytical model proposed by Vos et al. (2003) is based on the spectral bands of Sea-viewing Wide Field-of-view Sensor (SeaWiFS). The model estimates $a_{phy}(670)$ by establishing relationship between several parameters (see Table 1.2). This model is classified as semi-analytical since $a(\lambda)$ is not required to estimate $a_{phy}(670)$. A quasi-analytical model reparametrized by Li et al. (2013) estimated the total absorption without the water absorption ($a_{t\text{-}w}$) from a relationship between R_{rs}, b_b, and a_w at 709 nm, $b_b(\lambda)$, and $a_w(\lambda)$. The model assumed that at 665 nm the total absorption is entirely caused by phytoplankton, and therefore, $a_{t\text{-}w}$ can be used to estimate chl-a concentration.

1.4.2 Performance of Bio-optical Models

Several studies have been published in recent years comparing the performance of existing semi-empirical bio-optical models, mainly for the estimation of chl-a concentration in inland waters. Gurlin et al. (2011) compared the performance of two-band (Gitelson, 1992) and three-band NIR-red models (Dall'Olmo et al., 2003; Dall'Olmo and Gitelson, 2005) using data from several productive lakes in Nebraska at MERIS spectral channels (Moses

TABLE 1.2 Examples of the Five Types of Bio-Optical Models used for chl-*a* Estimation

Model type	Reference	Sensor	Model structure
Empirical	Allan et al. (2015)	Landsat/TM	$Chl - a \cong \left(865.17 \bullet \left(\frac{B_2 + B_3}{2} \right)^2 \right) + \left(19.8 \bullet \left(\frac{B_2 + B_3}{2} \right) \right) + 0.24$
Semi-empirical	Mishra and Mishra (2012)	MERIS	$Chl - a \propto \dfrac{R_{rs}(B_9) - R_{rs}(B_7)}{R_{rs}(B_9) + R_{rs}(B_7)}$
Semi-analytical	Vos et al. (2003)	SeaWiFS	$a_{phy}(670) = \left(\dfrac{fb_b(670)}{R_{rs}(670)} \right) - (b_b(670) + g_{440}\bar{a}_{CDOM}(670) + a_w(670) + a^*_{TSM}(670)C_{TSM})$
Quasi-analytical	Li et al. (2013)	Ocean Optics in situ sensor (USB-4000)	$a_{t-w}(\lambda) = \dfrac{R_{rs}(709)b_b(\lambda)[a_w(709) + b_b(709)]}{R_{rs}(\lambda)b_b(709)} - b_b(\lambda) - a_w(\lambda)$
Analytical	Bricaud et al. (1995)	Spectrophotometer	From inversion of a R_{rs} model

Note: There are numerous examples exist in the literature under each category.

et al., 2009a,b). The study showed that the two-band model was more accurate in estimating chl-*a* than the three-band model. Augusto-Silva et al. (2014) compared the performance of two- and three-band NIR-red models and NDCI at close range in an inland tropical reservoir and found that NDCI was the most accurate among the three models. Beck et al. (2016) compared 12 bio-optical models used for the estimation of chl-*a* concentrations in Harsha Lake, OH, USA by simulating reflectance at spectral bands of different sensors. The authors concluded that NDCI is the most widely applicable model and performs well for most of the sensors typically used in remote sensing for inland waters including MERIS, WorldView-2, MSI Sentinel-2, and OLCI Sentinel-3.

Odermatt et al. (2012) provided an overview of different semi-empirical and empirical bio-optical models for estimating water constituents using satellite data. The study evaluated different satellite sensors, such as Sea-Viewing Wide Field-of-View Sensor (SeaWiFS), Moderate Resolution Imaging Spectroradiometer (MODIS), MEdium Resolution Imaging Spectrometer (MERIS), Landsat, and Hyperspectral Imager for the Coastal Ocean (HICO) in order to compare the bio-optical models for chl-*a*, TSM and CDOM estimation (Figs. 1.5−1.7).

The Ocean Color (OC) series of models (O'Reilly et al., 2000) were able to estimate accurately the chl-*a* concentration in optically complex waters only for chl-*a* below $10 \, \text{mg/m}^3$ (Fig. 1.5). For higher concentrations $(10−100 \, \text{mg/m}^3)$, red-NIR based empirical and semi-empirical models were able to accurately estimate the chl-*a* concentration. More detailed insight into bio-optical modeling of chl-*a* (phytoplankton) and sun-induced chl-*a* fluorescence is provided in Chapter 6 and Chapter 7, respectively.

Fig. 1.6 shows a list of bio-optical models developed between 2006 and 2011 for TSM estimation. Although the study showed that different types of bio-optical models could be used to retrieve TSM, still there is a strong sensitivity of the models towards the concentration range at which these models perform accurately. None of the TSM models tested in this study performed accurately over a broad concentration range $(0−1000 \, \text{mg/m}^3)$. The overestimation is because of the effects of chl-*a* and CDOM which often contribute to turbidity. A more detailed discussion about bio-optical models involving TSM (or TSS) estimation is presented in Chapter 5.

CDOM estimation models for SeaWiFS, MERIS, and MODIS imagery were also reviewed by Odermatt et al. (2012) (Fig. 1.7). They reported a similar bias in the performance of the models toward the CDOM range because CDOM absorption wavelength range (blue range) is usually affected by other water constituents. Chapter 4 of this book presents a complete review of remote sensing and bio-optical modeling of CDOM.

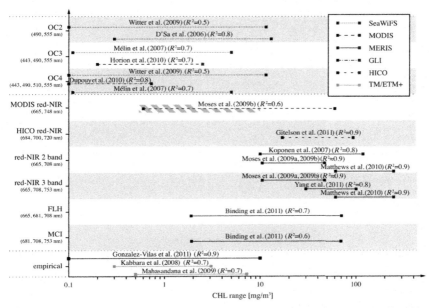

FIGURE 1.5 Performance summary of the semi-empirical and empirical models for chl-*a* estimation using various satellite sensors. *Source: Figure 1.1 from Odermatt et al. (2012).*

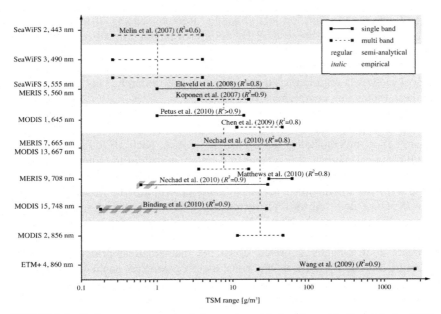

FIGURE 1.6 Performance summary of the semi-empirical and empirical models for TSM estimation using various satellite sensors. *Source: Figure 1.2 from Odermatt et al. (2012).*

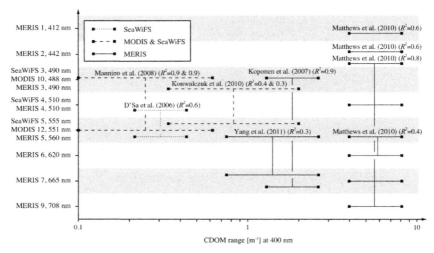

FIGURE 1.7 Performance summary of the semi-empirical and empirical models for CDOM estimation using various satellite sensors. *Source: Figure 1.3 from Odermatt et al. (2012).*

1.5 BOOK CONTENT

The book contains nine chapters covering key topics on remote sensing of inland waters. The topics were chosen based on a comprehensive review of remote sensing literature on inland waters. Remote sensing of inland waters is a complex problem because of the significant interaction between optically active constituents. Therefore, a fundamental understanding of the radiative transfer theory, optical properties of the water constituents, and existing bio-optical models is required to advance this field of research. This book covers all these topics in a comprehensive manner.

Chapter 2 provides an overview of radiative transfer theory describing the physical principles of interaction of light with constituents in the water column. For understanding the optical properties, such as IOPs and AOPs and bio-optical models, it is crucial to develop an insight to radiative transfer theory. The propagation of light through water and its absorption and scattering due to the interaction with constituents determines the radiant intensity or reflectance detected by a sensor deployed above or under water, be it in situ or airborne or space-borne. To decompose these composite reflectance spectra, one needs to understand the properties of the medium and the constituents in it. Chapter 2 breaks down the mathematical principles behind radiative transfer by IOPs, AOPs, constituents, geometric parameters, surface effects, and under water light field in a detail manner. It also provides a seamless transition to subsequent chapters, which deal with bio-optical modeling of optically active constituents.

Light received by an airborne or space borne remote sensor includes contributions due to its interactions with gaseous molecules and aerosol particles in the atmosphere in addition to the light reflected directly from the water body. The effects of atmospheric interference need to be adequately corrected for prior to applying field-based bio-optical models to remotely sensed data. Atmospheric correction of remotely sensed data is more challenging for inland waters than for open ocean waters due to a number of factors. Chapter 3 lays out the challenges and provides a comprehensive review of several existing atmospheric correction algorithms for inland waters, with brief discussions of the basic assumptions underlying each algorithm.

Atmospheric correction chapter is followed by Chapters (4–8) dealing with bio-optical modeling of water constituents starting with CDOM. CDOM is the optically active part of carbon. In remote sensing, it can be used as a proxy of dissolved organic carbon in an aquatic system. CDOM increases radiative heating of water bodies, limits the amount of light available for primary production, and protects aquatic life from excessive solar radiation. Accurate modeling and quantification of CDOM is important in studying the role of inland waters in global carbon cycle. In addition to providing a broad overview of the importance of CDOM in inland waters, Chapter 4 discusses in detail the optical properties of CDOM and architecture of existing empirical and semi-analytical bio-optical models.

TSS, a proxy for sediment flux, contaminants, and pollutants, plays an important role in controlling primary productivity in inland waters. Fortunately, TSS is one of the frequently studied and accurately modeled water constituents in both inland and coastal waters. Chapter 5 reviews the optical properties of TSS, mainly absorption and backscattering properties in several subalpine lakes as case studies. It also discusses a list of models including empirical and semi-analytical those have been used successfully before. In addition to field-based models, Chapter 5 also presents scaled-up model case studies from rivers to deep lakes involving airborne data, and Landsat-8 and MERIS images.

Similar to TSS, chl-a is another frequently studied water constituent using remote sensing techniques. Chl-a concentration, a proxy for phytoplankton biomass and productivity, is one of the most important water quality parameters because of its sensitivity and quick response to environmental and landscape changes. Chapter 6 discusses the theoretical basis of principles for estimating chl-a using remote sensing measurements via the optical pathways of absorption, fluorescence, and scattering. It also discusses the challenges in estimating chl-a in inland waters such as interference from other constituents and models to address them.

Chapter 7 deals with a different aspect of chl-a estimation using the sun-induced chl-a fluorescence. This chapter is focused on analyzing the relationship between fluorescence and phytoplankton abundance, CDOM absorption, and TSS concentration using a four component bio-optical model implemented on simulated and field remote sensing data. The chapter also discusses the use and limitations of fluorescence line height algorithms for estimating chlorophyll from various satellite data including MERIS and MODIS, which can be applicable also to Geostationary Ocean Color Imager and Sentinel-3's Ocean and Land Color Instrument. The detailed analysis of the models and remote sensing techniques presented in this chapter are useful for monitoring sun-induced chl-*a* fluorescence from water bodies over wide geographic regions and varied optical properties.

Although chl-*a* is commonly used for monitoring of algal blooms, remote estimation of cyanobacterial bloom is generally achieved through PC, a unique accessory pigment in cyanobacteria. Toxin producing cyanobacteria are a public health issue and frequent monitoring of these blooms may lead to targeted early warning systems, which would be extremely beneficial to human and animal health. Chapter 8 reviews the status of remote sensing of cyanobacteria using bio-optical models. It expands on the advantages and disadvantages of existing PC models and evaluates representative algorithms using a large field dataset. Discussions about the factors affecting the accuracy of the PC models will aid in developing scale-up techniques allowing their implementation on satellite data.

Submerged aquatic vegetation (SAV) or aquatic macrophytes grow in relatively clear water environments and provide suitable habitats for fish and zooplankton. They also influence the nutrient and energy cycling and limit phytoplankton growth. Developing remote sensing techniques to monitor their status and trend often leads to better water resource management. SAV can also act as a source of uncertainty for bio-optical models aimed at estimating water constituents by interfering with the remote sensing reflectance data. Therefore, isolating the SAV contribution in R_{rs} would lead to accurate estimation of the optically active water constituents. The final chapter of this book, Chapter 9 provides an overview of the current research on remote sensing of SAV. This chapter discuses several important topics, such as spectral properties of SAV, existing remote sensing models, application of airborne and space-borne data in mapping SAV biophysical properties, and need for regional assessment and global monitoring.

The content of this book summarizes the progress that has been made so far by remote sensing community worldwide on the broad topic of "remote sensing of inland waters." Although the bio-optical modeling of different water constituents is the main topic of this book, physical understanding of necessary theories to carry out bio-optical modeling, such as radiative transfer theory and atmospheric correction are also presented. Since it is a relatively new area of remote sensing research, there are several challenges that will not be easily solved, such as the development of robust and widely applicable radiative transfer models for inland waters. Nevertheless, the book discusses the theory and current state of the art in remote sensing of inland waters considering the availability of sensors with an adequate spatial, spectral, temporal, and radiometric resolution in future. The potential of these methods coupled with current and future satellite missions will enhance the opportunity to create an operational system or protocol for monitoring inland waters using remote sensing technologies.

REFERENCES

Allan, M.G., Hamilton, D.P., Hicks, B., Brabyn, L., 2015. Empirical and semi-analytical chlorophyll a models for multi-temporal monitoring of New Zealand lakes using Landsat. Environ. Monit. Assess. 187, 364. Available from: http://dx.doi.org/10.1007/s10661-015-4585-4.

Augusto-Silva, P.B., Ogashawara, I., Barbosa, C.C.F., de Carvalho, L.A.S., Jorge, D.S.F., Fornari, C.I., et al., 2014. Analysis of MERIS reflectance models for estimating chlorophyll-a concentration in a Brazilian Reservoir. Remote Sens. 6, 11689−117077.

Bastviken, D., Tranvik, L.J., Downing, J.A., Crill, P.M., Enrich-Prast, A., 2011. Freshwater methane emissions offset the continental carbon sink. Science 331, 50.

Bricaud, A., Babin, M., Morel, A., Claustre, H., 1995. Variability in chlorophyll-specific absorption coefficients of natural phytoplankton: analysis and parameterization. J. Geophys. Res. 100 (7), 13321−13332. Available from: http://dx.doi.org/10.1029/95JC00463.

Cox, C., Munk, W., 1954. Measurement of the roughness of the sea surface from photographs of the sun's glitter. J. Opt. Soc. Am. 44 (11), 838−850. Available from: http://dx.doi.org/10.1364/JOSA.44.000838.

Dall'Olmo, G., Gitelson, A.A., 2005. Effect of bio-optical parameter variability on the remote estimation of chlorophyll-a concentration in turbid productive waters: experimental results. Appl. Opt. 44 (3), 412−422. Available from: http://dx.doi.org/10.1364/AO.44.000412.

Dall'Olmo, G., Gitelson, A.A., Rundquist, D.C., 2003. Towards a unified approach for remote estimation of chlorophyll-a in both terrestrial vegetation and turbid productive waters. Geophys. Res. Lett. 30 (18), 1938. Available from: http://dx.doi.org/10.1029/2003GL018065.

Dekker, A.G., 1993. Detection of Optical Water Quality Parameters for Eutrophic Waters by High Resolution Remote Sensing. Ph.D. Thesis. Vrije Universiteit, Amsterdam, The Netherlands.

Duan, H., Ma, R., Xu, J., Zhang, Y., Zhang, B., 2010. Comparison of different semi-empirical models to estimate chlorophyll-a concentration in inland lake water. Environ. Monitor. Assess. 170, 231−244.

Giardino, C., Bresciani, M., Villa, P., Martinelli, A., 2010. Application of remote sensing in water resource management: the case study of Lake Trasimeno, Italy. Water Resour. Manag. 24, 3885−3899. Available from: http://dx.doi.org/10.1007/s11269-010-9639-3.

Gitelson, A., 1992. The peak near 700 nm on radiance spectra of algae and water: relationships of its magnitude and position with chlorophyll concentration. Int. J. Remote. Sens. 13, 3367−3373. Available from: http://dx.doi.org/10.1080/01431169208904125.

Gitelson, A.A., Gritz, U., Merzlyak, M.N., 2003. Relationships between leaf chlorophyll content and spectral reflectance and algorithms for non-destructive chlorophyll assessment in higher plant leaves. J. Plant. Physiol. 160 (3), 271−282.

Gitelson, A.A., Dall'Olmo, G., Moses, W., Rundquist, D.C., Barrow, T., Fisher, T.R., et al., 2008. A simple semi-analytical model for remote estimation of chlorophyll-a in turbid waters: Validation. Remote Sens. Environ. 112, 3582−3593. Available from: http://dx.doi.org/10.1016/j.rse.2008.04.015.

Gons, H.J., Auer, M.T., Effler, S.W., 2008. MERIS satellite chlorophyll mapping of oligotrophic and eutrophic waters in the Laurentian Great Lakes. Remote Sens. Environ. 112, 4098−4106. Available from: http://dx.doi.org/10.1016/j.rse.2007.06.029.

Gordon, H., Morel, A., 1983. Remote assessment of ocean color for interpretation of satellite visible imagery: a review. Lecture Notes on Coastal and Estuarine Studies. Springer Verlag, New York, NY, USA, p. 4.

Gordon, H.R., Brown, O.B., Jacobs, M.M., 1975. Computed relationships between the inherent and apparent optical properties of a flat homogeneous ocean. Appl. Opt. 14 (2), 417–427. Available from: http://dx.doi.org/10.1364/AO.14.000417.

Gordon, H.R., Brown, O.B., Evans, R.H., Brown, J.W., Smith, R.C., Baker, K.S., et al., 1988. A semianalytic radiance model of ocean color. J. Geophys. Res. 93 (D9), 10909–10924.

Gurlin, D., 2012. Near infrared-red models for the remote estimation of chlorophyll-α concentration in optically complex turbid productive waters: From in situ measurements to aerial imagery [doctoral dissertation]. Univ. of Nebraska-Lincoln.

Gurlin, D., Gitelson, A.A., Moses, W.J., 2011. Remote estimation of chl-a concentration in turbid productive waters—return to a simple two-band NIR-red model? Remote Sens. Environ. 115, 3479–3490. Available from: http://dx.doi.org/10.1016/j.rse.2011.08.011.

Hadjimitsis, D.G., Clayton, C., 2009. Assessment of temporal variations of water quality in inland water bodies using atmospheric corrected satellite remotely sensed image data. Environ. Monit. Assess. 159, 281–292.

Huang, C., Shi, K., Yang, H., Li, Y., Zhu, A., Sun, D., et al., 2015. Satellite observation of hourly dynamic characteristics of algae with Geostationary Ocean Color Imager (GOCI) data in Lake Taihu. Remote Sens. Environ. 159, 278–287. Available from: http://dx.doi.org/10.1016/j.rse.2014.12.016.

Jerlov, N.G., 1968. Optical oceanography, Elsevier Oceanographic Series, v. 5. Elsevier, Amsterdam, the Netherlands.

Jerlov, N.G., 1976. Marine optics, Elsevier Oceanographic Series, v. 14. Elsevier, Amsterdam, the Netherlands.

Kortum, G., 1969. Reflectance Spectroscopy. Principles, Methods, Applications. Springler-Verlag, New York Inc.

Kubelka, P., Munk, F., 1931. An article on optics of paint layers. Zeitschr. Fur techn. Physik 12, 593–601.

Kutser, T., Pierson, D.C., Tranvik, L., Reinart, A., Sobek, S., Kallio, K., 2005a. Using satellite remote sensing to estimate the colored dissolved organic matter absorption coefficient in lakes. Ecosystems 8, 709–720. Available from: http://dx.doi.org/10.1007/s10021-003-0148-6.

Kutser, T., Pierson, D., Kallio, K., Reinart, A., Sobek, S., 2005b. Mapping lake CDOM by satellite remote sensing. Remote Sens. Environ. 94, 535–540. Available from: http://dx.doi.org/10.1016/j.rse.2004.11.009.

Kutser, T., Paavel, B., Metsamaa, L., Vahtmae, E., 2009. Mapping coloured dissolved organic matter concentration in coastal waters. Int. J. Remote. Sens. 30, 5843–5849. Available from: http://dx.doi.org/10.1080/01431160902744837.

Kutser, T., Alikas, K., Kothawala, D.N., Köhler, S.J., 2015. Impact of iron associated to organic matter on remote sensing estimates of lake carbon content. Remote Sens. Environ. 156, 109–116. Available from: http://dx.doi.org/10.1016/j.rse.2014.10.002.

Le, C., Li, Y., Zha, Y., Sun, D., Huang, C., Lu, H., 2009. A four-band semi-analytical model for estimating chlorophyll a in highly turbid lakes: The case of Taihu Lake, China. Remote Sens. Environ. 113, 1175–1182. Available from: http://dx.doi.org/10.1016/j.rse.2009.02.005.

Lee, Z., Carder, K.L., Arnone, R.A., 2002. Deriving inherent optical properties from water color: a multiband quasi-analytical model for optically deep waters. Appl. Opt. 41 (27), 5755–5772. Available from: http://dx.doi.org/10.1364/AO.41.005755.

Li, L., Li, L., Song, K., Li, Y., Tedesco, L.P., Shi, K., et al., 2013. An inversion model for deriving inherent optical properties of inland waters: establishment, validation and application. Remote Sens. Environ. 135, 150–166. Available from: http://dx.doi.org/10.1016/j.rse.2013.03.031.

Matthews, M.W., Bernard, S., 2013. Characterizing the absorption properties for remote sensing of three small optically-diverse South African reservoirs. Remote Sens. 5, 4370−4404. Available from: http://dx.doi.org/10.3390/rs5094370.

Millennium Ecosystem Assessment (MEA), 2005. Ecosystems and Human Well-Being: Synthesis. Island Press, Washington, DC.

Mishra, S., Mishra, D.R., 2012. Normalized difference chlorophyll index: a novel model for remote estimation of chlorophyll-a concentration in turbid productive waters. Remote Sens. Environ. 117, 394−406. Available from: http://dx.doi.org/10.1016/j.rse.2011.10.016.

Mishra, S., Mishra, D.R., 2014. A novel remote sensing algorithm to quantify phycocyanin in cyanobacterial algal blooms. Environ. Res. Lett. 9, 114003. Available from: http://dx.doi.org/10.1088/1748-9326/9/11/114003.

Mishra, S., Mishra, D.R., Schluchter, W., 2009. A novel model for predicting phycocyanin concentrations in cyanobacteria: a proximal hyperspectral remote sensing approach. Remote Sensing 1 (4), 758−775. Available from: http://dx.doi.org/10.3390/rs1040758.

Mishra, S., Mishra, D.R., Lee, Z., Tucker, C.S., 2013. Quantifying cyanobacterial phycocyanin concentration in turbid productive waters: a quasi-analytical approach. Remote Sens. Environ. 133, 141−151. Available from: http://dx.doi.org/10.1016/j.rse.2013.02.004.

Mishra, S., Mishra, D.R., Lee, Z., 2014. Bio-optical inversion in highly turbid and cyanobacteria dominated waters. IEEE. Trans. Geosci. Remote. Sens. 52, 375−388. Available from: http://dx.doi.org/10.1109/TGRS.2013.2240462.

Mobley, C.D., 2001. Radiative transfer in the ocean. In: Steele, J.H. (Ed.), Encyclopedia of Ocean Sciences. Academic Press Elsevier, London, UK, pp. 2321−2330.

Mobley, C.D., Stramski, D., Bissett, W.P., Boss, E., 2004. Optical modeling of ocean waters: Is the Case 1-Case 2 classification still useful? Oceanography 17 (2), 60−67.

Morel, A., 2001. Bio-optical models. In: Steele, J.H. (Ed.), Encyclopedia of Ocean Sciences. Academic Press Elsevier, London, UK, pp. 317−326.

Morel, A., Gordon, H.R., 1980. Report of the working group on water color. Bound.-Layer Meteor. 18 (4), 343−355.

Morel, A., Prieur, L., 1977. Analysis of variation in ocean colour. Limnol. Oceanogr. 22, 709−722. Available from: http://dx.doi.org/10.4319/lo.1977.22.4.0709.

Moses, W.J., Gitelson, A.A., Berdnikov, S., Povazhnyy, V., 2009a. Estimation of chlorophyll-a concentration in case II waters using MODIS and MERIS data—successes and challenges. Environ. Res. Lett. 4/4, 045005. Available from: http://dx.doi.org/10.1088/1748-9326/4/4/045005.

Moses, W.J., Gitelson, A.A., Berdnikov, S., Povazhnyy, V., 2009b. Satellite estimation of chlorophyll-a concentration using the red and NIR bands of MERIS—the Azov Sea case study. IEEE Geosci. and Rem. Sens. Lett. 6 (4), 845−849.

National Research Council (NRC), 2002. The Drama of the Commons. National Academy Press, Washington, DC, USA.

O'Reilly, J.E., Maritorena, S., O'Brien, M.C., Siegel, D.A., Toole, D., Menzies, D., et al., 2000. SeaWiFS Postlaunch Calibration and Validation Analyses, Part 3, Volume 11. National Aeronautics and Space Administration, Washington, DC, USA.

Odermatt, D., Gitelson, A., Brando, V.E., Schaepman, M., 2012. Review of constituent retrieval in optically deep and complex waters from satellite imagery. Remote Sens. Environ. 118, 116−126. Available from: http://dx.doi.org/10.1016/j.rse.2011.11.013.

Ogashawara, I., 2015. Terminology and classification of bio-optical models. Remote Sens. Lett. 6 (8), 613−617. Available from: http://dx.doi.org/10.1080/2150704X.2015.1066523.

Ostrom, E., Stern, P.C., Dietz, T., 2003. Water rights in the commons. Water Resources IMPACT 5 (2), 9–12.

Peng, F., Effler, S.W., 2013. Spectral absorption properties of mineral particles in western Lake Erie: insights from individual particle analysis. Limnol. Oceanogr. 55 (5), 1775–1789. Available from: http://dx.doi.org/10.4319/lo.2013.58.5.1775.

Petzold, T.J., 1972. Volume Scattering Functions for Selected Ocean Waters (Ref. 72-28). Scripps Institute of Oceanography, University of California, San Diego, CA, USA.

Preisendorfer, R.W., 1976. Hydrologic Optics, Vol. II. Foundations. U.S. Dept. of Commerce, Washington, D.C., USA.

Schalles, J.F., Yacobi, Y.Z., 2000. Remote detection and seasonal patterns of phycocyanin, carotenoid, and chlorophyll pigments in eutrophic waters. Archi. Hydrobiol. 55, 153–168.

Smith, R.C., Baker, K.S., 1977. The Bio-Optical State if Ocean Waters and Remote Sensing. Scripps Institution of Oceanography, University of California, San Diego, CA, USA.

Simis, S.G.H., Peters, S.W.M., Gons, H.J., 2005. Remote sensing of the cyanobacterial pigment phycocyanin in turbid inland water. Limnol. Oceanogr. 50 (1), 237–245. Available from: http://dx.doi.org/10.4319/lo.2005.50.1.0237.

Song, K., Li, L., Tedesco, L., Clercin, N., Li, L., Shi, K., 2014. Spectral characterization of colored dissolved organic matter for productive inland waters and its source analysis. Chin. Geogra. Sci. 25 (3), 295–308. Available from: http://dx.doi.org/10.1007/s11769-014-0690-5.

Verpoorter, C., Kutser, T., Seekell, D.A., Tranvik, L.J., 2014. A global inventory of lakes based on high-resolution satellite imagery. Geophys. Res. Lett. 41, 6396–6402. Available from: http://dx.doi.org/10.1002/2014GL060641.

Vos, R.J., Hakvoort, J.H.M., Jordans, R.W.J., Ibelings, B.W., 2003. Multiplatform optical monitoring of eutrophication in temporally and spatially variable lakes. Sci. Total. Environ. 312, 221–243, http://dx.doi.org/10.1016/S0048-9697(03)00225-0.

Yacobi, Y., Gitelson, A., Mayo, M., 1995. Remote sensing of chlorophyll in Lake Kinneret using high spectral resolution radiometer and Landsat Thematic Mapper: spectral features of reflectance and algorithm development. J. Plankton Res. 17 (11), 1–19.

Chapter 2

Radiative Transfer Theory for Inland Waters

Peter Gege
Deutsches Zentrum für Luft- und Raumfahrt (DLR), Remote Sensing Technology Institute, Weßling, Germany

2.1 INTRODUCTION

When we enjoy the beauty of nature, it is the interaction of light with matter that is responsible for most of our perception. Light carries much of the information we utilize for recognizing and categorizing our world. Most objects in our environment do not emit light by themselves, but their perceived colors and contours are produced by reflected or scattered light originating from an external light source, for example the sun. The paths of light from the source to our eyes can be very complicated, and such can be the modification of the originally emitted radiation through the manifold interactions with matter before it reaches the eye. Radiative transfer (RT) theory is the physical approach to describe these paths and processes quantitatively.

In water, propagation of electromagnetic radiation is heavily hampered by absorption. Fig. 2.1 shows the absorption coefficient of pure water (Hale and Querry, 1973; Segelstein, 1981) and the related penetration depth specifying the distance after which 90% of incident radiation is absorbed. It can be seen that the penetration depth is very small in most spectral regions and exceeds 1 m only in the visible; the constituents present in natural waters decrease it further. Consequently, only this narrow range is relevant for biologic processes making use of electromagnetic radiation (vision and photosynthesis) and for remote sensing of objects below the water surface. For these reasons, this chapter restricts the discussion of RT theory and material properties to the visible, only a bit extended to the ultraviolet and infrared, and uses the terms *electromagnetic radiation* and *light* as synonyms. The term *optics* describes the properties and behavior of light and its interaction with matter.

RT theory describes the wavelength-dependent changes in direction and intensity of electromagnetic radiation as a function of density and physical properties of the interacting matter. The theory consists of a mathematical

Bio-optical Modeling and Remote Sensing of Inland Waters.
DOI: http://dx.doi.org/10.1016/B978-0-12-804644-9.00002-1

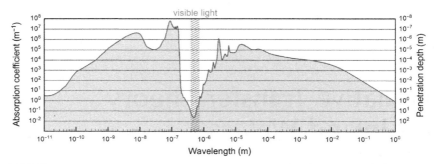

FIGURE 2.1 Absorption coefficient of pure water and related penetration depth for electromagnetic radiation.

framework and a set of material parameters. The mathematical formalism is known as radiative transfer equation (RTE), in which attenuation, scattering, and luminescence are the relevant physical processes linking the properties of the matter with the light field. The term luminescence summarizes emission processes like fluorescence and phosphorescence. Scattering and luminescence are responsible for changes in beam direction.

The comprehensive treatise of RT dates back to 1950, when Chandrasekhar first published his classical book "Radiative transfer" (Chandrasekhar, 1960). Chandrasekhar developed sophisticated mathematical methods to solve the RTE for planetary and stellar atmospheres, in particular to handle polarization, transfer in semi-infinite and finite atmospheres, and Rayleigh scattering. However, he did not consider inelastic processes like fluorescence and Raman scattering. While Chandrasekhar's methods are still in use in atmosphere physics and astrophysics, they have not found their way to aquatic sciences. A study by Klier (1972) is perhaps the only attempt to adapt Chandrasekhar's approach for the aquatic environment.

The systematic description of the RT processes in water begun in 1976 with the publication of Preisendorfer's six volumes of "Hydrologic Optics" (Preisendorfer, 1976) and Jerlov's book "Marine Optics" (Jerlov, 1976). These works established the physically sound backbone of oceanographic and limnologic optics. All nonempirical models and methods currently in use in these disciplines fit into these frameworks.

The RT in natural environments with the sun as light source is obviously a very complicated process. Thus, it is not surprising that the RTE has for most applications no analytical solution and requires approximations. Furthermore, the material parameters used in the RTE cannot easily be measured directly, and vice versa, many measurement quantities are not expressed explicitly in the RTE. Since each application has its specific challenges and can differ significantly from other disciplines concerning the relevant effects, material properties and measurement methods, the approximations and considered effects are application specific. This variety of

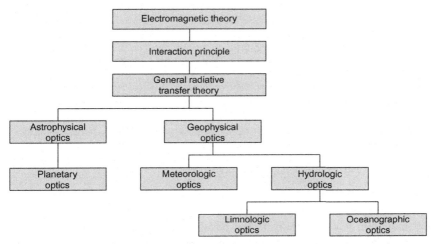

FIGURE 2.2 Preisendorfer's classification scheme of radiative transfer theory. *After Preisendorfer (1976).*

approaches to solve the RTE splits up the optics deduced from RT theory into different disciplines. A schematic overview of the descendance from electromagnetic theory to RT theory, and further down to the different applications is shown Fig. 2.2. While Chandrasekhar developed fundamental concepts for astrophysical, planetary and meteorologic optics, the works of Preisendorfer and Jerlov covered mainly hydrologic optics.

The chapter provides an overview about the physical principles of interaction of light with matter and the RT in water, and it describes the deduced bio-optical models forming the backbone of data analysis for optical in situ measurements and remote sensing. The reader interested in more details is referred to the textbooks about hydrologic optics such as Preisendorfer (1976), Jerlov (1976), Shifrin (1988), Zimmermann (1991), Mobley (1994), Bukata et al. (1995), Kirk (2011), and Mobley et al. (2017).

2.2   BASIC PRINCIPLES

2.2.1   Interaction of Light with Matter

The nature of light was discussed controversial for long time. The first plausible explanation was Huygens' wave theory from 1678, which explained the laws of optics by wave propagation. Newton published in 1704 his famous work "Opticks" with an alternate explanation of the same laws by propagation of small discrete particles. In the late 19th century, Maxwell explained light as the propagation of electromagnetic waves. Hertz verified in 1887 the validity of Maxwell's equations by experiment. This ambiguity of theories and observations initiated fruitful discussions about the concept of matter

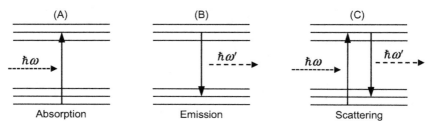

FIGURE 2.3 Term schemes illustrating the fundamental processes for the interaction of light with matter. The *y*-axis is energy. Horizontal lines represent the quantized energy states of matter, vertical arrows the transition of an electron. The horizontal arrows represent absorbed and emitted photons with energies of $\hbar\omega$ and $\hbar\omega'$, respectively.

and radiation, leading finally to the development of quantum mechanics in the early 20th century and a consistent theory for the interaction of matter with light. The elementary interaction processes are illustrated in Fig. 2.3.

Fundamental for the development of quantum mechanics was the hypothesis of Planck in 1900 that matter cannot change energy in any arbitrary way, but can absorb and emit only certain amounts (lat. *quantum*) of energy, i.e., the energy of material objects is quantized. Fig. 2.3 illustrates this hypothesis: the horizontal lines represent the allowed energies, and the gaps represent the energies that cannot be carried by the object. Einstein then published in 1905 the hypothesis that light itself is quantized. Both hypotheses have been validated since then by many experiments and form the basis for our current understanding of radiation and matter, and for their interaction. In modern quantum field theory, each fundamental type of interaction is described by a field, with particles (called *gauge bosons*) and waves being complementary descriptions of the force carriers. The particles associated to the light field, *photons*, are the gauge bosons of electromagnetism.

Each photon carries a certain amount of energy,

$$\varepsilon = \hbar\omega = hc/\lambda \,[\text{J}] \tag{2.1}$$

with ω frequency [s^{-1}], λ wavelength [m], $h = 6.6 \times 10^{-34}$ J s the Planck constant, $\hbar = h/2\pi$, and $c = 3 \times 10^{8}$ m s^{-1} the speed of light in vacuum. Light can only interact with matter if $\hbar\omega$ corresponds to an allowed energy difference of the object, and all electromagnetic radiation emitted by the object has energies $\hbar\omega'$ representing the differences of allowed energy levels. All interaction can be reduced to the processes absorption and emission (Fig. 2.3A,B). Emission of a photon is frequently stimulated by an absorption process (Fig. 2.3C). If absorbed and emitted photons have identical energy, the process is called elastic, otherwise inelastic. Examples for inelastic processes are fluorescence and Raman scattering. Many measurement quantities like reflectance, transmission, and attenuation represent second-order physical processes and can be lead back to the first-order processes of absorption and emission.

The term schemes of Fig. 2.3 illustrate in a simplified way the allowed energies of a molecule. The energy levels are grouped into two regions separated by a large gap. These groups represent the electronic ground state and the first electronically excited state. Each electronic state allows certain vibrational movements of the molecule, constituting the electronic fine structure. The Pauli Exclusion Principle permits only two electrons with opposite spin per energy level, thus transitions are only possible if the target level is not yet occupied. Photons in the near ultraviolet and visible have sufficient energy $\hbar\omega$ to raise an electron from the molecule's ground state to another electronic state (Fig. 2.3A). Transitions within the fine structure are linked to interaction with other molecules and to absorption and emission of infrared radiation, thus the population of the energy levels depends on density and temperature. Spectroscopy allows uncovering the current electronic configuration.

The equation behind Fig. 2.3C, which contains the complete information about two-photon interactions of an atom or molecule, is the Kramers—Heisenberg scattering tensor (Hassing and Nørby Svendsen, 2004; Siebert and Hildebrandt, 2008)

$$[\alpha_{if}]_{\rho\sigma} = \frac{1}{h}\sum_l \left(\frac{<f|M_\rho|l><l|M_\sigma|i>}{\omega_l - \omega' - \omega - i\Gamma_l} + \frac{<f|M_\sigma|l><l|M_\rho|i>}{\omega_l - \omega' + \omega + i\Gamma_l}\right). \quad (2.2)$$

It describes quantitatively the probability for absorption, scattering, and luminescence. The symbols i, l, and f refer to the initial, intermediate and final states of the molecule, ρ, σ to polarization, and M_ρ, M_σ are dipole moments. The bracketed terms in the numerators describe, in Dirac's notation, molecule specific probabilities for absorption and emission of photons. ω and ω' denote the frequencies of incident and scattered light, respectively, and ω_l the frequency of a molecular vibration. Γ_l is a damping constant that is related to the lifetime of the vibrational intermediate state l. The sum indicates that the scattering tensor is controlled by the transition probabilities involving all vibrational states.

Scattering consists of Rayleigh ($\omega = \omega'$, $i = f$) and Raman ($\omega \neq \omega'$, $i \neq f$) scattering. In both cases, the secondary photon is emitted prior to the dephasing of the intermediate state ($\rho = \sigma$), hence the intensity of scattered light depends on polarization of the incident beam. All light appearing after the dephasing of the intermediate state ($\rho \neq \sigma$) is termed *luminescence* and consists of "hot" luminescence when it originates from the intermediate state (resonance fluorescence), or from states not yet in thermal equilibrium. "Ordinary" luminescence (fluorescence, phosphorescence) occurs when light originates from thermally equilibrated states less energetic than the intermediate level (Champion and Albrecht, 1982).

The macroscopic material properties can be derived from the scattering tensor [Eq. (2.2)] by a suitable averaging over the contributions from the various molecules. This averaging takes different forms depending on whether we consider a coherent process like Rayleigh scattering or an incoherent process like Raman scattering. For coherent scattering the scattering tensor must be averaged with respect to molecular orientation, while for incoherent scattering, it is the square of the tensor, which must be averaged, since in that case each molecule scatters independently of the others (Hassing and Sonnich Mortensen, 1980).

2.2.2   Radiometric Quantities

Radiometry is the science of the measurement of electromagnetic energy (Mobley, 1994). All radiometric quantities can be led back to energies and fluxes of photons. The *radiant flux* is the energy of N photons per unit of time, t:

$$\phi(\lambda) = \frac{N\varepsilon(\lambda)}{t} [\text{J s}^{-1} = \text{W}]. \tag{2.3}$$

The energy of a photon depends on its wavelength λ according to Eq. (2.1), hence all radiometric quantities are wavelength-dependent. This λ dependence is omitted in many equations of Section 2.2 for brevity.

To define geometric relationships, it is convenient to use a spherical coordinate system with origin in an infinitesimal small areal element dA. The zenith angle θ and the azimuth angle φ of a beam's optical axis are defined according to Fig. 2.4. The *solid angle* Ω is the ratio of a small areal element S of a sphere's surface to the square of sphere radius r:

$$\Omega = \frac{S}{r^2} [\text{sr}]. \tag{2.4}$$

The radiant flux inside that solid angle,

$$I = \frac{d\phi}{d\Omega} [\text{W sr}^{-1}], \tag{2.5}$$

is called *radiant intensity*, and the radiant flux per areal element,

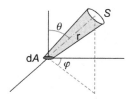

FIGURE 2.4   Definition of geometry parameters.

$$E = \frac{d\phi}{dA} [\text{W m}^{-2}], \quad (2.6)$$

is termed *irradiance*. The radiant intensity incident on dA from direction (θ, φ) or emitted in that direction is called *radiance*:

$$L(\theta, \varphi) = \frac{dI(\theta, \varphi)}{dA} = \frac{d^2\phi(\theta, \varphi)}{dA \, d\Omega} [\text{W m}^{-2}\text{sr}^{-1}]. \quad (2.7)$$

In order to have well defined geometries for relating E with L, two special surface geometries are commonly distinguished, i.e., plane and spherical surfaces, and the radiant flux is divided into upward and downward directions, indicated by the subscripts "*u*" and "*d*". For a plane surface, the flux per area is proportional to $|cos\theta|$, hence the hemispherical integrals of radiance are

$$E_u = \int_0^{2\pi} \int_{\pi/2}^{\pi} L(\theta, \varphi)|\cos \theta|\sin \theta \, d\theta \, d\varphi, \quad (2.8)$$

$$E_d = \int_0^{2\pi} \int_0^{\pi/2} L(\theta, \varphi)|\cos \theta|\sin \theta \, d\theta \, d\varphi. \quad (2.9)$$

These are called *plane irradiances*. For a spherical surface, the flux per area is independent from surface orientation, and the corresponding angular integrals of radiance are called *scalar irradiance*:

$$E_{0u} = \int_0^{2\pi} \int_{\pi/2}^{\pi} L(\theta, \varphi) \sin \theta \, d\theta \, d\varphi, \quad (2.10)$$

$$E_{0d} = \int_0^{2\pi} \int_0^{\pi/2} L(\theta, \varphi) \sin \theta \, d\theta \, d\varphi. \quad (2.11)$$

Plane irradiances are the appropriate quantities for describing the radiation incident on a horizontally extended layer, while the illumination of randomly oriented particles is better approximated by scalar irradiances.

2.2.3 Radiative Transfer Equation

A precise formulation of the RT in an absorbing and scattering medium is given by the classical RTE (Jerlov, 1976; Bukata et al., 1995; Kirk, 2011)

$$\frac{dL(\lambda, z, \theta, \varphi)}{dr} = -c(\lambda, z)L(\lambda, z, \theta, \varphi) + L^*(\lambda, z, \theta, \varphi) + L_S^*(\lambda, \lambda', z, \theta, \varphi). \quad (2.12)$$

It describes the change of radiance, dL, of a light beam of wavelength λ traveling a small distance dr in a medium at depth z in the direction (θ, φ). The first term on the right represents loss by attenuation, the second gain by elastic scattering, and the third gain by luminescence. In case of a plane parallel medium, it is $dr = dz/\cos \theta$, and the equation can be rewritten as

$$\cos \theta \frac{dL(\lambda, z, \theta, \varphi)}{dz} = -c(\lambda, z)L(\lambda, z, \theta, \varphi) + L^*(\lambda, z, \theta, \varphi) + L_S^*(\lambda, \lambda', z, \theta, \varphi).$$

(2.13)

The term $L^*(\lambda, z, \theta, \varphi)$, called the path function, involves every volume element in the medium as a source of scattering. It is given by

$$L^*(\lambda, z, \theta, \varphi) = \int_0^{2\pi} \int_0^\pi \beta(\lambda; \theta, \varphi; \theta', \varphi')L(\lambda, z, \theta', \varphi')\sin \theta' d\theta' d\varphi'.$$

(2.14)

The volume scattering function $\beta(\lambda; \theta, \varphi; \theta', \varphi')$ defines the probability that radiance will be scattered from its initial directions (θ', φ') to the direction (θ, φ) without change in wavelength. Transspectral processes like Raman scattering and fluorescence, and emission processes like bioluminescence are summarized in the source function $L_S^*(\lambda, \lambda', z, \theta, \varphi)$, which is given by (Mobley, 1994)

$$L_S^*(\lambda, \lambda', z, \theta, \varphi) = \int_0^{2\pi} \int_0^\pi \beta_S(\lambda, \lambda'; \theta, \varphi; \theta', \varphi')L(\lambda', z, \theta', \varphi')\sin \theta' d\theta' d\varphi'. \quad (2.15)$$

The volume scattering function for source processes, $\beta_S(\lambda, \lambda'; \theta, \varphi; \theta', \varphi')$, describes the gain of radiance of wavelength λ from different origin like excitation at wavelength λ' or conversion of nonradiant energy to light, and specifies its geometric distribution (θ, φ) for the sources emitting in the directions (θ', φ').

The RTE is an integro-differential equation and cannot be solved analytically, hence application specific approaches are necessary that are adapted to the illumination geometry and to the shape, consistency and composition of the considered medium. In meteorologic optics, the challenge lies in the parameterization of the atmosphere, while illumination geometry is well defined by the sun as light source. In hydrologic optics, however, a mathematically useful description of the medium water is only one of the challenges. The major challenge is here the illumination geometry. The incident direction is not just given by the sun position like in meteorologic optics, but the sky and eventually clouds illuminate the water surface hemispherical.

As challenging as is already modeling the radiance distribution of the sky, it is even more challenging to model the incident radiation field under water due to the dynamic water surface. Waves, ripples, and foam refract the rays from the sun, the sky, and eventually the clouds in ever changing directions, and the surface reflects furthermore a large portion of the underwater upwelling radiance back in downward directions. Since the details of atmospheric conditions and the water surface geometry are not known in practice, only approximate solutions of the RTE can be expected in hydrologic optics, and their computation requires coupled models of the atmosphere, the water surface, the water body, and in case of shallow waters, additionally of the bottom. Zhai et al. (2009) provide an overview of the developed methods to solve the RTE and the RT codes currently available for a coupled atmosphere-ocean system.

2.2.4 Inherent Optical Properties

Preisendorfer (1961) has introduced the terms inherent and apparent optical property (IOP, AOP) to distinguish between material properties which are independent (IOP) and dependent (AOP) on the illumination and observation geometries. The IOPs are defined with the help of an imaginary parallel layer of medium of infinitesimal thickness dz, illuminated at right angle by a parallel beam of light with incident radiant flux ϕ (Fig. 2.5). Only two things can happen to the photons inside the layer: they can be absorbed or scattered. The losses are quantified by the absorbed flux, $d\phi_a$, and the scattered flux, $d\phi_b$. The energy loss of the transmitted beam is $d\phi_a + d\phi_b$, i.e., the transmitted flux is $\phi_t = \phi - d\phi_a - d\phi_b$.

The *absorption coefficient* is defined as the absorbed flux per length relative to the incident flux:

$$a = \frac{1}{\phi}\frac{d\phi_a}{dz}\,[\mathrm{m}^{-1}]. \tag{2.16}$$

Similarly, the *scattering coefficient* is the scattered flux per length relative to the incident flux:

$$b = \frac{1}{\phi}\frac{d\phi_b}{dz}\,[\mathrm{m}^{-1}]. \tag{2.17}$$

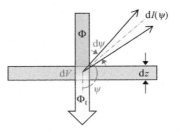

FIGURE 2.5 Interaction of a beam of light (radiant flux ϕ) with a thin layer of medium.

The sum of both is the *beam attenuation coefficient*:

$$c = a + b = \frac{1}{\phi}\frac{d\phi_a + d\phi_b}{dz} = \frac{1}{\phi}\frac{\phi - \phi_t}{dz} \, [\text{m}^{-1}]. \tag{2.18}$$

The ratio

$$\omega_0 = \frac{b}{c}, \tag{2.19}$$

known as *single-scattering albedo*, is the probability that a photon will be scattered rather than absorbed (Mobley, 1994).

The coefficients *a*, *b*, and *c* are sufficient to describe the material properties for the interaction of light in one dimension, i.e., for a parallel beam of light. However, scattering adds two more dimensions to the process, as photons can be scattered out of beam direction. The angle ψ between the directions of incident and scattered beam is called *scattering angle* (Fig. 2.5).

Let d*V* denote an infinitesimal small volume element illuminated with irradiance *E*. The radiant intensity d$I(\psi)$ scattered from that volume element into direction ψ, normalized by the incident irradiance *E*, is called *volume scattering function*:

$$\beta(\psi) = \frac{1}{E}\frac{dI(\psi)}{dV} \, [\text{m}^{-1}\text{sr}^{-1}]. \tag{2.20}$$

Integration of $\beta(\psi)$ over all scattering angles yields the *scattering coefficient*:

$$b = \int_0^{2\pi}\int_0^{\pi} \beta(\psi) \sin\psi \, d\psi \, d\varphi = 2\pi \int_0^{\pi} \beta(\psi) \sin\psi \, d\psi. \tag{2.21}$$

Since the scatterers in water are usually oriented randomly, scattering is symmetric in azimuth direction, i.e., β does not depend on the azimuth angle φ. The ratio

$$\tilde{\beta}(\psi) = \frac{\beta(\psi)}{b} \, [\text{sr}^{-1}] \tag{2.22}$$

is called *scattering phase function* (Mobley, 1994). Since it is a normalized function,

$$2\pi \int_0^{\pi} \tilde{\beta}(\psi)\sin\psi \, d\psi = 1, \tag{2.23}$$

it is a useful parameter for separating the volume scattering function into the intensity (*b*) and into the angular pattern ($\tilde{\beta}(\psi)$): $\beta(\psi) = b \cdot \tilde{\beta}(\psi)$. Its cosine weighted average over all scattering directions,

$$g = 2\pi \int_0^{\pi} \tilde{\beta}(\psi) \cos\psi \sin\psi \, d\psi, \tag{2.24}$$

called *asymmetry parameter*, is a convenient measure of its shape. Values of g can range from -1 for complete backward scattering to $+1$ for complete forward scattering, with values in the range of 0.8 to 0.95 being typical for oceanic waters (Mobley et al., 2017).

The RTE requires the knowledge of c and $\beta(\theta, \varphi; \theta', \varphi')$ to model the propagation of light. Illumination geometry and viewing direction determine the relationship between the scattering angle ψ and the directions of incidence, (θ', φ'), and observation, (θ, φ). Like any IOP, $\beta(\psi)$ is additive, i.e., the β's of water and the different suspended materials have to be added to obtain the effective $\beta(\psi)$ of the water body. Inelastic processes like fluorescence and Raman scattering require in addition the appropriate function $\beta_S(\lambda, \lambda', \psi)$ describing besides the geometric also the spectral re-distribution of radiance from excitation wavelengths λ' to emission wavelengths λ. Consistent definitions are obtained by replacing in Eq. (2.20) $dI(\psi)$ with the corresponding emission radiant intensity. The volume scattering for fluorescence, $\beta_F(\lambda, \lambda', \psi)$, and its relationship to the fluorescence radiance are outlined in Section 2.4.4, the Raman volume scattering function is discussed in Schroeder et al. (2003).

The bio-optical models described below make use of further IOPs. Most common is the *backscattering coefficient*, which is the integral of all photons scattered in backward directions, i.e., at angles between $\pi/2$ and π:

$$b_b = 2\pi \int_{\pi/2}^{\pi} \beta(\psi) \sin \psi \, d\psi. \tag{2.25}$$

Likewise, the *forward scattering coefficient* summarizes all photons scattered at angles between 0 and $\pi/2$:

$$b_f = 2\pi \int_{0}^{\pi/2} \beta(\psi) \sin \psi \, d\psi. \tag{2.26}$$

The sum of both is the scattering coefficient:

$$b = b_f + b_b. \tag{2.27}$$

It should be noted that in hydrologic optics literature some of the definitions (2.20) to (2.26) are sometimes expressed in terms of the zenith angle θ instead of the scattering angle ψ, in particular b_b in bio-optical models. This is motivated by the convenience of coordinate systems in which θ distinguishes the two hemispheres separated by the water surface. Such definition makes the parameter however to an AOP and requires correction for the angle of incidence (Sathyendranath and Platt, 1997).

2.2.5 From Microscopic to Macroscopic Material Parameters

The relationships between macroscopic material properties and microscopic molecule parameters are depicted in Fig. 2.6. This scheme shows how the

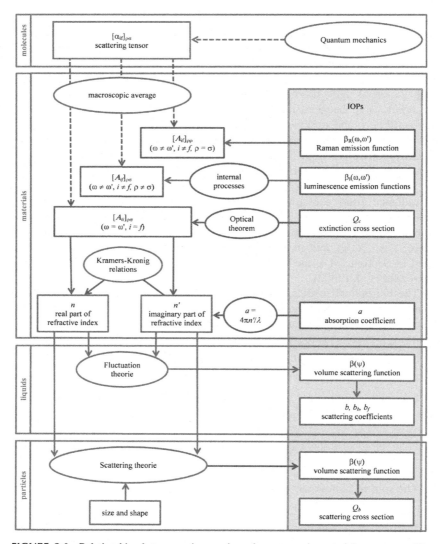

FIGURE 2.6 Relationships between microscopic and macroscopic material parameters. The parameters are represented by *boxes*, the models by *ellipses*, and the most common directions of obtaining a parameter by *arrows*. Viable steps are indicated by solid lines, generally impracticable computations by dashed lines.

different IOPs are connected with the scattering tensor $[\alpha_{if}]_{\rho\sigma}$ given by Eq. (2.2), and how they are related to each other. The numerical values of the tensor elements can be obtained from quantum mechanics with sufficient accuracy only for few very simple molecules.

Averaging the contributions of many molecules to obtain macroscopic material properties (denoted $[A_{if}]_{\rho\sigma}$ in Fig. 2.6) is difficult, since besides electronic energy transfer a number of further processes have to be taken into consideration, for example thermal excitations and collisions. Furthermore, surrounding molecules can alter the potential curve and geometry of the molecule and thus change the energy levels. This makes it necessary to determine macroscopic material properties experimentally.

Material properties responsible for inelastic scattering processes ($\omega \neq \omega'$) can be obtained from spectral measurements of Raman scattering, fluorescence, and phosphorescence, while those representing elastic processes ($\omega = \omega'$) can be determined by measuring extinction. If s_0 denotes a unit vector in the direction of the incident wave, $e = (e_\varrho, e_\sigma)$ the complex vector amplitude of the incident field, and $[A_{ii}]_{\rho\sigma}(s_0, s_0)$ the scattering tensor amplitude in the direction of the incident beam, the extinction cross section is given by (Born and Wolf, 1999; Lytle II et al., 2005)

$$Q_c = \frac{\omega}{2} Im \sum_{i,\varrho,\sigma} \frac{e_\varrho e_\sigma^*}{|e|^2} [A_{ii}]_{\rho\sigma}(s_0, s_0).$$ (2.28)

Im denotes the imaginary part. The sum term is called forward scattering amplitude (Born and Wolf, 1999). Eq. (2.28) is known as optical theorem.

The elementary macroscopic material property for elastic processes is the *complex refractive index*

$$m = n - in'.$$ (2.29)

Its real part, n, determines the phase lag of the wave traveling through the medium, changing the phase velocity of the light wave, c, to c/n in the medium (van de Hulst, 1957). Its imaginary part, n', describes the wave's decrease of intensity in forward direction. For stratified homogeneous media, n' is proportional to the absorption coefficient (van de Hulst, 1957),

$$a = \frac{4\pi n'}{\lambda}.$$ (2.30)

Complex numbers in physics always imply a strong relationship between the real and the imaginary part. This connection is here the scattering tensor of Eq. (2.2), which expresses absorption and emission as inverse processes from the perspective of a molecule. The theory of complex functions in conjunction with the principle of causality leads to fundamental relationships between the real and imaginary part of physical quantities, known as Kramers-Kronig (KK) relations. In particular, the following KK relation can be helpful to derive n from the much easier measurable parameter n' (Lucarini et al., 2005):

$$n(\omega) = 1 + \frac{2}{\pi} P \int_0^\infty \frac{\omega' \cdot n'(\omega')}{\omega^2 - \omega'^2} d\omega'.$$ (2.31)

P is the principal value of the integral, i.e., the contribution at the pole $\omega = \omega'$ is omitted. Note that no other parameter besides $n'(\omega)$ must be known to get $n(\omega)$. Cheng and Weng (2012) illustrate the usefulness of Eq. (2.31) by deriving the refractive index of water from absorption. However, Eq. (2.31) requires the knowledge of n' for all wavelengths to obtain exact results. If n is known for one or more wavelengths, this problem can be reduced using *subtracted* KK relations (Palmer et al., 1998).

The origin of light scattering in liquids was revealed between 1908 and 1910 by Einstein and Smoluchowski. Their model, known as fluctuation theory, makes density fluctuations responsible for the scattering processes, and provides the following parameterization of the volume scattering function (Morel, 1974; Zhang and Hu, 2009):

$$\beta(\psi) = \frac{2\pi^2}{\lambda^4} kTn^2 \frac{1}{\eta} \left(\frac{\partial n}{\partial p}\right)_T^2 \frac{6+6\delta}{6-7\delta} \left(1 + \frac{1-\delta}{1+\delta}\cos^2\psi\right). \tag{2.32}$$

k is the Boltzmann constant, T temperature, n the real part of the refractive index, η isothermal compressibility, p pressure, and δ the fraction of unpolarized light for scattering at $90°$. β is proportional to λ^{-4} only by approximation since n and δ are wavelength dependent. Many liquids are accurately described by this model, but in case of water, the measured temperature and pressure dependencies are not consistent to model predictions. Hence, empirical relationships are used for these.

Light scattering by particles depends on their size and geometry and can be very complicated, in particular if the particles or surface structures are in the order of the wavelength. In such cases, the scattered waves can show complex interference patterns. Their computation is based on Maxwell's equations. Exact solutions exist only for a few simple shapes like sphere, spheroid, cylinder and ellipsoid.

The particles in water are frequently approximated by small spheres. This reduces the complicated scattering theory to the simpler Mie theory (Hergert and Wriedt, 2012) and allows applying the analytical equations of van de Hulst's (1957) anomalous diffraction approximation for computation. In such way, fundamental material properties can be determined by combining Mie theory with measurement. For example, the extinction cross section of Eq. (2.28) is given in Mie theory for nonabsorbing particles by the simple equation

$$Q_c = 2 - \frac{4}{\rho}\sin\rho + \frac{4}{\rho^2}(1 - \cos\rho) \tag{2.33}$$

with

$$\rho = \frac{2\pi d}{\lambda}(n - 1). \tag{2.34}$$

The parameter ρ represents the phase shift of light with wavelength λ passing through a particle of diameter d relative to that passing around it (Bryant et al., 1969). These equations allow deriving the material parameter Q_c from measurements of d and n. Q_c is related to the beam attenuation coefficient c as follows (Morel and Bricaud, 1981):

$$c = \frac{N}{V} s Q_c. \tag{2.35}$$

N is the number of particles with geometric cross section s ($= \pi d^2/4$ for spheres) per volume V. The absorption and scattering coefficients a and b are related analogously to the corresponding material properties Q_a and Q_b.

2.3 BIO-OPTICAL MODELS

To acknowledge the fact that in many oceanic environments the optical properties of water bodies are essentially subordinated to the biological activity, and ultimately to phytoplankton and their derivatives, the expression "bio-optical state of ocean waters" was coined in 1978 (Morel, 2001). For this historical reason, the equations describing the optical properties of a water body are called *bio-optical models*.

2.3.1 Water Composition

Natural waters contain a huge number of molecules and particles differing in size, shape, and chemical composition. Besides inorganic components like salts, quartz sand, clay minerals, and metal oxides, a complete ecosystem of living organisms and dead organic material is observed in most open waters. Gas bubbles of various sizes are also part of the aquatic world.

The large variety of water constituents makes it necessary to define classes with comparable IOPs. Four classes are usually distinguished: total suspended matter (TSM), colored dissolved organic matter (CDOM), phytoplankton, and detritus, the nonalgal fraction of TSM. TSM dominates scattering, the other classes mainly affect absorption. The most relevant IOPs for remote sensing are the absorption coefficient, a, and the backscattering coefficient, b_b. These are frequently factorized in bio-optical models as follows:

$$a(\lambda) = a_w(\lambda) + C \cdot a^*_{phy}(\lambda) + Y \cdot a^*_{CDOM}(\lambda) + D \cdot a^*_d(\lambda), \tag{2.36}$$

$$b_b(\lambda) = b_{b,w}(\lambda) + X \cdot b^*_{b,TSM}(\lambda). \tag{2.37}$$

$a_w(\lambda)$ and $b_{b,w}(\lambda)$ are IOPs of pure water. The symbols C, Y, D, and X represent groups of components with similar optical properties, and the associated spectra $a^*_{phy}(\lambda), a^*_{CDOM}(\lambda), a^*_d(\lambda), b^*_{b,TSM}(\lambda)$ are class specific normalized IOP averages. The star symbol indicates normalization to concentration (common for phytoplankton and TSM) or wavelength (common for CDOM and detritus). Concentration normalized IOPs are called *specific inherent optical properties* (SIOPs).

Such grouping introduces natural variability to the SIOPs, since the IOPs of the individual components forming a class are generally not identical and their relative abundances can change significantly. Models can account for this SIOP variability by increasing the number of classes, by selecting site or season specific SIOPs from a database, or by using models relating the SIOPs to environmental conditions.

2.3.1.1 Phytoplankton

Being the first link in the aquatic food chain, phytoplankton is the most important water constituent for ecologically and biologically oriented aquatic sciences. Many disciplines use its "fingerprint molecule" chlorophyll-*a* (chl-*a*) as a measure of concentration since it is an inevitable molecule in the photosynthesis process and thus present in all species. Chl-*a* concentration ([chl-*a*]) was mapped in lakes already in 1974 from aircraft and satellite (Strong, 1974), and it was the first parameter derived quantitatively from ocean color satellite sensors. The correlation of [chl-*a*] with other water constituents is the basis of water type classification in remote sensing: case-I waters are those where phytoplankton and its degradation products dominate the optical properties (i.e., all constituents are correlated to [chl-*a*]), and case-II waters are the others (Morel and Prieur, 1977; Gordon and Morel, 1983). Open oceans are usually case-I, and coastal and inland waters frequently case-II.

The focus on chl-*a* continues to the present, but phytoplankton composition is receiving increasing attention. The ocean color community has realized that capturing the role of phytoplankton in the global cycles of major and minor elements in the ocean (e.g., carbon, nitrogen, sulfur, and iron) requires improved representation of their complex functionalities (IOCCG, 2014). In inland waters, the raising interest in phytoplankton classification is mainly motivated by the increasing number of blooms including toxic cyanobacteria affecting water usage for drinking, irrigation, and recreational purposes (Dörnhöfer and Oppelt, 2016).

The phytoplankton community of a typical lake consists of hundreds of species, most of them with similar optical properties. Perhaps the first optical classification for integration into a bio-optical model was made for Lake Constance (Gege, 1998). Four spectral classes could be distinguished (cryptophyta, diatoms, green algae, and dinoflagellates) and their concentrations determined if these were above certain thresholds. The chl-*a* specific absorption coefficients $a_{phy}^*(\lambda)$ of these classes were derived from measurements of pure cultures grown in the laboratory (Fig. 2.7). A problem for bio-optical modeling is the ability of cryptophyta to adapt their pigment inventory to

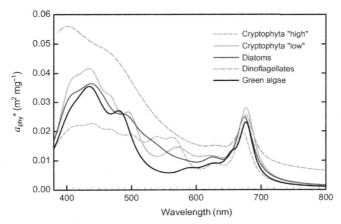

FIGURE 2.7 Chl-*a* specific absorption coefficients of phytoplankton. *Adapted from Gege (1998) with modifications described in Gege (2012b).*

environmental conditions, which makes their optical properties quite variable; it was tackled by representing the class with two absorption spectra. Also cyanobacteria have this ability, called chromatic adaption, making remote sensing of such classes challenging.

Due to the omnipresence of chl-*a*, bio-optical models commonly express phytoplankton concentration C in terms of chlorophyll-*a* concentration [chl-*a*]. Phytoplankton cells yet contain further light absorbing pigments, some unique to certain phytoplankton taxa and useable as unambiguous markers (Nair et al., 2008). These pigmentation differences are one reason for the variability of $a_{phy}^*(\lambda)$. Another reason is the packaging effect (Duysens, 1956) which leads to a flattening of the absorption spectrum with increasing cell size (Kirk 1975a, 1975b). Bio-optical modeling usually accounts for the natural variability of $a_{phy}^*(\lambda)$ by combining linearly the contributions of N spectra $a_{phy,i}^*(\lambda)$ from a database,

$$C \cdot a_{phy}^*(\lambda) = \sum_{i=1}^{N} C_i \cdot a_{phy,i}^*(\lambda), \tag{2.38}$$

with C_i the concentration of class i and $C = \Sigma\, C_i$ the total chl-*a* concentration. The "end members" $a_{phy,i}^*(\lambda)$ are however frequently not well known, and the linear approach cannot describe well the nonlinear packaging effect. Bricaud et al. (1995) have derived a nonlinear empirical equation for the changes of $a_{phy}^*(\lambda)$ with concentration C:

$$a_{phy}^*(\lambda) = A(\lambda) \cdot C^{-B(\lambda)}. \tag{2.39}$$

It is based on 815 $a_{phy}^*(\lambda)$ measurements from different regions of the world ocean, covering the [chl-a] range 0.02–25 mg m^{-3}. The author is not aware of a validation of Eq. (2.39) for inland waters or of alternate nonlinear approaches to tackle the variability of $a_{phy}^*(\lambda)$.

2.3.1.2 CDOM

While phytoplankton is usually the main absorber in the open ocean, absorption of inland and coastal waters is frequently dominated by CDOM. It is the colored fraction of the water constituents passing a filter with a pore size of about 0.2 μm (traditionally 0.4 μm). Historical names are, besides others, gelbstoff, and yellow matter. Most of its compounds are produced during the decay of plant matter and consist of various phenolhumic (humic) acids and carbohydrate humic (fulvic) acids or melanoids. Kirk (2011) provides an overview of CDOM concentration ranges (his Table 3.2): absorption at 440 nm ranges from ~0 to 0.16 m^{-1} in the open oceans, but from 0.004 to 3.82 m^{-1} in coastal areas and from 0.06 to 19.1 m^{-1} in inland waters. Many inland waters are dominated by the soluble humic fraction leached from the soils in the catchment area (Kirk, 2011).

Absorption of humic and fulvic substances is mainly caused by aromatic groups with various degrees and types of substitution (Korshin et al., 1997). The absorption maxima of these groups, called chromophores, lie in the UV, e.g., at 180, 203, and 253 nm for benzene (Korshin et al., 1997) which is a common aromatic ring structure in humic matter. Since these wavelengths are outside the range of interest for hydrologic optics, most measurements cover only the longwave tail of these peaks above 300 nm. This tail is usually approximated by an exponential function:

$$a_{CDOM}^*(\lambda) = \exp\{-S\cdot(\lambda - \lambda_0)\}. \tag{2.40}$$

Such normalization at a reference wavelength λ_0 defines CDOM concentration in units of absorption at λ_0:

$$Y = a_{CDOM}(\lambda_0)[\text{m}^{-1}]. \tag{2.41}$$

The slope S depends on CDOM composition. Values ranging from 0.004 to 0.053 nm^{-1} were observed (Aas, 2000), but in most cases S is in the range between 0.010 nm^{-1} for humic acid dominated waters and 0.020 nm^{-1} when fulfic acids prevail (Bricaud et al., 1981; Zepp and Schlotzhauer, 1981; Carder et al., 1989; Dekker and Peters, 1993; Schwarz et al., 2002; Babin et al., 2003a,b; Laanen, 2007; Binding et al., 2008; Kirk, 2011).

It is frequently assumed that S is practically constant in the near UV and visible. Højerslev and Aas (2001) have reviewed the literature about this assumption and recap that there exist measurements supporting this assumption, but many other disprove it. They conclude that there seems to be a tendency for S to decrease with increasing wavelength, possibly by 10%–15%

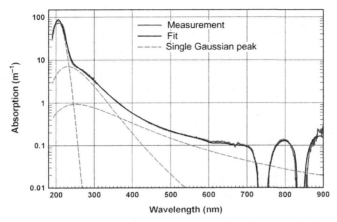

FIGURE 2.8 Typical CDOM absorption spectrum from Lake Constance. The fit curve (*red*, dark gray in print versions) is the sum of three Gaussians (*green*, light gray in print versions) after conversion from wavelengths to wavenumbers, plus a function accounting for temperature effects of pure water absorption. *From Gege (2000).*

from the UV to the visible spectral domain. Such S decrease is consistent to the observation that CDOM absorption peaks in the far UV, as the slope of a peaked function decreases with the distance from its maximum. In order to illustrate this peaked function, Fig. 2.8 shows an absorption measurement of CDOM for a wide wavelength range from the far UV (190 nm) to the near infrared (900 nm). A semilogarithmic plot was chosen to make clear that Eq. (2.40) is just an approximation for a restricted wavelength interval. From 300 to 700 nm, Eq. (2.40) represents a tangent to a convex curve at the chosen reference wavelength λ_0, thus it underestimates absorption at all other wavelengths in this range. Gege (2000) has derived an equation for the dependence of S on λ_0, and he determined the parameters of this $S(\lambda_0)$ model for Lake Constance. For this lake, the error of the approximation of Eq. (2.40) exceeds 10 % for $|\lambda - \lambda_0| > 60$ nm.

The $S(\lambda_0)$ equation of Gege (2000) is based on modeling CDOM absorption as the sum of two Gaussians centered at 233 ± 5 nm and 251 ± 11 nm with line widths of 6979 ± 575 cm^{-1} and 10650 ± 1916 cm^{-1}, respectively. It reproduced very well the range 270–900 nm of his measurements, while a third peak at 204.7 ± 1.0 nm was required for the range from 190 to 270 nm. However, as pointed out by Sipelgas et al. (2003), the 6 parameters of the two Gaussians cannot be determined reliably if the spectral range of the measurement does not capture the peaks. They suggest a hyperexponential function, $\exp(-\alpha\lambda^\eta)$, which is both a generalization of Eq. (2.40) ($\alpha = S$, $\eta = 1$) and of a Gaussian curve ($\eta = 2$). It fitted well their measurements in 7 Finnish and 13 Estonian lakes in the range 300–700 nm, while Eq. (2.40) led to a large dependence of S on the chosen spectral range.

As CDOM dominates absorption in many inland waters, an accurate spectral model of $a^*_{CDOM}(\lambda)$ is here arguably more important than for the open ocean, where Eq. (2.40) represents the established bio-optical model of CDOM absorption. Scattering of CDOM is small and commonly ignored, yet fluorescence may be significant and should be accounted for in RT models at high concentrations (Pozdnyakov et al., 2002). CDOM fluorescence is however not well studied in inland waters, and a bio-optical model is lacking.

2.3.1.3 Total Suspended Matter

All natural waters contain high numbers of particles: mineral particles derived from the land or from bottom sediments, phytoplankton, bacteria, dead cells and fragments of cells, etc. (Kirk, 2011). The sum of these components is called TSM or seston. TSM without phytoplankton is known as detritus, and the inorganic fraction as tripton.

The size distribution of TSM is roughly hyperbolic (Bader, 1970), i.e., the number of particles with a diameter greater than D is proportional to $D^{-\gamma}$, where γ varies widely from 0.7 to 6 in different water bodies (Jerlov, 1976). Although small particles are numerous, their low scattering efficiency makes scattering in natural waters frequently dominated by particles with $D > 2\ \mu m$ (Jerlov, 1976). The wavelength dependency of scattering of such particle distributions follows approximately the law of Angström:

$$b_{TSM}(\lambda) = b^*_{TSM} \cdot \left(\frac{\lambda}{\lambda_s}\right)^{-n}. \tag{2.42}$$

The Angström exponent n is typically between 0 and 1 in the open ocean and in clear coastal waters (Morel and Maritorena, 2001; Blondeau-Patissier et al., 2009), and around zero in shallow and inland waters (Babin et al., 2003a; Chami et al., 2005). Specific scattering coefficients b^*_{TSM} are on the order of 0.5 to 1.0 $m^2\ g^{-1}$ at $\lambda_S = 555$ nm (Babin et al., 2003a). The specific backscattering coefficient of Eq. (2.37) can be expressed in terms of $b_{TSM}(\lambda)$ as follows:

$$b^*_{b,TSM}(\lambda) = \tilde{b}_b \cdot b_{TSM}(\lambda). \tag{2.43}$$

The ratio of backscattering to total scattering, $\tilde{b}_b = b_b/b$, ranges from about 0.2% to 3% (Chami et al., 2005; Antoine et al., 2011). It is frequently assumed constant over the visible range (Ulloa et al. 1994), but some studies suggest a wavelength dependence (McKee and Cunningham, 2005). The power law of scattering and backscattering is only valid for a hyperbolic size distribution of nonabsorbing particles. Due to a lack of measurements, it is difficult to estimate potential deviations for inland waters.

While scattering of TSM depends more on particle size and concentration than on type, absorption differs significantly for the individual components. Phytoplankton absorption is characterized by a number of peaks attributed to specific pigments (Fig. 2.7). Detritus and tripton absorption can be approximated by an exponential equation alike CDOM absorption,

$$a_d^*(\lambda) = \exp\{-S_d \cdot (\lambda - \lambda_0)\}, \tag{2.44}$$

with a spectral slope typically less than for CDOM. Studies in coastal waters report S_d values of $0.0123 \pm 0.0013 \, \text{nm}^{-1}$ (Babin et al., 2003b) and $0.011 \, \text{nm}^{-1}$ (D'Sa et al., 2006). Detritus absorption has sometimes shoulders due to the breakdown products of photosynthetic pigments (Kirk, 2011). Tripton may include particles of various origins, such as continental dust deposited on the water by winds or volcanic deposits, thus its spectral shape can change significantly.

2.3.2   Apparent Optical Properties

The AOPs describe bulk material properties under natural illumination conditions in the field. They are defined in terms of ratios or depth derivatives of radiometric quantities, since these are much less dependent on the light field as the radiometric quantities themselves.

Light extinction by a water body is characterized by a number of attenuation coefficients, defined as the rates of change with depth of radiometric quantities. The *diffuse attenuation coefficient of irradiance* is defined as the depth derivative of irradiance:

$$K = -\frac{1}{E}\frac{dE}{dz}\,[\text{m}^{-1}], \tag{2.45}$$

and the *diffuse attenuation coefficient of radiance* is analogously given by

$$k(\theta, \varphi) = -\frac{1}{L(\theta, \varphi)}\frac{dL(\theta, \varphi)}{dz}[\text{m}^{-1}]. \tag{2.46}$$

Common variants of K are K_d, K_u, K_{Od} and K_{Ou}, in which E of Eq. (2.45) is replaced by the respective irradiances E_d, E_u, E_{Od}, and E_{Ou} defined in Section 2.2.2; and common variants of k are k_u and k_d with L of Eq. (2.46) being measured in nadir ($\theta = 0$) and zenith ($\theta = \pi$) direction, respectively.

Crude but useful one-parameter measures of the directional structures of the light fields are the *average cosines*

$$\overline{\mu}_d = \frac{E_d}{E_{Od}}, \quad \overline{\mu}_u = \frac{E_u}{E_{Ou}}, \quad \overline{\mu} = \frac{E_d - E_u}{E_{Od} + E_{Ou}}. \tag{2.47}$$

If the radiance distribution is isotropic, then $\overline{\mu}_d = \overline{\mu}_u = 1/2$, and if all radiance is incident from direction (θ_0, φ_0) in the upper hemisphere, then $\overline{\mu}_d = \cos \theta_0$ ($\overline{\mu}_u$ is undefined since there are no photons heading upwards). A further useful measure of the isotropy of the upwelling radiance is the *light distribution factor*

$$Q = \frac{E_u}{L_u} [\text{sr}], \qquad (2.48)$$

which is equal to π for an isotropic light field.

The intensity of light reflected from a surface depends on the object's material properties and surface structure and on the illumination and observation angles. To avoid systematic errors under hemispherical illumination conditions (i.e., all outdoor measurements), a proper definition of the term *reflectance* is necessary. Nine definitions are possible for a plane surface, corresponding to all possible ratios of the quantities ϕ, L, and E for incident and reflected light (Schaepman-Strub et al., 2006). The reflectance parameter relevant in the RTE, corresponding to the volume scattering function for scattering, is the directional-conical reflectance factor L_u/ϕ_d, which is usually approximated by the bidirectional reflectance distribution function (BRDF) ϕ_u/ϕ_d. Both are IOPs and cannot be measured in the field.

In hydrologic optics, the measured reflectance is composed of light reflected from surfaces (water surface and bottom substrates) and scattered from a volume (water body). Since reflection at the water surface is specular, while scattering in the water body and reflection at the bottom is diffuse, the reflectance parameters need to be even more specific than for land applications. The most common definitions are *remote sensing reflectance*

$$R_{rs} = \frac{L_w(0^+)}{E_d(0^+)} \left[\text{sr}^{-1}\right], \qquad (2.49)$$

remote sensing ratio (Mobley et al., 2017)

$$r_{rs}(z) = \frac{L_u(z)}{E_d(z)} \left[\text{sr}^{-1}\right], \qquad (2.50)$$

and *irradiance reflectance*

$$R(z) = \frac{E_u(z)}{E_d(z)}. \qquad (2.51)$$

R_{rs} is the ratio of water leaving radiance L_w to the downwelling planar irradiance E_d; the symbol 0^+ indicates a measurement just above the water surface. L_w cannot be measured directly but has to be derived from a measurement of upwelling radiance, $L_u = L_w + L_r$, by subtracting the radiance L_r reflected at the water surface. While R_{rs} is defined above the surface, r_{rs} and R are defined for underwater measurements, as indicated by the depth parameter z. The ratio L_u/E_d corresponds to the land community's hemispherical-conical reflectance factor, and E_u/E_d to bihemispherical reflectance, generally called *albedo* (Schaepman-Strub et al., 2006).

2.3.3 AOP Models

Analytical models have been developed for a number of AOPs. The focus is here on models for K_d, R, r_{rs}, and R_{rs}. See Hirata et al. (2009) for Q, and Talaulikar et al. (2014) for $\bar{\mu}_d$.

From the many attenuation coefficients, K_d is the most common and widely used one, as it describes the decrease of natural illumination under water. K_d has been studied extensively; see Bukata et al. (1995) for a literature overview. The following approximation is used frequently:

$$K_d(\lambda) = \kappa_0 \frac{a(\lambda) + b_b(\lambda)}{\cos \theta'_{sun}} \tag{2.52}$$

with θ'_{sun} the sun zenith angle in water. It originates from Gordon (1989), who has shown by Monte Carlo simulations for oceanic waters that this equation provides an accuracy of $\sim 3\%$ near the water surface for sun zenith angles below 60°. He determined a value of $\kappa_0 = 1.0395$ for oceanic waters, Albert and Mobley (2003) obtained $\kappa_0 = 1.0546$ for inland waters.

While Eq. (2.52) has been derived for oceanic waters with relatively low attenuation, Lee et al. (2005a) developed a semi-analytical model for both oceanic and coastal waters:

$$K_d(\lambda) = m_0 \cdot a(\lambda) + m_1 \left(1 - m_2 e^{-m_3 \cdot a(\lambda)}\right) \cdot b_b(\lambda). \tag{2.53}$$

The empirical constants are $m_0 \approx 1 + 0.005 \cdot \theta_{sun}$ with θ_{sun} the sun zenith angle in air, $m_1 = 4.18$, $m_2 = 0.52$, and $m_3 = 10.8$. Comparisons with field data show that Eq. (2.53) performs quite well even for water with high attenuation: the average absolute differences were $\sim 14\ \%$ at 490 nm and $\sim 11\%$ at 443 nm in a study of Lee et al. (2005b).

For modeling reflectance, Gordon et al. (1975) demonstrated the usefulness of the function

$$u(\lambda) = \frac{b_b(\lambda)}{a(\lambda) + b_b(\lambda)} \tag{2.54}$$

(X in their notation) by Monte Carlo simulations. They derived a polynomial of third order in $u(\lambda)$ for the depth dependency of irradiance reflectance of optically deep water. Its approximation for small z,

$$R^{deep}(\lambda) = f \cdot u(\lambda) \tag{2.55}$$

with $f = 1/3$, became a frequently used R model in ocean optics, and the corresponding equation

$$r_{rs}^{deep}(\lambda) = f_{rs} \cdot u(\lambda) \tag{2.56}$$

with $f_{rs} = f/Q$, became a common r_{rs} model. The superscript "deep" indicates a measurement of optically deep water, and Q is the light distribution factor defined by Eq. (2.48).

The factors f and f_{rs} are not really constant, but AOPs. Analytical models have been derived, e.g., by Lee et al. (1998, 1999),

$$f_{rs}(\lambda) = 0.084 + 0.0170 \cdot u(\lambda), \tag{2.57}$$

and by Albert (Albert and Mobley 2003, Albert 2004):

$$f(\lambda) = 0.1034 \cdot \left(1 + 3.3586 \cdot u(\lambda) - 6.5358 \cdot u(\lambda)^2 + 4.6638 \cdot u(\lambda)^3\right)$$
$$\cdot \left(1 + \frac{2.4121}{\cos \theta'_{sun}}\right) \cdot (1 - 0.0005 \ v_W), \tag{2.58}$$

$$f_{rs}(\lambda) = 0.0512 \cdot \left(1 + 4.6659 \cdot u(\lambda) - 7.8387 \cdot u(\lambda)^2 + 5.4571 \cdot u(\lambda)^3\right)$$
$$\cdot \left(1 + \frac{0.1098}{\cos \theta'_{sun}}\right) \cdot \left(1 + \frac{0.4021}{\cos \theta'_v}\right) \cdot (1 - 0.0044 \ v_W). \tag{2.59}$$

v_W is the wind speed in units of (m s^{-1}), θ'_{sun} the sun zenith angle in water, and θ'_v the viewing zenith angle in water. A combination of these models yields an analytical model for Q:

$$Q(\lambda) = \frac{f(\lambda)}{f_{rs}(\lambda)}. \tag{2.60}$$

For remote sensing, radiance reflectance models are more important than irradiance reflectance models since the sensors measure upwelling radiance. The most general r_{rs} models can be applied to both deep and shallow waters. Such models have been developed by Lee et al. (1998, 1999) and Albert (Albert and Mobley, 2003; Albert 2004) based on simulations with the RT program Hydrolight (Mobley et al., 1993). They used the same approach,

$$r_{rs}(\lambda) = r_{rs}^{deep}(\lambda) \cdot \left[1 - A_{rs,1} \cdot \exp\left\{-(K_d(\lambda) + k_{uW}(\lambda)) \cdot z_B\right\}\right]$$
$$+ A_{rs,2} \cdot R_{rs}^b(\lambda) \cdot \exp\left\{-(K_d(\lambda) + k_{uB}(\lambda)) \cdot z_B\right\}, \tag{2.61}$$

but different expressions for the model parameters. A comparison of the two parameterizations is given in Table 2.1. The Albert model accounts for illumination and viewing geometry, which is neglected in the Lee model, and it has been developed for significantly larger ranges of environmental parameters, including most of the high concentrations observed in inland waters.

Remote sensing reflectance R_{rs} is related to subsurface r_{rs} as follows (Mobley, 1994; Lee et al., 1998):

$$R_{rs}(\lambda) = \frac{\zeta \cdot r_{rs}(\lambda)}{1 - \Gamma \cdot r_{rs}(\lambda)} + R_{rs}^{surf}(\lambda). \tag{2.62}$$

$\zeta \approx 0.52$ is the water-to-air radiance divergence factor, the denominator with $\Gamma \approx 1.6$ accounts for the effects of internal reflection from water to air, and $R_{rs}^{surf}(\lambda)$ are the reflections at the water surface discussed in

TABLE 2.1 Comparison of the Shallow Water Models of Lee and Albert

Parameter	Lee et al. (1998, 1999)	Albert and Mobley (2003), Albert (2004)
Input to Hydrolight Runs		
C (mg m^{-3})	0.4, 1.0, 2.0, 5.0	0.5, 1.0, 1.5, 2, 3, 5, 7, 10, 20, 40, 60, 80, 100
Y (m^{-1})	0.05, 0.1, 0.3	0.05, 0.10, 0.15, 0.2, 0.3, 0.4, 0.5, 0.6, 0.7, 0.8, 0.9, 1.0, 1.3, 2.5, 4.0, 5.0
X (g m^{-3})	3 values of a parameter related to C, corresponds roughly X range $0.04 - 3.2$	0.5, 1.0, 1.5, 2, 3, 4, 5, 6, 7, 9, 10, 30, 50
$R^b(\lambda)$	0, 0.1, 0.3, 1.0	sand, sediment, green algae, macrophyte
z_B (m)	0.5, 1, 3, 8, 16, 32, infinite	1, 3, 6, 10, 20, 30, infinite
Wind speed (m/s)	5	0, 5, 10
θ_{sun} (°)	0, 30, 60	11, 19, 28, 37, 48, 57, 65, 73
λ (nm)	400−700, steps of 20	400−750, steps of 5
Resulting Parameterization for $r_{rs}(\lambda)$		
$K_d(\lambda)$	Eq. (2.52)	Eq. (2.52)
$f_{rs}(\lambda)$	Eq. (2.57)	Eq. (2.59)
$k_{uW}(\lambda)$	$1.03(1+2.4u(\lambda))^{0.5}\cdot[a(\lambda)+b_b(\lambda)]$	$\dfrac{a(\lambda)+b_b(\lambda)}{\cos\theta_v}\cdot[1+u(\lambda)]^{3.54}\cdot\left[1-\dfrac{0.279}{\cos\theta'_{sun}}\right]$
$k_{uB}(\lambda)$	$1.04(1+5.4u(\lambda))^{0.5}\cdot[a(\lambda)+b_b(\lambda)]$	$\dfrac{a(\lambda)+b_b(\lambda)}{\cos\theta_v}\cdot[1+u(\lambda)]^{2.27}\cdot\left[1+\dfrac{0.058}{\cos\theta'_{sun}}\right]$
$A_{rs,1}$	1	1.1576
$A_{rs,2}$	1	1.0389
κ_0	1	1.0546

Section 2.4.2. Fig. 2.9 illustrates the application of an AOP model for simulating measurements of $R_{rs}(\lambda)$ for different water depths z_B.

The reflectance properties of the sea floor are relevant when $r_{rs}(\lambda, z)$ is affected by the light reflected at bottom substrates. Waters with such influence up to the surface ($z = 0$) are called optically shallow. The irradiance reflectance of the bottom, $R^b(\lambda)$, is the weighted sum of the contributions of the different substrates:

$$R^b(\lambda) = \sum_{n=1}^{N} f_n \cdot A_n(\lambda). \tag{2.63}$$

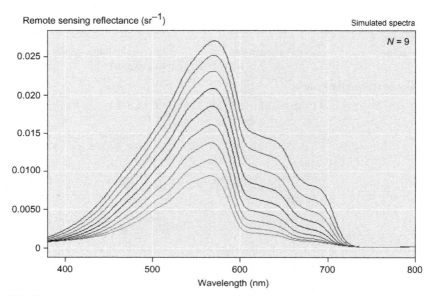

FIGURE 2.9 Simulated R_{rs} measurements for water depths between 1 m (uppermost curve) and 5 m (lowermost curve). The calculations were made by applying the shallow water model of Albert (Albert and Mobley, 2003; Albert, 2004) to a bottom albedo spectrum of sand for the model parameters $C = 1$ mg m^{-3}, $X = 1$ g m^{-3}, $Y = 0.3$ m^{-1}, $S = 0.014$ nm^{-1}, $\theta_{sun} = 40°$. Screenshot of WASI (Gege, 2017).

Irradiance reflectance of a surface is called *albedo*, thus the symbol $A_n(\lambda)$ is used to specify the albedo of substrate number n. The weights f_n are the areal fractions of the substrates within the sensor's field of view ($\Sigma f_n = 1$). When the upwelling radiation is measured by a radiance sensor, the remote sensing reflectance of the bottom can be expressed correspondingly as follows:

$$R_{rs}^b(\lambda) = \sum_{n=1}^{N} f_n \cdot B_n \cdot A_n(\lambda). \tag{2.64}$$

B_n is the proportion of radiation, which is reflected towards the sensor. $B_n = 1/\pi = 0.318$ sr^{-1} represents isotropic reflection of so-called Lambertian surfaces.

2.4 LIGHT FIELD MODELS

If the spatial distribution and the optical properties of all absorbing and scattering components of the atmosphere, the water and the sea floor are known, and also the geometry of the water surface, numerical models based directly on the RT equation (2.12) allow an accurate calculation of the light fields in air and water. Zhai et al. (2009) give a review of such models. These models are valuable for exemplary simulation of measurements and for theoretical studies about the dependency of the light field on certain parameters (sensitivity studies). They are however not suited for data analysis since their high computational expense makes them by several orders of magnitude too slow for remote sensing applications. Since detailed knowledge of the environmental conditions is anyway not available in practice, all data analysis software is based on computationally faster methods that have been developed using a "numerically exact" method as reference. Examples of such inversion methods are neural networks, look-up tables, principal component inversion, and analytical approximations. Since the latter make the underlying physical laws directly visible, these are used to discuss the light field topic.

2.4.1 Incident Radiation

Apart from artificial light sources and bioluminescence, the origin of all light in hydrologic optics is the sun. Its disk with $0.53°$ diameter in the sky provides for an observer at the Earth surface the direct component of downwelling irradiance, $E_{dd}(\lambda)$. This irradiance depends on the extraterrestrial solar irradiance, $F_0(\lambda)$, on the path length through the atmosphere (called air mass), and on the optical properties of the atmosphere. It can be approximated in the visible and near infrared by the following equation (Gregg and Carder, 1990):

$$E_{dd}(\lambda) = \cos \theta_{sun} F_0(\lambda) T_r(\lambda) T_{aa}(\lambda) T_{as}(\lambda) T_{oz}(\lambda) T_o(\lambda) T_{wv}(\lambda). \qquad (2.65)$$

The relative air mass is given by $\cos \theta_{sun}$. The T_i are the transmission coefficients after scattering or absorption of atmospheric component i (T_r: Rayleigh scattering, T_{aa}: aerosol absorption, T_{as}: aerosol scattering, T_{oz}: ozone absorption, T_o: oxygen absorption, and T_{wv} water vapor absorption). Gregg and Carder (1990) have developed a parameterization of the $T_i(\lambda)$ for the wavelength range $400-700$ nm.

The Earth atmosphere and clouds scatter many photons out of the beam from the sun before the sunlight reaches the surface. Such photons constitute the major part of the diffuse component of downwelling irradiance, $E_{ds}(\lambda)$. A minor part is light from the Earth that is scattered in downward directions by the atmosphere or clouds. This adjacency effect is commonly ignored in analytical E_{ds} models. The model of Gregg and Carder (1990) distinguishes between the Rayleigh (E_{dsr}) and aerosol (E_{dsa}) components of E_{ds}:

$$E_{dsr}(\lambda) = \tfrac{1}{2} \cos \theta_{sun} F_0(\lambda) \left[1 - T_r(\lambda)^{0.95}\right] T_{aa}(\lambda) T_{oz}(\lambda) T_o(\lambda) T_{wv}(\lambda), \qquad (2.66)$$

$$E_{dsa}(\lambda) = \cos \theta_{sun} F_0(\lambda) T_r(\lambda)^{1.5} T_{aa}(\lambda) T_{oz}(\lambda) T_o(\lambda) T_{wv}(\lambda) [1 - T_{as}(\lambda)] F_a. \qquad (2.67)$$

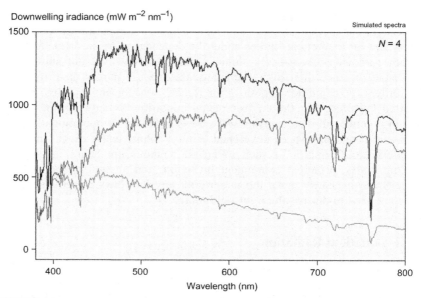

FIGURE 2.10 Simulated irradiances E_d (upper curve), E_{dd} (mid curve), and E_{ds} (lower curve) using the adopted model of Gregg and Carder (1990). Sun zenith angle: 40°, ozone scale height: 0.3 cm, water vapor scale height: 2.5 cm, Angström exponent of aerosol scattering: 1.32, turbidity coefficient: 0.26. Screenshot of WASI (Gege, 2017).

F_a is the aerosol forward scattering probability. The downwelling irradiance is the sum of the direct and diffuse components:

$$E_d(\lambda) = E_{dd}(\lambda) + E_{dsr}(\lambda) + E_{dsa}(\lambda). \qquad (2.68)$$

Software tools like Hydrolight (Mobley et al., 1993) and WASI (Gege, 2012a) use the Gregg and Carder model (updated and extended to the range 300−1000 nm) for the simulation of $E_d(\lambda)$. Fig. 2.10 illustrates such a simulation.

$E_d(\lambda)$ represents well the illumination for horizontally extended, diffuse reflecting objects like the water body and level sea floors. It provides however not the correct angular weighting of the incident radiation for specularly reflecting or tilted surfaces like the water surface, individual parts of water plants or inclined sea floor. A proper description of the radiation incident on such surfaces requires a model of the angular dependency of sky radiance, $L_{sky}(\theta, \varphi)$. A mathematical relationship has been proposed already in 1929 by Pokrovski for clear sky conditions. Harrison and Coombes (1988) have derived an empirical analytical model on the basis of more than 3000 scans of clear skies above Calgary in Canada.

2.4.2 Water Surface Effects

The water surface acts like a mirror and reflects a small part of the sky in the viewing direction (θ_v, φ_v) of a downward looking observer. For a plane surface, the reflected piece of sky is centered at $(\pi - \theta_v, \varphi_v)$ and has the size of the observer's field of view. The reflected radiance is proportional to the radiance L_{sky} of that piece of sky:

$$L_{r,sky}(\lambda, \theta_v, \varphi_v) = \rho_L(\theta_v) \cdot L_{sky}(\lambda, \pi - \theta_v, \varphi_v). \tag{2.69}$$

$L_{r,sky}$ is called *sky glint*. The reflectance factor ρ_L is obtained from the Fresnel equation, which reads for unpolarized light (Jerlov, 1976):

$$\rho_L(\theta) = \frac{1}{2} \left| \frac{\sin^2(\theta - \theta')}{\sin^2(\theta + \theta')} + \frac{\tan^2(\theta - \theta')}{\tan^2(\theta + \theta')} \right|. \tag{2.70}$$

θ' is the angle of refraction and related to the angle of incidence, θ, by Snell's law

$$\sin \theta = n_w \sin \theta', \tag{2.71}$$

with $n_w \approx 1.33$ the refractive index of water. For viewing angles near nadir, $\rho_L \approx 0.02$. The sky radiance $L_{sky}(\lambda, \pi - \theta_v, \varphi_v)$ can be obtained either from measurement during field campaigns, or from combining a spectral model (Gregg and Carder, 1990) with an angular model (Pokrovski, 1929; Harrison and Coombes, 1988).

The water surface is rarely perfectly flat, but waves alter steadily its geometry. If a facet of the water surface is tilted in the direction (β, γ) relative to the normal, it reflects a piece of sky located at $(\pi - \theta_v - \beta, \varphi_v + \gamma)$ towards the observer. The tilted facets increase the reflected sky area proportional to the probability distribution function $p(\beta, \gamma)$ of the surface slopes. In this way, waves can reflect light from the sun disk towards the sensor. The radiance of this *sun glint* is given by (Kay et al., 2009)

$$L_{r,sun}(\lambda) = \rho_L(\omega) \cdot \frac{E_{dd}(\lambda) \cdot p(z_x, z_y)}{4 \cos^4 \beta \cdot \cos \theta_v}. \tag{2.72}$$

ω is the angle between the normal to a facet and sun direction, and $z_x = \partial z / \partial x = \tan \beta \cos \gamma$ and $z_y = \partial z / \partial y = \tan \beta \sin \gamma$ are the facet slopes. z_x, z_y, ω and β are functions of the solar and sensor geometry. Cox and Munk (1954a, 1954b, 1956) have derived from aerial photographs empirical equations relating $p(z_x, z_y)$ to wind speed and wind direction. The sun glint correction algorithms of most ocean color instruments are based on these equations. Recent studies have suggested some minor modifications but they have left unchanged the main features (Kay et al., 2009).

The wave spectrum is ranging spatially from millimeters to tens of meters (Elfouhaily et al., 1997) and temporally from milliseconds to tens of seconds

(Darecki et al., 2011), thus statistical models of the surface geometry smooth out the effects of individual waves only for areas above approximately 100×100 m^2 or integration times above 100 s. This makes the Cox and Munk model unsuited for most measurements in inland waters. Furthermore, the $p(z_x, z_y)$ model is based implicitly on an undisturbed interaction of wind and water for many kilometers during which the wind direction is stable. Since these criteria are frequently not fulfilled, the applicability of the $p(z_x, z_y)$ model to inland waters is questionable.

For these reasons, modeling and correction of sun glint demands for an alternate approach for inland waters based on radiometry rather than geometry. A number of models have been developed utilizing signals in the near-infrared (Hochberg et al., 2003; Hedley et al., 2005; Ruddick et al., 2006; Goodman et al., 2008; Kutser et al., 2009; Lee et al., 2010; Kutser et al., 2013; Simis and Olsson, 2013), but these are not yet working reliably under all conditions (Kay et al., 2009; Martinez-Vicente et al., 2013).

Mobley (1999) has shown that sun glint and sky glint are minimal under most conditions for an observation geometry of $\theta_v = 40°$, $\varphi_v = 135°$. This geometry is thus frequently chosen during field measurements, supplemented by an $L_{sky}(\lambda, \pi - \theta_v, \varphi_v)$ measurement. It has, however, the drawback to provide the water leaving radiance for an angle that differs considerably from nadir, making an anisotropy model necessary for comparison with remote sensing data. Furthermore, it extends the path length of the water layer for measurements in shallow water, making it more difficult to retrieve information from the bottom.

2.4.3 Underwater Light Field

The downwelling irradiance in water consists of radiation transmitted through the air-water interface, and of a fraction ρ_u of *upwelling* irradiance E_u that is reflected at the surface in downward directions. Expressing the transmitted radiation by the direct and diffuse irradiance components E_{dd} and E_{ds} from Section 2.4.1, the downwelling plane irradiance just below the water surface (symbol 0^-) is given by (Gregg and Carder, 1990; Baker and Smith, 1990; Mobley, 1994; Zheng et al. 2002)

$$E_d(\lambda, 0^-) = (1 - \rho_{dd})E_{dd}(\lambda) + (1 - \rho_{ds})E_{ds}(\lambda) + \rho_u E_u(\lambda, 0^-). \qquad (2.73)$$

Not all incident radiation penetrates into the water, but the surface reflects a portion back to the upper hemisphere. The radiance reflection factors given by the Fresnel equation [Eq. (2.70)] average to irradiance reflection factors depending on the incident radiance distribution, slope distribution of the water surface facets, and solar elevation. For clear sky conditions they are in the order of $\rho_{dd} \approx 0.02 - 0.03$ for the direct component

(Jerlov, 1976; Preisendorfer and Mobley, 1986), $\rho_{ds} \approx 0.06-0.07$ for the diffuse component (Gregg and Carder, 1990), and $\rho_u \approx 0.50-0.57$ for the upwelling irradiance (Jerome et al., 1990; Mobley, 1999).

The relationship between the plane downwelling irradiances above (0^+) and just below (0^-) the surface can be expressed in terms of a transmittance function $t(\lambda, \theta_{sun})$ as follows (Zheng et al., 2002):

$$E_d(\lambda, 0^-) = \frac{t(\lambda, \theta_{sun})}{1 - \rho_u R(\lambda)} E_d(\lambda, 0^+), \qquad (2.74)$$

with $R(\lambda)$ irradiance reflectance. $t(\lambda, \theta_{sun})$ depends on wavelength and on the conditions of the water surface and the atmosphere (Baker and Smith, 1990).

The fluctuating water surface can induce strong short-term variability to under water measurements of E_d (Fig. 2.11). The undulations alter steadily the refraction angle of the incident rays, leading to focusing and defocusing effects (Walker, 1994; Zanefeld et al., 2001). E_d varies typically by 20%–40% in the upper few meters (Dera and Stramski, 1993; Hofmann et al., 2008), but flashes can be an order of magnitude above average (Hieronymi et al., 2012).

Since E_d is used to derive AOPs like K_d, R, or r_{rs}, the wave focusing effect can introduce large errors to underwater measurements of these parameters. The errors can be reduced by averaging sufficient measurements, or by applying a spectral model of the induced E_d changes for correction (Gege and Pinnel, 2011; Gege, 2012a). Geometry based models are useful for simulation, but inapt for data analysis since the actual water surface geometry cannot be determined with sufficient accuracy.

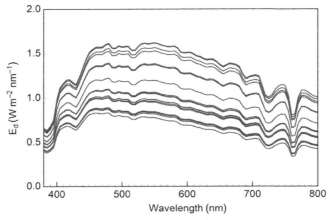

FIGURE 2.11 Illustration of variability of downwelling irradiance in water. The measurements were made within 100 s just below the water surface. *Adopted from Gege and Pinnel (2011).*

The decrease of E_d with depth z is usually approximated by the Lambert–Beer law

$$E_d(\lambda, z) = E_d(\lambda, 0^-) \cdot \exp\{- K_d(\lambda) \cdot z\}, \tag{2.75}$$

with $K_d(\lambda)$ the diffuse attenuation coefficient. However, the effective path lengths of direct and diffuse irradiance can be quite different, and such differences cannot be included correctly in the factorization of Eq. (2.75). Mainly for this reason $K_d(\lambda)$ is an AOP. Gege (2012a) has shown that separate treatment of E_{dd} and E_{ds} allows approximating the depth dependency of E_d in terms of the IOP $a(\lambda) + b_b(\lambda)$ instead of the AOP $K_d(\lambda)$. Models based on this separation not only allow improved simulation of the depth dependency of E_d, they can even be used to determine the concentration of absorbing water constituents (phytoplankton, CDOM) from single E_d measurements affected by wave focusing (Gege, 2012a,b; Linnemann et al., 2013).

Other common radiometric quantities describing the light field under water are the upwelling radiance, $L_u(\lambda, z)$, and the upwelling irradiance, $E_u(\lambda, z)$. These can be simulated by applying Eq. (2.50) or Eq. (2.51), i.e. by multiplying $E_d(\lambda, z)$ with a modeled spectrum $r_{rs}(\lambda)$ or $R(\lambda)$. Fig. 2.12 illustrates the spectral and depth dependency of $L_u(\lambda, z)$ in shallow water.

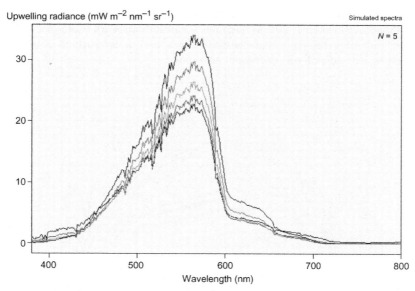

FIGURE 2.12 Simulated L_u measurements for a water depth of 5 m and sensor depths between 0 m (lowermost curve) and 4 m (uppermost curve). The calculations were made by applying the shallow water model of Albert (Albert and Mobley, 2003; Albert, 2004) to a bottom albedo spectrum of sand for the model parameters $C = 1$ mg m^{-3}, $X = 1$ g m^{-3}, $Y = 0.3$ m^{-1}, $S = 0.014$ nm^{-1}, $\theta_{sun} = 40°$. Screenshot of WASI (Gege, 2017).

2.4.4   Fluorescence

Fluorescence is an inelastic process where a fraction η_F of photons absorbed at wavelength λ' in a volume dV is emitted at longer wavelengths λ. To include it in bio-optical models, a model for the *fluorescence radiance* L_F is required. Maritorena et al. (2000) and Huot et al. (2005) have derived analytical equations in photon number units (mol m^{-2} s^{-1} sr^{-1} nm^{-1}) for optically deep water. To be consistent with the energy units used in this chapter (W m^{-2} nm^{-1} sr^{-1}), and to be applicable to shallow waters as well, an alternate derivation is provided.

Since the process is similar to scattering, except for a change of wavelength, the basic IOP of fluorescence, the *volume fluorescence scattering function* β_F, is defined analogously to the volume scattering function of Eq. (2.20), but now relative to a small excitation wavelength interval $d\lambda'$:

$$\beta_F(\lambda, \lambda', \psi) = \frac{1}{E(\lambda')}\frac{dI_F(\lambda, \psi)}{dVd\lambda'}\left[\text{m}^{-1}\text{sr}^{-1}\text{nm}^{-1}\right]. \tag{2.76}$$

I_F is the *fluorescence radiant intensity*, i.e., the emitted photon flux per solid angle element in accordance to Eq. (2.5). Integration over all angles yields the *volume fluorescence scattering coefficient* (Mobley, 1994):

$$b_F(\lambda, \lambda') = \int_0^{2\pi}\int_0^{\pi}\beta_F(\lambda, \lambda', \psi)\sin\psi\, d\psi\, d\varphi\left[\text{m}^{-1}\text{nm}^{-1}\right]. \tag{2.77}$$

The ratio

$$\eta_F(\lambda, \lambda') = \frac{N}{N'd\lambda} = \frac{b_F(\lambda, \lambda')}{a(\lambda')}\cdot\frac{\lambda}{\lambda'}\left[\text{nm}^{-1}\right] \tag{2.78}$$

is called *spectral fluorescence quantum efficiency* (Mobley, 1994), where N and N' denote the number of photons at wavelength intervals $[\lambda, \lambda + d\lambda]$ and $[\lambda', \lambda' + d\lambda']$, respectively, and $a(\lambda')$ [m^{-1}] is the absorption coefficient of the component in dV which is responsible for emission (Gordon, 1979).

Assuming isotropic fluorescence, Eq. (2.77) can be solved:

$$b_F(\lambda, \lambda') = 4\pi\beta_F(\lambda, \lambda'). \tag{2.79}$$

Inserting Eq. (2.76) and using $dV = dA\, dz'$ leads to

$$b_F(\lambda, \lambda') = 4\pi\frac{dI_F(\lambda)}{E(\lambda')dAdz'd\lambda'} = 4\pi\frac{L_F}{E(\lambda')dz'd\lambda'} \tag{2.80}$$

with the fluorescence radiance $L_F = dI_F/dA$ according to Eq. (2.7). Eq. (2.78) allows for replacing $b_F(\lambda, \lambda')$ and expressing L_F in terms of the incident irradiance E:

$$L_F(\lambda, \lambda', z') = \frac{1}{4\pi}\eta_F(\lambda, \lambda')\cdot E(\lambda', z')\cdot a(\lambda')\cdot dz'\cdot\frac{\lambda'}{\lambda}\cdot d\lambda'. \tag{2.81}$$

$L_F(\lambda, \lambda', z')$ represents the fluorescence radiance emitted in depth z' by a layer of thickness dz'. For a downward looking sensor in depth z, the total fluorescence signal is the sum of contributions from layers at depths $z' > z$ down to the bottom (depth z_B), weighted by the transmission of the water column between sensor and each layer. The transmission is obtained from the Lambert–Beer law using the attenuation coefficient for upwelling radiance, k_u. Hence,

$$L_F(\lambda, \lambda', z) = \frac{1}{4\pi} \int\limits_z^{z_B} \eta_F(\lambda, \lambda') \cdot E(\lambda', z') \cdot a(\lambda') \cdot \frac{\lambda'}{\lambda} \cdot d\lambda' \cdot \exp\{-k_u(\lambda) \cdot (z' - z)\} dz'.$$

(2.82)

As the incident irradiance at depth z' is related to that right below the water surface ($z' = 0^-$) also by the Lambert–Beer law:

$$E(\lambda', z') = E(\lambda', 0^-) \cdot \exp\{-K(\lambda') \cdot z'\},$$

(2.83)

but now in terms of K, the depth-averaged diffuse attenuation coefficient of irradiance, Eq. (2.82) can be expressed in terms of $E(\lambda', 0^-)$:

$$L_F(\lambda, \lambda', z) = \frac{1}{4\pi} \eta_F(\lambda, \lambda') \cdot E(\lambda', 0^-) \cdot a(\lambda') \cdot \frac{\lambda'}{\lambda} \cdot d\lambda'$$
$$\cdot \int\limits_z^{z_B} \exp\{-k_u(\lambda) \cdot (z' - z) - K(\lambda') \cdot z'\} dz'.$$

(2.84)

η_F and a are assumed independent of z' for this step in order to solve the integral analytically. Eq. (2.84) is valid for excitation in a narrow wavelength interval $d\lambda'$. In case of broadband excitation, the integral must be taken for the wavelength interval $[\lambda_u, \lambda_o]$ that excites fluorescence at wavelength λ:

$$L_F(\lambda, z) = \frac{1}{4\pi} \int\limits_{\lambda_u}^{\lambda_o < \lambda} \eta_F(\lambda, \lambda') \cdot E(\lambda', 0^-) \cdot a(\lambda') \cdot \frac{\lambda'}{\lambda}$$
$$\cdot \frac{\exp\{-K(\lambda') \cdot z\} - \exp\{-K(\lambda') \cdot z_B - k_u(\lambda) \cdot (z_B - z)\}}{K(\lambda') + k_u(\lambda)} d\lambda'.$$

(2.85)

Eq. (2.85) is similar to the equations derived by Maritorena et al. (2000) and Huot et al. (2005). Major differences are the lack of the factor λ'/λ for photon number units, and of water depth z_B for optically deep water. The other differences are due to parameterizations of $a(\lambda')$ and $\eta_F(\lambda, \lambda')$ for simulating phytoplankton fluorescence.

2.4.5   Polarization

All equations presented in this chapter neglect polarization. Accurate RT calculations should however include polarization effects (Plass et al., 1981; Kattawar and Adams, 1989; Adams and Kattawar, 1993; Mishchenko et al., 1994; Lacis et al., 1998). In addition, polarization properties could enhance the performance of inverse algorithms for ocean color purposes (Chami and Platel, 2007; Chami, 2007; Gordon et al., 1988), improve the contrast of underwater viewing systems (Cariou et al., 1990) and help understanding the behavior of marine organisms (Waterman, 1954). Many vector RT models have been proposed to predict the radiance and degree of polarization of the light; most of these are however not publicly available. See Chami et al. (2015) and the literature cited therein for a discussion of the polarization topic.

2.5   CONCLUSIONS

This chapter gave a brief overview about the complex interaction processes of light with matter in the aquatic environment. The physical principles are understood for a long time, and many methods and computer programs have been developed in the last decades for modeling the RT in the atmosphere-water system (overview in Zhai et al., 2009). However, the vast majority of accurate RT models have been implemented by scientists in their own software environment, while little software is publicly available for such calculations. A single software tool, Hydrolight (Mobley et al., 1993), is de facto the public "gold standard" for exact forward calculations in hydrologic optics.

An important application of the presented models is remote sensing of coastal, inland, and shallow waters. However, due to the high complexity of these water types, remote sensing cannot be considered operational so far (Odermatt et al., 2012; Palmer et al., 2014), except for some regions with site-specific knowledge. Certain research groups and companies have the know-how and software for deriving a number of parameters accurately and with known uncertainty, but many satellite and airborne data are not processed with such optimized algorithms and expert knowledge. The applied generic algorithms can introduce large and unknown errors if not validated independently. Major challenges for data analysis are the following:

- *Reflections at the water surface.* The light reflected at the surface can be as intense as that from the water itself, or even higher, thus above-water measurements require an accurate correction of sun glint and sky glint. A reliable method for correction does not yet exist.
- *Complex composition.* The large variety of type and concentration of water constituents and bottom substrates of the lakes on Earth makes it difficult to develop a generic model that can be applied to all water types. Lake and season specific data of the concentration ranges and

SIOPs can improve model performance, but adequate databases for inland waters like LIMNADES (Globolakes, 2014) are still in the evolving phase.

- *Spectral variability of components.* The water constituents and bottom types cannot be characterized by a constant set of specific inherent optical properties (SIOPs), as the SIOPs can change regionally and temporally. Only little knowledge exists about this variability in inland waters.
- *Dissolved organic matter.* As absorption of many inland waters is dominated by dissolved substances including fluorescing components, bio-optical models of such waters require accurate spectral models of CDOM absorption and fluorescence. The usual exponential approximation for absorption may be too coarse at high concentration, and fluorescence is not well studied.
- *Spectral ambiguities.* The diversity of composition and SIOPs introduces a fundamental numerical problem to inversion: different combinations of water constituents can lead to indistinguishable reflectance spectra (Defoin-Platel and Chami, 2007). It can be handled by utilizing correlations between parameters or by restricting the ranges of some parameters. Both measures require site-specific knowledge and are difficult to implement in software for processing of global data sets.
- *Atmosphere correction.* A challenge for remote sensing of inland waters is atmosphere correction which is described in detail in Chapter 3. An inland water specific problem is the adjacency effect, where scattering in the atmosphere contaminates water pixels with radiation from the surrounding land. The algorithms applied by the data providers are frequently not optimized for inland waters, and publicly available software that can be tuned by the user does not exist.

While such methodological challenges are relevant for algorithm developers, the availability of ready data products and software is the crux for many users (Dörnhöfer and Oppelt, 2016). Many scientists have developed their own codes for data analysis (comparison in Dekker et al., 2011), but publicly available software is rare. For satellite data, SeaDAS (NASA, 2016) and the Sentinel-3 Toolbox (ESA, 2017) provide visualization, analysis and processing tools for the exploitation of a number of satellite missions. For airborne data, BOMBER (Giardino et al., 2012) and WASI-2D (Gege, 2014, 2017) allow inversion of multi- and hyperspectral images for user-defined SIOPs. Both tools are tailored for inland waters, but they require atmospherically corrected data as input. In situ measurements can be simulated and inverted using WASI (Gege, 2004, 2017).

Currently a new generation of satellite sensors is upcoming that bears much potential for monitoring of inland waters (e.g., Landsat-8, Sentinel-2, Sentinel-3, HyspIRI, and EnMAP). It can perhaps trigger facing the mentioned challenges and push the development of software and infrastructure for operational remote sensing of inland waters. This book, with its systematic compilation of the actual state-of-the-art, can eventually also foster such advancement.

ACKNOWLEDGMENTS

Many thanks to my colleagues Anna Göritz and Sebastian Riedel for helpful feedback to an earlier version of the chapter. An anonymous reviewer is acknowledged for very detailed comments on the submitted chapter, which led to significant improvements in some parts of the text.

REFERENCES

Aas, E., 2000. Spectral slope of yellow substance: problems caused by small particles. Proc. Ocean Optics XV Conference, October 16–20, 2000, Monaco.

Adams, C.N., Kattawar, G.W., 1993. Effect of volume-scattering function on the errors induced when polarization is neglected in radiance calculations in an atmosphere-ocean system. Appl. Opt. 32 (24), 4610–4617.

Albert, A., 2004. Inversion technique for optical remote sensing in shallow water. Ph.D. Dissertation. Universität Hamburg, Hamburg, Germany, p. 188.

Albert, A., Mobley, C.D., 2003. An analytical model for subsurface irradiance and remote sensing reflectance in deep and shallow case-2 waters. Opt. Express. 11, 2873–2890.

Antoine, D., Siegel, D.A., Kostadinov, T., Maritorena, S., Nelson, N.B., Gentili, B., et al., 2011. Variability in optical particle backscattering in three contrasting bio-optical oceanic regimes. Limnol. Oceanogr. 56 (3), 955–973.

Babin, M., Morel, A., Fournier-Sicre, V., Fell, F., Stramski, D., 2003a. Light scattering properties of marine particles in coastal and open ocean waters as related to the particle mass concentration. Limnol. Oceanogr. 48, 843–859.

Babin, M., Stramski, D., Ferrari, G.M., Claustre, H., Bricaud, A., Obolensky, G., et al., 2003b. Variations in the light absorption coefficients of phytoplankton, nonalgal particles, and dissolved organic matter in coastal waters around Europe. J. Geophys. Res. 108, C7.

Bader, H., 1970. The hyperbolic distribution of particle sizes. J. Geophys. Res. 75, 2822–2830.

Baker, K.S., Smith, R.C., 1990. Irradiance transmittance through the air/water interface. Spinrad, R.W. (Ed.), Proc. SPIE 1302, Ocean Optics X, 556–565.

Binding, C.E., Jerome, J.H., Bukata, R.P., Booty, W.G., 2008. Spectral absorption properties of dissolved and particulate matter in Lake Erie. Remote Sens. Environ. 112, 1702–1711.

Blondeau-Patissier, D., Brando, V.E., Oubelkheir, K., Dekker, A.G., Clementson, L.A., Daniel, P., 2009. Bio-optical variability of the absorption and scattering properties of the Queensland inshore and reef waters, Australia. J. Geophys. Res. 114, C05003.

Born, M., Wolf, E., 1999. Principles of Optics, seventh ed. Cambridge University Press, Cambridge.

Bricaud, A., Morel, A., Prieur, L., 1981. Absorption by dissolved organic matter of the sea (yellow substance) in the UV and visible domains. Limnol. Oceanogr. 26, 43–53.

Bricaud, A., Babin, M., Morel, A., Claustre, H., 1995. Variability in the chlorophyll specific absorption coefficients of natural phytoplankton: analysis and parameterization. J. Geophys. Res. 100, 13,321–13,332.

Bryant, F.D., Seiber, B.A., Latimer, P., 1969. Absolute optical cross sections of cells and chloroplasts. Arch. Biochem. Biophys. 135 (1), 97−108.

Bukata, R.P., Jerome, J.H., Kondratyev, K.Y., Pozdnyakov, D.V., 1995. Optical Properties and Remote Sensing of Inland and Coastal Waters. CRC Press, Boca Raton, New York, London, Tokyo.

Carder, K.L., Harvey, G.R., Ortner, P.B., 1989. Marine humic and fulvic acids: their effects on remote sensing of ocean chlorophyll. Limnol. Oceanogr. 34, 68−81.

Cariou, J., Jeune, B.L., Lotrian, J., Guern, Y., 1990. Polarization effects of seawater and underwater targets. Appl. Opt. 29 (11), 1689−1695.

Chami, M., 2007. Importance of the polarization in the retrieval of oceanic constituents from the remote sensing reflectance. J. Geophys. Res. Oceans 112 (C5), C05026.

Chami, M., Platel, M.D., 2007. Sensitivity of the retrieval of the inherent optical properties of marine particles in coastal waters to the directional variations and the polarization of the reflectance. J. Geophys. Res. 112 (C5), C05037.

Chami, M., Shybanov, E.B., Churilova, T.Y., Khomenko, G.A., Lee, M.E.-G., Martynov, O.V., et al., 2005. Optical properties of the particles in the Crimea coastal waters (Black Sea). J. Geophys. Res. 110, C11020.

Chami, M., Lafrance, B., Fougnie, B., Chowdhary, J., Harmel, T., Waquet, F., 2015. OSOAA: a vector radiative transfer model of coupled atmosphere-ocean system for a rough sea surface application to the estimates of the directional variations of the water leaving reflectance to better process multi-angular satellite sensors data over the ocean. Opt. Express. 23, 27829−27852.

Champion, P.M., Albrecht, A.C., 1982. Resonance Raman scattering: the multimode problem and transform methods. Ann. Rev. Phys. Chem. 33, 353−376.

Chandrasekhar, S., 1960. Radiative Transfer. Dover, New York.

Cheng, M., Weng, F., 2012. Kramers-Kronig analysis of leaf refractive index with the PROSPECT leaf optical property model. J. Geophys. Res. 117, D18106.

Cox, C., Munk, W., 1954a. Statistics of the Sea surface derived from Sun glitter. J. Mar. Res. 13, 198−227.

Cox, C., Munk, W., 1954b. Measurement of the roughness of the Sea surface from photographs of the Suns glitter. J. Opt. Soc. Am. 44, 838−850.

Cox, C., Munk, W., 1956. Slopes of the Sea surface deduced from photographs of Sun glitter. Scripps Inst. Oceanogr. Bull. 6, 401−488.

Darecki, M., Stramski, D., Sokolski, M., 2011. Measurements of high-frequency light fluctuations induced by ocean surface waves with an underwater porcupine radiometer system. J. Geophys. Res. 116, C00H09.

Defoin-Platel, M., Chami, M., 2007. How ambiguous is the inverse problem of ocean color in coastal waters? J. Geophys. Res. 112, C03004.

Dekker, A.G., Peters, S.W.M., 1993. The use of the Thematic Mapper for the analysis of eutrophic lakes: a case study in The Netherlands. Int. J. Remote Sensing 14, 799−822.

Dekker, A.G., Phinn, S.R., Anstee, J., Bissett, P., Brando, V.E., Casey, B., et al., 2011. Intercomparison of shallow water bathymetry, hydro-optics, and benthos mapping techniques in Australian and Caribbean coastal environments. Limnol. Oceanogr. Methods 9, 396−425.

Dera, J., Stramski, D., 1993. Focusing of sunlight by sea surface waves: new results from the Black Sea. Oceanologia 34, 13−25.

Dörnhöfer, K., Oppelt, N., 2016. Remote sensing for lake research and monitoring−recent advances. Ecol. Indic. 64, 105−122.

D'Sa, E.J., Miller, R.L., Del Castillo, C., 2006. Bio-optical properties and ocean color algorithms for coastal waters influenced by the Mississippi River during a cold front. Appl. Opt. 45, 7410–7428.

Duysens, L.N.M., 1956. The flattening of the absorption spectrum of suspensions as compared to that of solutions. Biochim. Biophys. Acta. 19, 1–12.

Elfouhaily, T., Chapron, B., Katsaros, K., Vandemark, D., 1997. A unified directional spectrum for long and short winddriven waves. J. Geophys. Res. 102 (C7), 15781–15796.

ESA, 2017. The Sentinel-3 Toolbox. <https://sentinel.esa.int/web/sentinel/toolboxes/sentinel-3> (accessed 28.01.17).

Gege, P., 1998. Characterization of the phytoplankton Lake Constance for classification by remote sensing. In: E., Gaedke, U. (Eds.) Lake Constance, Characterization of an Ecosystem in Transition, Bäuerle, Arch. Hydrobiol. Spec. Issues Advanc. Limnol. 53: 179–193.

Gege, P., 2000. Gaussian model for yellow substance absorption spectra. Proc. Ocean Optics XV Conference, October 16–20, 2000, Monaco.

Gege, P., 2004. The water colour simulator WASI: an integrating software tool for analysis and simulation of optical in-situ spectra. Comput. Geosci. 30, 523–532.

Gege, P., 2012a. Analytic model for the direct and diffuse components of downwelling spectral irradiance in water. Appl. Opt. 51, 1407–1419.

Gege, P., 2012b. Estimation of phytoplankton concentration from downwelling irradiance measurements in water. Israel J. Plant Sci. 60, 193–207.

Gege, P., 2014. WASI-2D: A software tool for regionally optimized analysis of imaging spectrometer data from deep and shallow waters. Comput. Geosci. 62, 208–215.

Gege, P., 2017. WASI (Water Colour Simulator). <http://www.ioccg.org/data/software.html> (accessed 28.01.2017).

Gege, P., Pinnel, N., 2011. Sources of variance of downwelling irradiance in water. Appl. Opt. 50, 2192–2203.

Giardino, C., Candiani, G., Bresciani, M., Lee, Z.P., Gagliano, S., Pepe, M., 2012. BOMBER: a tool for estimating water quality and bottom properties from remote sensing images. Comput. Geosci. 45, 313–318.

Globolakes, 2014. <http://www.globolakes.ac.uk/limnades/> (accessed 28.01.17).

Goodman, J.A., Lee, Z., Ustin, S.L., 2008. Influence of atmospheric and sea-surface corrections on retrieval of bottom depth and reflectance using a semi-analytical model: a case study in Kaneohe Bay, Hawaii. Appl. Opt. 2008 (47), F1–F11.

Gordon, H.R., 1979. Diffuse reflectance of the ocean: the theory of its augmentation by chlorophyll a fluorescence. Appl. Opt. 21, 2489–2492.

Gordon, H.R., 1989. Can the Lambert-Beer law be applied to the diffuse attenuation coefficient of ocean water? Limnol. Oceanogr. 34, 1389–1409.

Gordon, H.R., Morel, A., 1983. Remote Assessment of Ocean Color for Interpretation of Satellite Visible Imagery. A Review. Springer, New York, p. 144.

Gordon, H.R., Brown, O.B., Jacobs, M.M., 1975. Computed relationships between the inherent and apparent optical properties of a flat, homogeneous ocean. Appl. Optics 14, 417–427.

Gordon, H.R., Brown, J.W., Evans, R.H., 1988. Exact Rayleigh scattering calculations for use with the Nimbus-7 coastal zone color scanner. Appl. Opt. 27 (5), 862–871.

Gregg, W.W., Carder, K.L., 1990. A simple spectral solar irradiance model for cloudless maritime atmospheres. Limnol. Oceanogr. 35, 1657–1675.

Hale, G.M., Querry, M.R., 1973. Optical constants of water in the 200 nm to 200 µm wavelength region. Appl. Opt. 12, 555–563.

Harrison, A.W., Coombes, C.A., 1988. Angular distribution of clear sky short wavelength radiance. Sol. Energy 40, 57–63.

Hassing, S., Sonnich Mortensen, O., 1980. Kramers-Kronig relations and resonance Raman scattering. J. Chem. Phys. 73, 1078–1083.

Hassing, S., Nørby Svendsen, E., 2004. The correct form of the Raman scattering tensor. J. Raman Spectrosc. 35, 87–90.

Hedley, J., Harborne, A., Mumby, P., 2005. Simple and robust removal of Sun glint for mapping shallow-water benthos. Int. J. Remote. Sens. 26, 2107–2112.

Hergert, W., Wriedt, T., 2012. The Mie Theory. Springer-Verlag, Berlin, Heidelberg.

Hieronymi, M., Macke, A., Zielinski, O., 2012. Modeling of wave-induced irradiance variability in the upper ocean mixed layer. Ocean Sci. 8, 103–120.

Hirata, T., Hardman-Mountford, N., Aiken, J., Fishwick, J., 2009. Relationship between the distribution function of ocean nadir radiance and inherent optical properties for oceanic waters. Appl. Opt. 48, 3130–3139.

Hochberg, E., Andrefouet, S., Tyler, M., 2003. Sea surface correction of high spatial resolution Ikonos images to improve bottom mapping in near-shore environments. IEEE. Trans. Geosci. Remote. Sens. 41, 1724–1729.

Hofmann, H., Lorke, A., Peeters, F., 2008. Wave-induced variability of the underwater light climate in the littoral zone. Verh. Internat. Verein. Limnol. 30, 627–632.

Højerslev, N.K., Aas, E., 2001. Spectral light absorption by yellow substance in the Kattegat–Skagerrak area. Oceanologia 43, 39–60.

Huot, Y., Brown, C.A., Cullen, J.J., 2005. New algorithms for MODIS sun-induced chlorophyll fluorescence and a comparison with present data products. Limnol. Oceanogr. Methods 3, 108–130.

IOCCG, 2014. Phytoplankton functional types from space. In: Sathyendranath, S. (Ed.), Reports of the International Ocean-Colour Coordinating Group, No. 15, IOCCG. Dartmouth, Canada.

Jerlov, N.G., 1976. Marine Optics, second ed Elsevier, Amsterdam – Oxford – New York.

Jerome, J.H., Bukata, R.P., Bruton, J.E., 1990. Determination of available subsurface light for photochemical and photobiological activity. J. Great. Lakes. Res. 16, 436–443.

Kattawar, G.W., Adams, C.N., 1989. Stokes vector calculations of the submarine light field in an atmosphere–ocean with scattering according to a Rayleigh phase matrix: effect of interface refractive index on radiance and polarization. Limnol. Oceanogr. 34 (8), 1453–1472.

Kay, S., Hedley, J.D., Lavender, S., 2009. Sun glint correction of high and low spatial resolution images of aquatic scenes: a review of methods for visible and near-Infrared wavelengths. Remote Sens. 1, 697–730.

Kirk, J.T.O., 1975a. A theoretical analysis of the contribution of algal cells to the attenuation of light within natural waters I. General treatment of suspensions of pigmented cells. New. Phytol. 75, 11–20.

Kirk, J.T.O., 1975b. A theoretical analysis of the contribution of algal cells to the attenuation of light within natural waters II. Spherical cells. New. Phytol. 75, 21–36.

Kirk, J.T.O., 2011. Light and Photosynthesis in Aquatic Ecosystems, third ed. Cambridge University Press, New York.

Klier, K., 1972. Absorption and scattering in plane parallel turbid media. J. Opt. Soc. Am. 62, 882–885.

Korshin, G.V., Chi-Wang, L., Benjamin, M.M., 1997. Monitoring the properties of natural organic matter through UV spectroscopy: a consistent theory. Water. Res. 31, 1787–1795.

Kutser, T., Vahtmäe, E., Praks, J.A., 2009. Sun glint correction method for hyperspectral imagery containing areas with non-negligible water leaving NIR signal. Remote Sens. Environ. 113, 2267–2274.

Kutser, T., Vahtmäe, E., Paavel, B., Kauer, T., 2013. Removing glint effects from field radiometry data measured in optically complex coastal and inland waters. Remote Sens. Environ. 133, 85–89.

Laanen, M., 2007. Yellow Matters–improving the remote sensing of Coloured Dissolved Organic Matter in inland freshwaters. PhD thesis. Vrije Universiteit, Amsterdam, p. 267.

Lacis, A., Chowdhary, J., Mishenko, M.I., Cairns, B., 1998. Modeling errors in diffuse-sky radiation: vector vs scalar treatment. Geophys. Res. Lett. 25 (2), 135–138.

Lee, Z.-P., Carder, K.L., Mobley, C.D., Steward, R.G., Patch, J.S., 1998. Hyperspectral remote sensing for shallow waters: 1. A semi-analytical model. Appl. Opt. 37, 6329–6338.

Lee, Z.-P., Carder, K.L., Mobley, C.D., Steward, R.G., Patch, J.S., 1999. Hyperspectral remote sensing for shallow waters: 2. Deriving bottom depths and water properties by optimization. Appl. Opt. 38, 3831–3843.

Lee, Z.-P., Du, K.-P., Arnone, R., 2005a. A model for the diffuse attenuation coefficient of downwelling irradiance. J. Geophys. Res. 110, C02016.

Lee, Z.-P., Darecki, M., Carder, K.L., Davis, C.O., Stramski, D., Rhea, W.J., 2005b. Diffuse attenuation coefficient of downwelling irradiance: an evaluation of remote sensing methods. J. Geophys. Res. 110, C02017.

Lee, Z.-P., Ahn, Y.-H., Mobley, C.D., Arnone, R., 2010. Removal of surface-reflected light for the measurement of remote-sensing reflectance from an above-surface platform. Opt. Express. 18, 171–182.

Linnemann, K., Gege, P., Rößler, S., Schneider, T., Melzer, A., 2013. CDOM retrieval using measurements of downwelling irradiance. Proc. SPIE Remote Sensing, September 23–26, 2013. Dresden, Germany.

Lucarini, V., Saarinen, J.J., Peiponen, K.-E., Vartiainen, E.M., 2005. Kramers-Kronig relations and sum rules in linear optics. Kramers-Kronig Relations in Optical Materials Research. Springer, Heidelberg, Germany, pp. 27–48.

Lytle II, D.R., Carney, P.S., Schotland, J.C., Wolf, E., 2005. Generalized optical theorem for reflection, transmission, and extinction of power for electromagnetic fields. Phys. Rev. E 71, 056610.

Maritorena, S., Morel, A., Gentili, B., 2000. Determination of the fluorescence quantum yield by oceanic phytoplankton in their natural habitat. Appl. Optics 39, 6725–6737.

Martinez-Vicente, V., Simis, S.G.H., Alegre, R., Land, P.E., Groom, S.B., 2013. Above-water reflectance for the evaluation of adjacency effects in earth observation data: Initial results and methods comparison for near coastal waters in the western channel, UK. J. Eur. Optical Soc. 8, 13060.

McKee, D., Cunningham, A., 2005. Evidence for wavelength dependence of the scattering phase function and its implication for modeling radiance transfer in shelf seas. Appl. Opt. 44, 126–135.

Mishchenko, M.I., Lacis, A.A., Travis, L.D., 1994. Errors induced by the neglect of polarization in radiance calculations for Rayleigh-scattering atmospheres. J. Quant. Spectrosc. Radiat. Transf. 51 (3), 491–510.

Mobley, C.D., 1994. Light and Water. Academic press, San Diego.

Mobley, C.D., 1999. Estimation of the remote-sensing reflectance from above-surface measurements. Appl. Opt. 38, 7442–7455.

Mobley, C.D., Gentili, B., Gordon, H.R., Jin, Z., Kattawar, G.W., Morel, A., et al., 1993. Comparison of numerical models for computing underwater light fields. Appl. Opt. 32, 7484–7504.

Mobley, C.D., Boss, E., Roesler, C., 2017. Ocean Optics Web Book. < http://www.oceanoptics-book.info/ > (accessed 28.01.17).

Morel, A., 1974. Optical properties of pure water and pure Sea water. In: Jerlov, N.G., Steemann Nielsen, E. (Eds.), Optical Aspects of Oceanography. Academic press, London, pp. 1–24.

Morel, A., 2001. Bio-optical models. In: Thorpe, S.A., Turekian, K.K. (Eds.), Encyclopedia of Ocean Sciences. Academic press, pp. 317–326. . Available from: http://dx.doi.org/10.1006/rwos.2001.0407

Morel, A., Bricaud, A., 1981. Theoretical results concerning light absorption in a discrete medium, and application to specific absorption of phytoplankton. Deep-Sea Res. 28A (11), 1375–1393.

Morel, A., Maritorena, S., 2001. Bio-optical properties of oceanic waters: a reappraisal. J. Geophys. Res. 106, 7163–7180.

Morel, A., Prieur, L., 1977. Analysis of variations in ocean color. Limnol. Oceanogr. 22, 709–722.

Nair, A., Sathyendranath, S., Platt, T., Morales, J., Stuart, V., Forget, M.-H., et al., 2008. Remote sensing of phytoplankton functional types. Remote Sens. Environ. 112, 3366–3375.

NASA, 2016. OceanColor SeaDAS, < http://seadas.gsfc.nasa.gov/ > (accessed 28.01.17).

Odermatt, D., Gitelson, A., Brando, V.E., Schaepman, M., 2012. Review of constituent retrieval in optically deep and complex waters from satellite imagery. Remote Sens. Environ. 118, 116–126.

Palmer, K.F., Williams, M.Z., Budde, B.A., 1998. Multiply subtractive Kramers-Kronig analysis of optical data. Appl. Opt. 37, 2660–2673.

Palmer, S.C.J., Kutser, T., Hunter, P.D., 2014. Remote sensing of inland waters: challenges, progress and future directions. Remote Sens. Environ. 157, 1–8.

Plass, G.N., Humphreys, T.J., Kattawar, G.W., 1981. Ocean-atmosphere interface: its influence on radiation. Appl. Opt. 20 (6), 917–931.

Pokrovski, G.I., 1929. Über einen scheinbaren Mie-Effekt und seine mögliche Rolle in der Atmosphärenoptik. Z. Phys. 53, 67–71.

Pozdnyakov, D., Lyaskovsky, A., Grassl, H., Pettersson, L., 2002. Numerical modelling of trans-spectral processes in natural waters: implications for remote sensing. Int. J. Remote Sensing 22, 1581–1607.

Preisendorfer, R.W., 1961. Application of radiative transfer theory to light measurements in the sea. Union Geod. Geophys. Inst. Monogr. 10, 11–30.

Preisendorfer, R.W., 1976. Hydrologic Optics, 6 volumes. U.S. Dept. of Commerce, Washington.

Preisendorfer, R.W., Mobley, C.D., 1986. Albedos and glitter patterns of a wind-roughened sea surface. J. Phys. Ocean. 16, 1293–1316.

Ruddick, K.G., De Cauwer, V., Park, Y.-J., Moore, G., 2006. Seaborne measurements of near infrared water leaving reflectance: the similarity spectrum for turbid waters. Limnol. Oceanogr. 51, 1167–1179.

Sathyendranath, S., Platt, T., 1997. Analytic model of ocean color. Appl. Opt. 36, 2620–2629.

Schaepman-Strub, G., Schaepman, M.E., Painter, T.H., Dangel, S., Martonchik, J.V., 2006. Reflectance quantities in optical remote sensing—definitions and case studies. Remote Sens. Environ. 103, 27–42.

Schroeder, M., Barth, H., Reuter, R., 2003. Effect of inelastic scattering on underwater daylight in the ocean: model evaluation, validation, and first results. Appl. Opt. 42 (21), 4244−4260.

Schwarz, J.N., Kowalczuk, P., Kaczmarek, S., Cota, G.F., Mitchell, B.G., Kahru, M., et al., 2002. Two models for absorption by coloured dissolved organic matter (CDOM). Oceanologia 44, 209−241.

Segelstein, D.J., 1981. The Complex Refractive Index of Water. M.S. Thesis. University of Missouri-Kansas City, USA, p. 167.

Shifrin, K., 1988. Physical Optics of Ocean Water. American Institute of Physics, New York.

Siebert, F., Hildebrandt, P., 2008. Vibrational Spectroscopy in Life Science. Wiley-VCH Verlag GmbH & Co. KGaA, Weinheim, Germany.

Simis, S.G., Olsson, J., 2013. Unattended processing of shipborne hyperspectral reflectance measurements. Remote Sens. Environ. 135, 202−212.

Sipelgas, L., Arst, H., Kallio, K., Erm, A., Oja, P., Soomere, T., 2003. Optical properties of dissolved organic matter in Finnish and Estonian lakes. Nord. Hydrol. 34, 361−386.

Strong, A.E., 1974. Remote sensing of algal blooms by aircraft and satellite in Lake Erie and Utah Lake. Remote Sens. Environ. 3, 99−107.

Talaulikar, M., Suresh, T., Desa, E., Inamdar, A., 2014. An empirical algorithm to estimate spectral average cosine of underwater light field from remote sensing data in coastal oceanic waters. Limnol. Oceanogr. Methods 12, 74−85.

Ulloa, O., Sathyendranath, S., Platt, T., 1994. Effect of particle-size distribution of the backscattering ratio in seawater. Appl. Opt. 33, 7070−7077.

van de Hulst, H.C., 1957. Light Scattering by Small Particles. Dover, New York.

Walker, R.E., 1994. Marine Light Field Statistics. Wiley, New York.

Waterman, T.H., 1954. Polarization patterns in submarine illumination. Science 120 (3127), 927−932.

Zanefeld, J.R.V., Boss, E., Barnard, A., 2001. Influence of surface waves on measured and modeled irradiance profiles. Appl. Opt. 40, 1442−1449.

Zepp, R.G., Schlotzhauer, P.F., 1981. Comparison of photochemical behaviour of various humic substances in water: III. Spectroscopic properties of humic substances. Chemosphere. 10, 479−486.

Zhai, P.-W., Hu, Y., Trepte, C.R., Lucker, P.L., 2009. A vector radiative transfer model for coupled atmosphere and ocean systems based on successive order of scattering method. Opt. Express. 17, 2057−2079.

Zhang, X., Hu, L., 2009. Estimating scattering of pure water from density fluctuation of the refractive index. Opt. Express. 17, 1671−1678.

Zheng, X., Dickey, T., Chang, G., 2002. Variability of the downwelling diffuse attenuation coefficient with consideration of inelastic scattering. Appl. Opt. 41, 6477−6488.

Zimmermann, G., 1991. Fernerkundung des Ozeans. Akademie Verlag, Berlin.

Chapter 3

Atmospheric Correction for Inland Waters

Wesley J. Moses[1], Sindy Sterckx[2], Marcos J. Montes[1],
Liesbeth De Keukelaere[2], and Els Knaeps[2]
[1]U.S. Naval Research Laboratory, Washington, DC, United States, [2]Flemish Institute for Technological Research (VITO), Mol, Belgium

3.1 INTRODUCTION

Light received by a passive Earth-observing remote sensor goes through the Earth's atmosphere twice—from the Sun to the Earth's surface and from the surface to the sensor—before it reaches the sensor. As such, the light received at the sensor is invariably affected by absorption and scattering by gaseous molecules and particulate matter in the atmosphere. The process of correcting for the atmospheric effects and retrieving the reflectance of a target on the Earth's surface is called atmospheric correction. The atmospheric effect on the radiance received by a remote sensor is significantly large over water bodies because water is highly absorptive and contributes to only 20% or less of the total at-sensor radiance (e.g., Hovis and Leung, 1977). Correcting for these atmospheric effects is an essential prerequisite to retrieving accurate estimates of water-leaving radiance, which is the basis for deriving quantitative estimates of biophysical parameters from remotely sensed data.

Since the launch of the first ocean color sensor, the Coastal Zone Color Scanner (CZCS), in 1978, numerous algorithms have been developed to correct for atmospheric effects on remotely sensed data from water. Robust atmospheric correction algorithms have been developed for open ocean waters (e.g., Gordon and Wang, 1994), which have proven successful for correcting data from multispectral ocean color sensors such as CZCS, the MODerate resolution Imaging Spectroradiometer (MODIS), the Sea-viewing Wide Field-of-view Sensor (SeaWiFS), and the Visible Infrared Imaging Radiometer Suite (VIIRS). However, extending these algorithms to inland waters is not a simple, straightforward transition due to a number of factors

Bio-optical Modeling and Remote Sensing of Inland Waters.
DOI: http://dx.doi.org/10.1016/B978-0-12-804644-9.00003-3

69

that invalidate some basic assumptions inherent in these algorithms. The factors include:

- proximity to terrestrial sources of atmospheric pollution, which results in an optically heterogeneous atmosphere that is difficult to model;
- adjacency effects from neighboring land pixels, which is especially significant in cases of a raised, undulating topography around the water body;
- non-negligible reflectance of water in the near-infrared region due to high sediment concentrations in inland waters (caused by, for example, agricultural and industrial discharge from terrestrial sources, surface and sub-surface runoff, wind-driven resuspension of sediments, and sediment influx from landslides and shoreline erosion), which makes it difficult to accurately estimate and remove the effect of atmospheric aerosol contribution on the received signal; and
- variations in the altitude of inland water surface from the Mean Sea Level (MSL), which introduces uncertainties in the estimates of aerosol content in the atmospheric column above the water.

This chapter provides a discussion of some of the main challenges in atmospheric correction of remotely sensed data from inland waters, followed by a brief presentation of a few atmospheric correction algorithms that have been developed and used for inland waters in spite of the challenges and some recommendations to fill the gap in technology and algorithm development for atmospheric correction of remotely sensed data from inland waters.

3.2 CHALLENGES

The basic radiative transfer equation for atmospheric correction of remotely sensed data from a water body can be expressed as follows (radiance terms for contribution from adjacent terrestrial areas are omitted here for the sake of simplicity, assuming a large inland water body):

$$L_{at\text{-}sens} = L_{atm} + L_{spec}t_{spec} + L_w t_w \tag{3.1}$$

where $L_{at\text{-}sens}$ is the radiance received at the sensor, L_{atm} is the atmospheric path radiance, L_{spec} is the specular radiance from the water surface, t_{spec} is the transmittance of the specular radiance through the atmosphere, L_w is the water-leaving radiance (its radiative transfer through the water column has been discussed in Chapter 2), and t_w is the transmittance of the radiance from the water body reaching the sensor (Fig. 3.1). L_{atm} and $L_{spec}t_{spec}$ represent the non-water component of the radiance reaching the sensor and can be replaced by a single term, $L_0(= L_{atm} + L_{spec}t_{spec})$. The quantity of interest is the non-dimensional reflectance of water, ρ_w, which is expressed as,

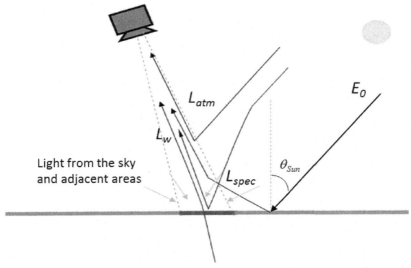

FIGURE 3.1 Components of the radiation reaching the sensor from a water target (modified from Sterckx et al. (2011)). L_w is the water-leaving radiance; L_{atm} is the atmospheric path radiance; L_{spec} is the specular radiance from the water surface; E_0 is the scalar solar irradiance; and θ_{sun} is the solar zenith angle.

$$\rho_w = \frac{\pi L_w}{\cos{(\theta_{sun})}E_0 t_{sun}} \qquad (3.2)$$

where E_0 is the scalar solar irradiance, θ_{sun} is the solar zenith angle, and t_{sun} is the transmittance of the solar radiation reaching the water body. Multiple reflections between the water body and the atmosphere are approximated by multiplying the reflectance term by $(1-s\rho_w)^{-1}$, where s is the spherical albedo of the atmosphere (Chandrasekhar, 1960).

Thus, Eq. (3.1) becomes

$$L_{at\text{-}sens} = L_0 + \frac{\rho_w \cos(\theta_{sun})E_0 t_{sun} t_w}{\pi(1 - s\rho_w)} \qquad (3.3)$$

The goal of atmospheric correction is to retrieve ρ_w from $L_{at\text{-}sens}$ after factoring out the other components. This is done by retrieving atmospheric properties from the received signal by modeling the scattering and absorption properties of the atmosphere, using known values of atmospheric parameters where possible, and employing assumptions about the reflectance properties of water. The at-surface remote sensing reflectance of water, R_{rs}^w, is calculated by dividing ρ_w by π. Atmospheric correction is particularly challenging for inland waters due to a number of factors that relate to the physical and bio-optical properties of inland waters and the inherent difficulties in accurately modeling the atmosphere above inland waters due to the optical complexity caused by influence from terrestrial sources of atmospheric pollution and the need to reconsider basic assumptions in atmospheric correction models that were originally developed for open ocean waters.

3.2.1 Challenges Due to Physical and Bio-optical Properties

High or anomalous reflectance from inland waters due to high turbidity, caused by influx of terrigenous particles from the surrounding land and wind-driven resuspension of sediments, the presence of floating objects on the water surface, and the influence of light reflected from adjacent land on the radiance received by the sensor present challenges for atmospheric correction.

3.2.1.1 High Turbidity and Floating Objects

Aerosol scattering contributes significantly to at-sensor radiance. Accurately retrieving the aerosol contribution and removing it from the remotely sensed data is one of the most challenging aspects of atmospheric correction. While there are different ways of correcting for aerosol contribution, the most common method involves estimating aerosol contribution using the at-sensor radiance at wavelengths where water is highly absorptive such that the received at-sensor radiance at those wavelengths can be fully attributed to aerosol contribution. This approach was originally implemented for atmospherically correcting data from the first ocean color sensor, CZCS (Gordon, 1978; Gordon and Clark, 1981), and subsequently the ocean color sensors that followed, such as SeaWiFS and MODIS (Gordon and Wang, 1994), wherein it was assumed that the water-leaving radiance was negligible in the near-infrared (NIR) region. This assumption, frequently referred to as the black-pixel assumption, holds true for clear ocean waters with very low concentrations of pigments (Gordon and Clark, 1981) and particulate matter. However, at moderate-to-high concentrations of particulate matter, there is appreciable scattering in the NIR region due to particles in water, resulting in non-negligible water-leaving radiance, and thus invalidating the black-pixel assumption (e.g., Doerffer and Fischer, 1994; Ruddick et al., 2000; Siegel et al., 2000; Hu et al., 2000). In such cases, atmospherically correcting the data using the black-pixel assumption would lead to overestimation of the aerosol contribution in the NIR region (as scattering from particles in the water will be erroneously treated as atmospheric aerosol scattering) and over-correction of atmospheric effects, resulting in very low, and often, negative reflectances at short wavelengths in the visible region (e.g., Hu et al., 2000; Ruddick et al., 2000; Moses et al., 2009).

Inland waters often contain significant concentrations of suspended particulate matter (SPM). Very high particulate concentrations, even on the order of more than a thousand grams per cubic meter, have been observed in many inland and estuarine waters (e.g., Doxaran et al., 2003; Wang et al., 2009).

High concentrations of SPM cause significant scattering of light in the NIR region. For example, Knaeps et al. (2012) collected measurements of reflectance and SPM concentration in Scheldt River, where SPM concentrations ranged from 15 to 402 g m^{-3}, and reported that reflectance in the NIR region was significantly non-zero. They observed a significant increase in reflectance between 950 and 1150 nm, with a peak around 1071 nm that had a strong positive correlation with SPM concentration. Thus, the NIR black-pixel assumption does not apply for turbid inland waters.

Several approaches have been developed, with limited success, to deal with the non-negligible NIR reflectance of turbid waters (see more details in Section 3.3). The high scattering of SPM in the NIR region necessitates alternate approaches, including the use of wavelengths in the longer, short wave infrared (SWIR) region, for retrieving atmospheric aerosol contribution. Reflectances at 1240, 1640, and 2130 nm have been recommended for retrieving aerosol contribution (e.g., Wang and Shi, 2005). Even within the SWIR region, non-negligible reflectance has been observed at shorter SWIR wavelengths (up to 1240 nm) in extremely turbid waters, such as the La Plata Estuary (Shi and Wang, 2009), Lake Taihu (Wang et al., 2011), and the Scheldt River (Knaeps et al., 2012). This has implications for applying SWIR-based atmospheric correction for very turbid waters from sensors such as the recently launched Ocean Land Colour Instrument, which has a single SWIR band at 1020 nm. Nevertheless, at longer SWIR bands (e.g., 1640 and 2130 nm), even highly turbid waters have virtually zero water-leaving radiance. Therefore, the black-pixel assumption would be valid for data in the SWIR bands for even very turbid waters. However, the noise level in relation to the signal is very high in the SWIR region, resulting in a very low signal-to-noise ratio (SNR), especially for SWIR bands designed for terrestrial applications, such as the SWIR bands of MODIS, VIIRS, and the Operational Land Imager (OLI) on Landsat-8. In order to achieve a high enough SNR while maintaining a spectral resolution that is fine enough to resolve atmospheric aerosol features, the spatial resolution will have to be relaxed. In the science definition team report for the National Aeronautics and Space Administration's (NASA) upcoming satellite mission, Plankton, Aerosol, Cloud, ocean Ecosystem (PACE), the authors recommend an SNR of ∼200−300 at 1240 and 1640 nm and ∼100 at 2130 nm, with spectral resolutions of 20, 40, and 50 nm, respectively, at those three bands, for robust atmospheric correction (PACE SDT, 2012). The PACE mission, which is designed for open ocean waters, has a threshold requirement of 1 km for the spatial resolution. However, a much finer spatial resolution, perhaps on the order of 100 m, is required for monitoring inland waters, considering the spatial extent of most inland waters and the spatial scales of variability of optically detectable biophysical phenomena in inland waters, which are expected to be similar to or smaller than the spatial scales of variability in coastal waters (Moses et al., 2016). Achieving the prescribed SNRs in the SWIR bands at a fine spatial resolution will be very challenging, especially for a spaceborne sensor.

Inland waters often contain algal scums, floating macrophytes such as water lilies, and floating debris that cover significant portions of the water surface and produce reflectance signals that are uncharacteristic of water and are very difficult to correct for atmospheric effects. Surface scums of algae and floating vegetation can produce a very high reflectance in the NIR region that resembles the reflectance from terrestrial vegetation (e.g., Kutser, 2004). Floating debris vary in their composition, size, and shape, and cannot be represented by a single reflectance feature. The anomalous reflectance produced by floating objects on the water surface causes atmospheric correction to fail. Unless the spectral properties of floating objects are known with sufficient accuracy that their spectral contribution can be explicitly isolated and removed through spectral modeling, image pixels with floating objects are usually flagged and not subjected to atmospheric correction.

3.2.1.2 Adjacency Effect

Adjacency effect is caused when light from objects surrounding the target area corresponding to a pixel gets reflected into the target-sensor path through atmospheric scattering and reaches the sensor. Adjacency effect, thus, modifies the at-sensor radiance recorded over the target and decreases the spectral contrast between the target and the surrounding region. Adjacency effect is more pronounced in cases of a dark target with a bright surrounding, such as an inland water body surrounded by land. The atmospheric aerosol optical depth (AOD), viewing and illumination geometries, and the topography are major factors that determine the adjacency effect.

Several studies based on airborne and spaceborne data have reported abnormally high NIR reflectance over inland and coastal waters due to adjacency effects (e.g., Reinersman and Carder, 1995; Van Mol and Ruddick, 2005). Adjacency effect is highest in images acquired with oblique viewing angles over a low-lying water body surrounded by raised topography in an optically turbid atmosphere. The extent of the adjacency effect depends significantly on the atmospheric AOD and the aerosol vertical distribution (Tanre et al., 1981; Santer and Schmechtig, 2000). Therefore, in order to reliably account for the adjacency effect, it is necessary to accurately estimate the optical properties and vertical distribution of aerosols, which, for reasons described in this chapter, is challenging for inland waters. Moreover, the vertical distribution of aerosols cannot be retrieved from remotely sensed visible-NIR (VNIR) data. Unless measurements of the aerosol vertical profile are available from airborne lidar or instruments deployed *in situ*, the aerosol distribution has to be modeled. Large discrepancies between the actual vertical distribution and the modeled distribution produce significant inaccuracies in the correction of adjacency effect. Conversely, adjacency effect in the spectral bands used for determining the aerosol contribution also adds uncertainty to the retrieval of aerosol properties (e.g., Santer and Schmechtig, 2000). In order to overcome these competing limitations, reference water pixels that are far enough from the land to be free from adjacency effects are used to determine the aerosol properties, with the assumption that the aerosol properties retrieved for the reference pixels can be carried over to the pixels with adjacency effects. Such an approach might work reasonably well in a coastal scenario with a fairly homogeneous atmosphere, where reference pixels may be found in offshore waters. However, for inland waters, depending on the spatial resolution of the remote sensor, the topography, and the size of the water body, the chances of finding such reference pixels might be slim.

3.2.2 Challenges Due to Difficulties in Atmospheric Modeling

3.2.2.1 Optical Heterogeneity Due to Terrestrial Influence

Certain atmospheric correction algorithms utilize physical models of the atmosphere in order to calculate the path radiance, transmission, etc. This involves knowing (or assuming) the spectral absorption and spectral phase function (angular scattering description) due to aerosols at each altitude in a modeled atmosphere. For this class of algorithms, the typical inputs required for these calculations include the optical properties (real and imaginary indices of refractions) of the aerosol particles and the size distributions (such as the volume size distribution). The vertical aerosol distribution is also required. Many atmospheric correction algorithms have a selection of aerosol models (combinations of optical properties, size distributions, and vertical distributions) that can be chosen, with the total amount of aerosol loading chosen based on the AOD at a reference wavelength. Additionally, some algorithms also allow the use of parameters derived from measurements obtained using handheld sun photometers, an Aerosol Robotic Network (AERONET) station (Holben et al., 1998), or even more complicated distributions.

The spatial and temporal variations of aerosol properties are important factors in atmospheric modeling. Indeed, on land, there are numerous aerosol sources due to human activities. Even though airborne pollutants are regulated at various levels around the world, many anthropogenic sources of aerosols, such as factories, vehicles, and dust (arising from work sites), are still present. In addition to anthropogenic sources, there are numerous natural sources, including natural fires, dust events, and volcanoes. Wind-blown pollen, which is a biogenic source of particulates, can affect remote sensing measurements (Noh et al., 2013) locally in both space and time, contributing up to $\sim 30\%$ of the AOD. Trees also emit chemicals that can form aerosols (Zhang et al., 2009), including the conspicuous "blue haze" of some mountain forests. Indeed, in the presence of so many sources, one can ask about the homogeneity of the aerosol distribution at a particular time and place, and how it may affect airborne and spaceborne VNIR-SWIR remote sensing measurements.

In trying to understand the effects of aerosol variability/heterogeneity on remote sensing observations, it is important to determine the spatial scales of importance in the problem. Spatial scales of importance include (i) the size of the water body, (ii) distances from point sources of aerosol, (iii) distances from nearby aerosol monitoring sensors, if any, and (iv) the image pixel size. One method of dealing with optical heterogeneity is to select observing dates that avoid, or at least minimize, anthropogenic and natural aerosols. Another method, which is really a standard "best practice," is to measure as much about the aerosols as possible during the sensor overpass, and to additionally use field spectrometers to measure large areas, covering multiple satellite pixel footprints on the land (natural and man-made targets) and water, if possible. However, *in situ* measurements are often not available to support opportunistic observations of interesting features from routinely collected remote sensing data

When no data are available from a nearby AERONET site or other ground measurements, AODs derived from other remotely sensed data can be used. For small inland water bodies, the best estimates are likely from nearby determinations over land. Kokhanovsky et al. (2007) compared AERONET-derived aerosol properties, such as the AOD, with those estimated from multispectral satellites for a single location in central Europe, with observations from several satellites with differing spatial, spectral, and angular coverage. There are obvious difficulties in comparing point measurements from ground-based AERONET sites to determinations made from satellites using many different algorithms. From the available suite of sensors and algorithms, it was found that there was a large dispersion of satellite-derived AODs compared to AODs measured at AERONET sites, but estimates from the MEdium Resolution Imaging Spectrometer (MERIS) were in best agreement with AERONET values.

3.2.2.2 Breakdown of Basic Assumptions

Atmospheric correction models originally developed for ocean color remote sensing are based on several implicit assumptions about the atmosphere and the target environment that are often not applicable for inland waters. For example, calculations of the atmospheric molecular and particulate loading are typically based on an assumption of water surface at MSL, whereas most inland waters are not typically at MSL. For inland waters at altitudes significantly higher than MSL, the atmosphere above would be thinner than it would be for a water body at MSL with similar environmental conditions. This can cause overestimation of atmospheric contribution for inland waters at high altitudes. Complex topography surrounding inland waters can cause additional scattering of sky light and thereby invalidate assumptions about clear sky illumination.

When modeling the aerosols in the atmosphere and their effects on the signal received by a sensor, one needs to know the surface elevation because the path radiance and transmission depend on the amount of atmosphere between the sensor and the surface. Algorithms that utilize Look-Up-Tables (LUTs) will clearly need values of atmospheric properties calculated over the required range of surface elevations in order to be useful. While this is not difficult, it does mean repeating calculations for many different surface elevations. Other algorithms that perform calculations as needed, such as the Second Simulation of the Satellite Signal in the Solar Spectrum (6S) (Vermote et al., 1997) and algorithms related to it, will not have problems in this regard.

Another quantity of interest is the aerosol scale height, which is typically a required input for modeling the effects of aerosols. A global study of AODs retrieved from the Cloud-Aerosol Lidar and Infrared Pathfinder Satellite Observations (CALIPSO) data showed that clear-sky aerosol scale heights were generally ≤3 km worldwide and typically <1.5 km over the ocean (Yu et al., 2010). Yu et al. (2010) show that within inland regions there is no appreciable difference in aerosol scale height for mountainous regions compared to neighboring areas. AERONET stations and measurements of visibility can be used to derive estimates of aerosol scale heights in a manner similar to that of Wong et al. (2009).

By definition, inland water bodies are surrounded mostly by land. Because of this, the mean reflectance of the surrounding environment is important in deriving corrected reflectance of the water body. In addition to the environmental adjacency effect, another effect, called the "topographic" adjacency effect, arises when large portions of the sky are blocked by the land surface (possibly covered by vegetation), such as with lakes surrounded by mountain slopes. While the direct solar beam is not affected (when the sun is not behind the hills/mountains), the diffuse illumination would be different than if the ground were flat.

One aspect of remote sensing of water bodies that may occur in both oceans and inland waters is the presence of sun glint. In order to understand the sub-surface signal, both atmospheric and glint corrections are necessary. Water surfaces typically have a superposition of waves with different wavelengths, amplitudes, and directions, but the wavelengths and amplitudes (depths/heights) are in general limited by the size and depth of water bodies. Additionally, since wind is responsible for most waves, smaller water bodies surrounded by raised topography may be protected from wind.

In the open ocean, there has been success in using the Cox-Munk (Cox and Munk, 1954) formalism in order to characterize the distribution of wind-formed capillary waves (≤2 cm wavelength) and determine the statistical effects of small facets reflecting sunlight directly from the sun into the sensor. Data from sensors such as MODIS and SeaWiFS, with 1 km pixels, are corrected using this particular statistical description, as part of the atmospheric correction procedure. However, when pixels are small enough that the mean surface within the pixel is no longer flat, as would be the case for high-resolution data from inland waters, then the Cox-Munk formalism is no longer sufficient. It can frequently be the case that a statistical treatment of glint is insufficient or unjustified in the remote sensing of inland waters. There can be many reasons for this: (i) the surface of shallow rivers frequently have their own non-flat topography due to the currents moving over the non-flat bottom, (ii) persistent surface features related to river in-flows and out-flows in lakes, ponds, and other inland water bodies, (iii) small pixel sizes used in remote sensing of small water bodies, and (iv) anthropogenic sources of wave generation (e.g., boats). In general, statistical solutions do not work well for small pixels. In this case, image-based solutions for glint removal become necessary, several of which were reviewed by Kay et al. (2009). Research continues to be active for both small pixels (Patterson, 2014; Martin et al., 2016) and large pixels (Mobley, 2015) towards developing methods of glint correction that account more completely for the true surface shape and for the polarized light signal. Because many of these small-pixel methods are image-driven, typically there is no need to know or model the wind speed. Martin et al. (2016) report results from an image-driven glint correction approach for inland waters. Such image-driven methods may work well even in sheltered water bodies with little or no waves.

3.3 EXISTING ALGORITHMS

In spite of the aforementioned challenges, several methods have been developed to account for atmospheric and adjacency effects on remotely sensed data and retrieve surface reflectance. This section provides brief descriptions of a few atmospheric correction and adjacency correction algorithms and a case study demonstrating combined atmospheric and adjacency correction

3.3.1 Atmospheric Correction Algorithms

Several approaches have been developed to atmospherically correct remotely sensed data from turbid productive waters, with various strategies to deal with non-negligible reflectance in the NIR region (Table 3.1).

3.3.1.1 Algorithms Deriving Aerosol Information from Clear Water Pixels in the Image, Assuming Spatial Homogeneity

In this approach, the aerosol type is retrieved from clear water pixels in the image and applied to the rest of the pixels, assuming spatial homogeneity of aerosol properties (Hu et al., 2000, Ruddick et al., 2000). The algorithm of Ruddick et al. (2000), often referred to as the MUMM algorithm (after the

TABLE 3.1 Overview of Atmospheric Correction Approaches for Inland Waters

Algorithmic Approach	Some References
Spatial extension of aerosol information from nearby clear water pixels	Ruddick et al. (2000)
SWIR black-pixel algorithms	Wang et al. (2011); Vanhellemont and Ruddick (2015); Gao et al. (2000); Wang et al. (2011)
Spatial extension of aerosol information from nearby land	Vidot and Santer (2005); Guanter et al. (2010)
Simultaneous retrieval of atmospheric and water components	Doerffer and Schiller (2007, 2008); Schroeder et al. (2007); Steinmetz et al. (2011); Kuchinke et al. (2009)
Image-based methods	Chavez (1996); Bernstein et al. (2005a,b)

first author's host institute, the Management Unit of the North Sea Mathematical Models, Belgium), was originally developed for the correction of SeaWiFS imagery over coastal and inland waters. The MUMM algorithm is based on two major assumptions. The first assumption is that the aerosol type does not vary within the image (or at least within a sub-image of interest) and therefore the ratio of the aerosol reflectances at two NIR bands, $\varepsilon(\lambda_{NIR1}, \lambda_{NIR2}) = \rho_a(\lambda_{NIR1})/\rho_a(\lambda_{NIR2})$, is constant. The aerosol reflectance ratio is extracted from clear water pixels in the image for which the water reflectance is assumed to be zero. In clear waters, the aerosol reflectance can be approximated by the Rayleigh-corrected reflectance. In the original work of Ruddick et al. (2000), clear water pixels were identified by visually inspecting the scatterplot of Rayleigh-corrected reflectance at the two NIR bands. A cluster of points in the lower left corner of the scatterplot, i.e., points corresponding to low Rayleigh-corrected reflectance at both bands, defined the aerosol reflectance ratio $\varepsilon(\lambda_{NIR1}, \lambda_{NIR2})$. Jiang and Wang (2014) noted that such an approach is not realistic in an operational processing environment. More automatic approaches to select the clear water pixels have been proposed by Vanhellemont and Ruddick (2015) and Goyens et al. (2013). For these automatically detected clear water pixels, the $\varepsilon(\lambda_{NIR1}, \lambda_{NIR2})$ value can be easily derived from the slope of the regression line or from the median ratio of the Rayleigh-corrected reflectances at the two NIR bands.

The second assumption in the MUMM algorithm is that for all water pixels in the image the ratio of the water reflectance in the two NIR bands, $\alpha(\lambda_{NIR1}, \lambda_{NIR2}) = \rho_w(\lambda_{NIR1})/\rho_w(\lambda_{NIR2})$, is constant. This assumption is based on the theoretical approximation that the absorption in the NIR region is determined entirely by pure water absorption and the backscattering has only weak spectral dependence. Therefore, the spectral shape of ρ_w in the NIR region is considered constant. This constant spectral shape of $\rho_w(NIR)$ has been referred to as the NIR similarity spectrum for turbid waters (Ruddick et al., 2006). Ruddick et al. (2006) have tabulated values of the similarity spectrum in the NIR region normalized by the reflectance at 780 nm, at 2.5 nm intervals, allowing the calculation of $\alpha(\lambda_{NIR1}, \lambda_{NIR2})$ for any NIR band pair.

The accuracy of the MUMM algorithm is highly depended on the validity of the two assumptions for a given image. Jaelani et al. (2013) reported varying performances of the MUMM algorithm when applied to MERIS images acquired from the extremely turbid Kasumigaura Lake in Japan. Poor performances were achieved when one or both of the assumptions were not met, while acceptable accuracies were obtained when both assumptions were valid. It is well documented that the validity of the second assumption is violated in highly turbid waters since the spectral relationship between the marine reflectances at two NIR bands is no longer linear (Doron et al., 2011; Goyens et al., 2013). Therefore, for extremely turbid waters, Goyens et al. (2013) proposed replacing the constant NIR reflectance ratio with a polynomial spectral relationship. The implementation of a polynomial spectral relationship reduces the error in the retrieved water reflectance but makes the algorithm more sensitive to errors in the retrieval of the aerosol model.

Additional complexity arises for inland waters. An entire inland water body might be turbid, with no clear water pixels (Vidot and Santer, 2005). Adjacency effects might contaminate the observed radiance, especially in the NIR region, and should be corrected for before applying the algorithm. Finally, in cases where the inland water body is dominated by floating vegetation or macrophytes, the spectral dependency of the reflectance in the NIR region cannot be considered constant, thereby invalidating the NIR similarity spectrum.

3.3.1.2 Algorithms Based on Extending the "Black-Pixel" Approach to the SWIR Region

This approach is conceptually similar to the standard Gordon and Wang (1994) approach. However, instead of bands in the NIR region, the aerosol contribution is retrieved using bands in the SWIR region, where the reflectance from water is virtually null even in turbid conditions due to the high absorption of pure water in this spectral region (Fig. 3.2).

Methods based on the SWIR black-pixel assumption have the advantage that no other assumptions have to be made regarding the optical properties of water. SWIR-based atmospheric correction schemes have been developed for ocean color sensors with SWIR bands, like MODIS and VIIRS, and

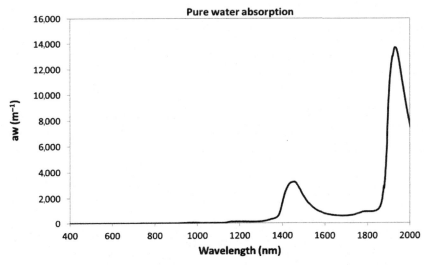

FIGURE 3.2   Pure water absorption coefficient (Kou et al., 1993; Pope and Fry, 1997).

also for sensors with higher spatial resolution, such as Landsat 8 OLI (Vanhellemont and Ruddick, 2015) and Sentinel-2. SWIR-based approaches were originally developed for coastal turbid waters, but they are in principle also applicable to inland waters (Wang et al., 2011).

The basic approach of SWIR-based methods is as follows: First, the spectral ratio of Rayleigh-corrected reflectances at two SWIR bands is used to deduce the aerosol model or aerosol spectral shape, as represented by the epsilon factor $\varepsilon(\lambda_s, \lambda_l) = \rho_a(\lambda_s)/\rho_a(\lambda_l)$, where λ_s and λ_l represent the shorter and longer SWIR bands, respectively, and ρ_a is the aerosol reflectance ratio, which can be approximated by the Rayleigh-corrected reflectance under the SWIR black-pixel assumption.

The retrieved ε values are compared against pre-computed values tabulated for a suite of aerosol models to select the appropriate aerosol model. As ρ_w is assumed to be zero at both λ_s and λ_l, there is no need to select clear water pixels. The sensitivity of ε and, consequently, the accuracy of the retrieved aerosol optical properties increase with increasing spectral distance between λ_s and λ_l. For example, aerosol optical properties retrieved using the 1240 nm and 2130 nm bands of MODIS are theoretically more accurate than those retrieved using the 1640 and 2130 nm bands (Wang and Liu, 2013). ε can be derived pixel-by-pixel or as a single value per image using the median value or regression slope of the reflectance ratio. The latter approach is often desirable as a pixel-by-pixel retrieval of ε might introduce additional noise in the estimates of aerosol optical properties due to the typically low SNR in the SWIR bands.

Once the aerosol model is determined, the AOD can be derived from the aerosol reflectance at a single SWIR band using a LUT containing aerosol reflectance as a function of AOD. Finally, the atmospheric correction parameters for all visible and NIR bands can be derived from the retrieved AOD and aerosol model through a simple exponential extrapolation of the spectral ε (Vanhellemont and Ruddick, 2015) or using a pre-computed LUT.

Algorithms based on the SWIR black-pixel assumption have been integrated into the SeaWiFS Data Analysis System (SeaDAS) development environment for processing ocean color data and the Atmospheric Correction for OLI "lite" (ACOLITE) processor (Vanhellemont and Ruddick, 2015), which has been designed for atmospherically correcting data from high-resolution sensors like the OLI sensor on Landsat 8 and the MultiSpectral Instrument on Sentinel-2. However, it should be noted that SWIR-based algorithms are not new and have been used in the hyperspectral community for many years. Gao et al. (2000), for instance, proposed the retrieval of aerosol parameters from hyperspectral data on the basis of a spectrum-matching technique using several bands beyond 1000 nm, and implemented a version for MODIS (Gao et al., 2007). This approach has been implemented in the Tafkaa atmospheric correction code (Montes et al., 2001). While Tafkaa owes a part of its heritage to the development of the ATmospheric REMoval (ATREM) algorithm (Gao et al., 1993), it uses a pre-calculated LUT to calculate scattering effects in the atmosphere (Gao et al., 2000). Additionally, sea-surface reflections are taken into account statistically by utilizing standard Cox-Munk calculations, which requires pixels with a mean flat/level orientation. Currently, the Tafkaa LUTs assume a lower boundary condition at MSL for the altitude. Because of both of these limitations, the use of Tafkaa for inland waters is currently limited to large water bodies at altitudes near MSL. Tafkaa offers the option of determining the solution pixel-by-pixel for the entire image or from one part of the image and applying the retrieved aerosol properties over the entire image.

While several successful applications of SWIR-based atmospheric correction schemes for estuarine and inland waters have been published in recent years (Wang et al., 2011; Dogliotti et al., 2011; He and Chen, 2014; Brando et al., 2015; Wang et al., 2012), there are also some limitations to be considered. Many sensors lack SWIR bands, which necessitates alternative methods. As stated earlier (Section 3.2.1.1), the black-pixel assumption may not be valid in extremely turbid waters at the shorter end of the SWIR region. Moreover, SWIR bands are prone to high levels of sensor noise. To reduce the effect of sensor noise in SWIR bands, Wang and Shi (2012) proposed spatially smoothing the at-sensor SWIR radiance using a simple moving box average approach. This approach, however, did not sufficiently remove noise when applied to Landsat-8 OLI data (Vanhellemont and Ruddick, 2015) and revealed spatial features caused by sensor artifacts. The use of a fixed ε per OLI image (or a subimage) resulted in improved reduction of noise.

An alternative method of addressing noise issues in the SWIR bands is the application of a NIR-SWIR switching approach, where pixels identified as turbid are processed with the SWIR-based method and non-turbid pixels with a NIR-based method. Pixels are identified as turbid when both a turbidity index, calculated based on reflectances at two bands (e.g., 748 and 1240 nm bands for MODIS), and the normalized water leaving radiance at a reference band (e.g., 869 nm) exceed certain thresholds. Due to the effect of sensor noise and the spectral distance across which the retrieved aerosol properties are extrapolated, which can be a particularly significant issue when absorbing aerosols are present, with everything else the same, uncertainties in SWIR-based retrievals are generally slightly higher than those from NIR-based retrievals. Therefore, NIR-based retrieval is preferred as long as the black-pixel assumption is valid at the NIR bands. When a SWIR-based retrieval is necessary, Oo et al. (2008) recommend constraining the SWIR-based aerosol model retrieval using reflectance from a very short visible band (e.g., the 412 nm band of MODIS) that can be assumed to be quasi-black for highly productive waters with high Colored Dissolved Organic Matter (CDOM). He et al. (2011) have suggested the use of ultra-violet (UV) bands for atmospheric correction in highly turbid waters, especially for sensors lacking SWIR bands.

3.3.1.3   Algorithms Based on Spatial Extension of Aerosol Information Retrieved from Nearby Land

In this approach, aerosol information retrieved from nearby land pixels using a land-based method is extended to water pixels, assuming spatial invariability of aerosols across the land and water. This approach does not rely on assumptions about water quality (i.e., clear or turbid) or the reflectance properties of water and is generally free from adjacency effects on aerosol retrieval. For aerosol retrieval over land, the Dense Dark Vegetation (DDV) method, originally developed by Kaufman and Sendra (1988), is often used. Slightly different implementations of the DDV method exist. They rely on the assumption that the surface reflectance of dark vegetation is either constant for a given wavelength or that its ratio at two wavelengths—one short (in the blue or red regions) and the other long (in the SWIR region)—is constant. Vidot and Santer (2005) corrected SeaWiFS data acquired over Lake Balaton and Lake Constance by assuming pre-determined, constant values for reflectance from DDV in blue and red bands. They used this assumption to retrieve AOD and aerosol type over DDV pixels. DDV pixels were selected based on a spectral index, the Atmospherically Resistant Vegetation Index (ARVI) (Kaufman and Tanre, 1992). The ARVI uses blue, red, and NIR reflectances corrected for Rayleigh scattering and gaseous absorption. Pixels with an ARVI value larger than a threshold value are identified as DDV. The DDV approach was also used by Campbell et al. (2011) for atmospheric correction of MERIS imagery from freshwater impoundments. However, instead of determining both AOD and aerosol type, only the AOD was retrieved from DDV pixels, while the aerosol type was selected *a priori* based on the wind conditions and location of the water body. Many researchers have reported difficulties in reliably retrieving the aerosol model over land (e.g. Ramon and Santer, 2005; Grey et al., 2006). A major issue with the DDV method is that pixels satisfying the criteria for DDV are sparse. For instance, only less than 1% of the land pixels in Western Europe can be considered as pure DDV pixels (Borde et al., 2003). Therefore, the aerosol type is often not derived from the image itself. The appropriate aerosol type can be selected based on climatology and/or local information of the site. Often, a fixed continental/rural type of aerosol model is used (Béal et al., 2007; Guanter et al., 2010).

A different land-based aerosol retrieval approach was proposed by Guanter et al. (2010) for the correction of MERIS data over inland waters, which does not rely on the presence of DDV pixels. The approach, which is implemented in the atmospheric processor SCAPE-M (Self-Contained Atmospheric Parameters Estimation for MERIS data) (Guanter et al., 2010) and the sensor-independent atmospheric correction scheme OPERA (OPERational Atmospheric correction algorithm) (Sterckx et al., 2015b), estimates the AOD at each macro-pixel, typically 30 km by 30 km, through a multi-parameter inversion of the at-sensor radiances at five reference pixels. These five reference pixels are land pixels with high spectral contrast (Guanter et al., 2007). It is assumed that these reference pixels can be represented by a linear combination of two endmember spectra, a vegetation spectrum and a soil spectrum. In the inversion step, the endmember abundances and the AOD are retrieved concurrently. Guanter et al. (2010) obtained a good correspondence between ρ_w retrieved through SCAPE-M and measured *in situ* from turbid waters. However, poor results were obtained over clear waters. SCAPE-M results presented in Jaelani et al (2013) showed that although the magnitude of the retrieved ρ_w was too high the spectral shape was consistent.

In general, the main advantage of the application of land-based aerosol retrieval methods for atmospheric correction of inland waters is that it is applicable to waters with various complexities, such as highly turbid waters, shallow waters, macrophyte-dominated waters, and highly eutrophic waters, where most of the water-based aerosol retrievals fail. It is also important to note that these atmospheric correction processors, originating from the land community, typically take into account various terrain altitudes in the LUTs, thus allowing proper correction of Rayleigh scattering for lakes situated at high altitudes. The difficulties in retrieving information on the aerosol type and uncertainties in the assumption of spatial homogeneity of aerosols over large water bodies are the main disadvantages of this approach.

3.3.1.4 Simultaneous Retrieval of Atmospheric and Water Components

Several methods have been developed to simultaneously retrieve atmospheric parameters, ρ_w, and biophysical properties of water, such as constituent concentrations, by taking into account the entire VNIR at-sensor radiance spectrum. Because of non-linear interactions in the radiative transfer, a nonlinear inversion method is required to correct for atmospheric effects and retrieve ρ_w from the at-sensor radiance. The nonlinear inversion is done through neural network or spectral optimization techniques.

Several artificial neural network approaches, such as the Case-2 Regional (C2R) processor (Doerffer and Schiller, 2007), the Free University Berlin (FUB) processor (Schroeder et al., 2007), the Eutrophic Lakes processor (Doerffer and Schiller, 2008), and the Boreal Lakes processor (Doerffer and Schiller, 2008) have been designed specifically for application to MERIS data and are included in the **Basic ERS & ENVISAT (A)ATSR and MERIS (BEAM)** toolbox. The C2R, Eutrophic Lakes, and Boreal Lakes processors follow the same coupled inverse/forward neural network structure, but they differ in the bio-optical properties of the data used to train the forward neural network. C2R was trained using optical properties and constituent concentrations from optically complex coastal waters, whereas the Eutrophic Lakes processor was trained with bio-optical data from eutrophic Spanish lakes, spanning a large range of chlorophyll-*a* (chl-*a*) concentrations ([chl-*a*]), and the Boreal Lakes processor was trained with data from boreal Finnish lakes with high CDOM absorption. The FUB processor consists of four neural networks. One network performs the atmospheric correction, while the other three networks directly retrieve the concentrations of constituents in water from the measured at-sensor radiance (Schroeder et al., 2007). The neural networks were trained with radiative transfer calculations performed with the Matrix Operator Method (MOMO) code for a range of atmospheric and in-water constituent concentrations. An overview of the ranges of training parameters for the different MERIS neural network based processors is given by Palmer et al. (2015). The main limitation of these neural network approaches is that the performance of these algorithms depends on whether the bio-optical properties of inland waters in a given image are within the ranges of parameters used to train the network because neural networks cannot accurately extrapolate beyond the ranges of the training parameters. For instance, Feng et al. (2014) attributed errors in the magnitude and spectral shape of ρ_w retrieved through the neural network processors included in the BEAM toolbox for Lake Poyang in China to the drastic differences between the optical properties of the Chinese lakes and the European lakes that were used for training the neural networks. Palmer et al. (2015), therefore, argued that more rigorous validation of the neural network processors is needed to better understand the uncertainty and performance limitations of the processors and select the most optimal processor for a given lake.

The POLYnomial based algorithm applied to MERIS (POLYMER), developed by Steinmetz et al. (2011), employs a spectral optimization technique to retrieve atmospheric and in-water parameters simultaneously from all available spectral bands. It combines a water reflectance model that depends on two parameters, namely, [chl-a] and particle backscattering, and a polynomial atmospheric reflectance model with three spectral components, which models the reflectance of the atmosphere, including aerosols and contamination by sun glint. The spectral optimization technique consists of optimizing the different parameters (five in total) of the atmospheric and water reflectance models in order to obtain the best spectral fit between measured and modeled at-sensor reflectances. POLYMER has been proven successful in retrieving ocean color information from MERIS data even under high sun glint conditions and has therefore been selected as the baseline MERIS processor for the Ocean Colour Climate Change Initiative. However, to our knowledge, there has been no report of applying POLYMER to inland water data.

A non-linear spectral optimization approach has also been proposed by Kuchinke et al. (2009) to simultaneously determine the parameters of the aerosol and bio-optical models for Case-2 waters. This algorithm assumes that some parameters that can be used to constrain the bio-optical model are known *a priori*, for example, from *in situ* observations, thereby making the algorithm very site-specific. Furthermore, it should be noted that because the parameters of the aerosol and bio-optical models are retrieved simultaneously an error in the coefficients of any parameter in the bio-optical model will impact the retrieval of all parameters, including the aerosol parameters. In spectral optimization algorithms, the parameters to be retrieved are often constrained by pre-defined lower and upper bounds to facilitate fast convergence of the solution. For example, in the algorithm of Kuchinke et al. (2009), the upper bound for [chl-a] is 50 mg m^{-3}. Therefore, the algorithm will probably fail under conditions with higher [chl-a], as would often be in inland waters with algal blooms.

3.3.1.5 Image-Based Algorithms

Image-based algorithms are solely based on the information retrieved from the image to convert at-sensor radiance to surface reflectance. They require no radiative transfer codes or ancillary information. Image-based methods are particularly popular for the correction of historical remote sensing data for which ancillary data are not available or data from sensors that may not be properly calibrated, as image-based algorithms are relatively insensitive to radiometric calibration inaccuracies. Furthermore, image-based approaches allow for the retrieval of approximate surface reflectance when the solar illumination intensity is unknown, for example under the presence of clouds.

The most simple image-based atmospheric correction is the Dark Object Subtraction (DOS) approach. The DOS method assumes that the image contains at least a few dark pixels. For these dark pixels, the DOS method assumes that the at-sensor radiance can be completely attributed to atmospheric contribution and specular reflection from the water surface. The DOS method searches for the lowest radiance in each spectral band for the entire image. This minimum radiance, taken as the dark pixel radiance, is then subtracted from the at-sensor radiance for all pixels. Gong et al. (2008) found the DOS approach to be not very effective in removing atmospheric effects over inland waters. One of the reasons for the poor performance is that the DOS method only corrects for the additive scattering component caused by path scattering and not for the multiplicative effect of transmittance (Chavez, 1996). Chavez (1996) extended the DOS method with a correction for the multiplicative effect of transmittance by approximating the downward atmospheric transmittance with the cosine of the solar zenith angle. This method, called COST or COS(TZ) as an abbreviation for the cosine of the solar zenith angle, has been used in various studies for atmospheric correction over inland waters, with slight modifications in some cases (e.g., Osińska-Skotak, 2005; Mancino et al., 2009; Zhou et al., 2006). Nevertheless, when meteorological data are available, an atmospheric correction algorithm based on a radiative transfer model is preferred (e.g., Allan et al., 2011).

The QUick Atmospheric Correction (QUAC) algorithm (Bernstein et al., 2005a,b) is a more sophisticated image-based approach. QUAC, which is available as an add-on to the image processing software ENVI (ENvironment for Visualizing Images), is based on three assumptions: (i) there are at least 10 spectrally diverse pixels (i.e. endmembers) in the image, (ii) the spectral standard deviation of the reflectance of the chosen set of diverse pixels is a wavelength-independent constant (i.e., spectrally flat), and (iii) there are sufficiently dark pixels in the image for a baseline atmospheric correction. Similar to the other image-based algorithms, the performance of QUAC depends strongly on the extent to which the aforementioned assumptions are met in a given image. Moses et al. (2012) applied the QUAC approach to hyperspectral data acquired by the Airborne Imaging Spectrometer for Applications (AISA) over turbid lakes in Nebraska, USA. They concluded that QUAC can be a reasonably viable atmospheric correction procedure capable of deriving smooth reflectance spectra, preserving spectral features related to chl-a, if the basic assumptions are satisfied. Zeng et al. (2013), however, compared four different atmospheric correction algorithms, including QUAC, for the correction of HJ-1A/1B images from the Poyang Lake in China and found that the accuracy of QUAC was lower than that of the three other algorithms, namely, the Fast Line-of-sight Atmospheric Analysis of Spectral Hypercubes (FLAASH), 6S, and COST.

3.3.2   Adjacency Correction Algorithms

While full 3D Monte-Carlo simulations allow accurate modeling of adjacency effects, simplified formulations are needed for implementation of an operationally fast correction scheme. One approach of correcting for adjacency effects involves the use of an atmospheric Point Spread Function (PSF), which allows calculation of contributions from neighboring pixels to the at-sensor radiance from a target pixel. In order to calculate the atmospheric PSF in a time-efficient manner, several methods rely on the single scattering approximation (Santer and Schmechtig, 2000). This primary scattering assumption is used in the sensor-independent adjacency correction algorithm developed by Kiselev et al. (2015). The adjacency correction module is integrated within the Modular Inversion and Processing (MIP) system (Heege and Fischer, 2004). MIP, which can be used for data from a variety of sensors, includes different physics-based modules (both atmospheric and bio-optical models) that are applied successively to derive biophysical parameters from the measured at-sensor radiance. Also, the Improved Contrast between Ocean and Land (ICOL) processor (Santer et al., 2007; Santer and Zagolski, 2009) developed specifically for MERIS and available via the BEAM toolbox relies on the primary scattering assumption. ICOL includes a simplified formalism for the influence of bright clouds and a land mask for the Fresnel reflection in order to account for the fact that the vicinity of land might reduce the direct to diffuse term of Fresnel reflectance. ICOL can be run as a preprocessor to correct the at-sensor radiance for adjacency effects, after which the data can be further processed with a user-selected atmospheric correction algorithm. Several studies (e.g. Odermatt et al., 2010; Kratzer and Vinterhav, 2010) have indicated an improvement in the retrieval of reflectance and derived biophysical parameters when the data were pre-processed with ICOL. Also Beltrán-Abaunza et al. (2014) showed that ICOL pre-processing improved the retrieval of reflectance from MERIS data with four atmospheric correction processors. Matthews et al. (2012), however, reported some unusual and unexpected spectral shapes following ICOL correction and Majozi et al. (2012) observed an overestimation of water reflectance when ICOL was applied. Differences in the performance might be attributed to the different versions of ICOL, which have been released over the years and/or changes in the operational processors used for MERIS (Beltrán-Abaunza et al., 2014).

A more simple approach for adjacency correction is to assume that the average background radiance can be calculated from surrounding pixels using an n x n low-pass filter. This approach has been integrated in the coastal Waters and Ocean MODTRAN-4 Based Atmospheric correction (c-Wombat-c) code (Brando and Dekker, 2003), which is based on the MODerate resolution TRANsmission (MODTRAN) (Berk et al., 1999) radiative transfer model. The spatial extent of the low-pass filter is either chosen arbitrarily (Brando et al., 2009) or determined iteratively by comparing with *in situ* reflectance spectra (Campbell et al., 2011).

Sterckx et al. (2015a) presented a sensor-independent adjacency correction method, the SIMilarity Environment Correction (SIMEC). The SIMEC approach, which has been integrated into OPERA (Sterckx et al., 2015b), was first proposed by Sterckx et al. (2011) for the correction of airborne hyperspectral imagery over inland and coastal waters. The correction algorithm estimates the contribution of the background iteratively by checking the correspondence of the retrieved water reflectance with the NIR similarity spectrum defined by Ruddick et al. (2006). The correspondence is checked at an NIR band (referred to as the "test" band) after normalization by the retrieved ρ_w at the spectral band nearest to 780 nm (i.e., the reference band). Therefore, no assumption is made on the exact value of the NIR reflectance (i.e., no black-pixel assumption); only its spectral shape in the NIR region is assumed to be invariable. An essential requirement for SIMEC is that the aerosol information must be obtained from sun photometer data or through a land-based aerosol retrieval method and not retrieved through a standard pixel-by-pixel, "water-based" method.

Sterckx et al. (2011) defined some basic rules for selecting the optimal test band: (i) the selected test band should be influenced minimally by gaseous absorption, (ii) it should not be located below 690 nm because of the large standard deviation of the NIR similarity spectrum, and (iii) it should preferably be located in the red-edge region of the spectrum because around these wavelengths the normalized ρ_w is mostly affected by insufficient correction of adjacency effects from the surrounding vegetated land. In the NIR region, the vegetation spectrum shows a sharp increase in reflectance (the red-edge). This high reflectance in the NIR region is in sharp contrast with the low $\rho_w(NIR)$. When the surrounding vegetation contributes to the at-sensor radiance from a water pixel through scattering in the atmosphere, it not only increases the retrieved $\rho_w(NIR)$ for the water pixel but also alters the spectral shape to something different than the NIR similarity spectrum. An *a priori* assumption in SIMEC is the validity of the NIR similarity spectrum. This assumption might be violated in extremely turbid waters (Doron et al., 2011), waters with macrophyte growth or algal blooms, and waters that are optically shallow in the NIR region.

3.3.3 Case Study: Combined Atmospheric and Adjacency Correction

The following is a case study showing the successful application of the atmospheric correction scheme OPERA in combination with the adjacency correction algorithm SIMEC.

OPERA is a sensor-independent atmospheric correction scheme that is based on MODTRAN-5 (Berk et al., 2006) and follows the radiative transfer modeling described by De Haan and Kokke (1996). A similar formalism is also used in the c-Wombat-c code developed by Brando and Dekker (2003). It is important to note that this formalism includes a correction for sky glint over water, which is generally not considered in land-based approaches like SCAPE-M. OPERA includes adjacency correction over water based on the SIMEC approach. For operational processing, OPERA separates land and water pixels in an image using a land-sea mask and retrieves the AOD from the land pixels following the approach described by Guanter et al. (2007). Alternatively, for more *ad hoc* studies, the AOD can be set by the user based on data from other sources, such as sun photometer data acquired concurrently with the sensor overpass. The OPERA atmospheric LUTs are generated with MODTRAN-5 for various surface elevations. This allows correction for lakes at high altitudes by taking into account the terrain elevation derived on a pixel-by-pixel basis from the digital elevation model available within OPERA. OPERA is currently being integrated into the Sentinel Application Platform (SNAP), which is the successor of the BEAM tool box, capable of processing data from multiple satellite missions.

OPERA was applied to a MERIS image acquired over Lake Geneva, Switzerland on September 10, 2007 (Fig. 3.3). Lake Geneva is a mesotrophic lake located at an altitude of approximately 372 m. Hand-held spectrometers RAMSES-ARC and RAMSES-ACC were used to collect *in situ* reflectance measurements (Odermatt et al., 2010). Fig. 3.3 shows a comparison between remote sensing reflectance measured *in situ* and reflectance retrieved through OPERA with and without the SIMEC adjacency correction for a near-shore location. When the data were not pre-processed with SIMEC, the retrieved remote sensing reflectance showed anomalously high reflectance between 709 and 753 nm, which cannot be explained by the reflectance properties of

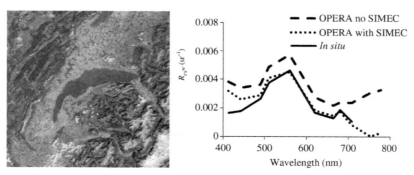

FIGURE 3.3 Left: A true color composite of a MERIS image acquired over Lake Geneva, Switzerland, with the *in situ* measurement location marked with a *red* (light gray in print versions) dot (*in situ* data courtesy of Daniel Odermatt). Right: Comparison between reflectance measured *in situ* and reflectance retrieved from MERIS data using OPERA with and without the SIMEC adjacency correction.

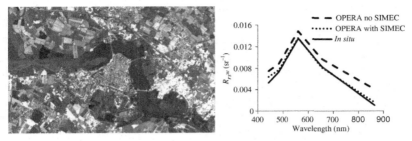

FIGURE 3.4 Left: A true color composite of a Landsat-8 OLI image acquired over Lake Mantua, Italy (courtesy of U.S. Geological Survey), with the *in situ* measurement location marked with a *red* (light gray in print versions) dot (*in situ* data courtesy of the National Research Council of Italy). Right: Comparison between reflectance measured *in situ* and reflectance retrieved from Landsat-8 OLI data using OPERA with and without the SIMEC adjacency correction.

water or the spectral dependence of aerosol scattering in case of an error in the aerosol retrieval. This increase in reflectance was caused by adjacency effects from nearby land that have a larger impact at 753 nm than at 709 nm, which is consistent with a typical vegetation spectrum that has a sharp increase in reflectance in the red-edge region, relatively low reflectance at 709 nm, and high reflectance in the NIR region. When the data were pre-processed with SIMEC, the OPERA-corrected reflectance showed a dramatic decrease in the reflectance between 709 and 753 nm, consistent with typical water reflectance spectra. Also, a slight decrease was observed in the reflectance in the visible region as well, resulting in improved comparison with the *in situ* measurements.

A similar comparison was also made for a Landsat-8 OLI image acquired over the Mantua Lakes system in Italy (Fig. 3.4) on September 23, 2010. The OLI radiance was enhanced using gain factors published by Pahlevan et al. (2014). Without pre-processing with SIMEC, remote sensing reflectance retrieved from OPERA was higher than the reflectance measured *in situ*, especially in the OLI NIR band. Application of SIMEC decreased the retrieved remote sensing reflectance, resulting in a better match with the *in situ* reflectance.

3.4 CONCLUSION

We have provided a brief discussion on the main challenges in the atmospheric correction of remotely sensed data for inland waters and some of the algorithms that have been developed to address those challenges. Several other studies have provided comparisons of different atmospheric correction algorithms for inland (e.g., Guanter et al., 2010; Jaelani et al., 2013) and coastal waters (e.g., Jamet et al., 2011). A general observation from such studies is that no single algorithm performs consistently better than others in all inland waters. A case in point is the QUAC algorithm, which was shown to perform better than FLAASH when applied to airborne hyperspectral data from several lakes in Nebraska, USA (Moses et al., 2012) but performed worse than FLAASH when applied to spaceborne hyperspectral data from Poyang Lake, China (Zeng et al., 2013). The performance of an atmospheric correction algorithm depends not only on its ability to handle various atmospheric and environmental conditions but also its sensitivity to the radiometric calibration of the data. Therefore, the same algorithm could perform differently for data from different sensors collected under the same atmospheric and environmental conditions. Moreover, results from comparison studies are also sensitive to the instruments and methods used for collecting the *in situ* "ground truth" data used for validating atmospheric correction. Therefore, inferences drawn from a comparative study of atmospheric correction algorithms are generally limited in their scope to the particular set of atmospheric and environmental conditions encountered and instruments used in that study. A comprehensive comparative study would require a large dataset collected using the same set of instruments and measurement protocols from inland waters from various geographic locations around the world, covering a wide range of atmospheric, environmental, and water quality conditions. We have not attempted to provide such a comprehensive analysis in this chapter. Instead, we have discussed a few atmospheric correction approaches that are particularly suitable for inland waters, with brief descriptions of a number of algorithms for each approach. The algorithms discussed have achieved varying levels of success, each generally performing well when conditions match the assumptions built into the algorithm and not so well when conditions violate the assumptions.

The optical complexities of inland waters and the atmosphere above, in addition to spatial limitations and interferences from surrounding topography, make atmospheric correction for inland waters very challenging. The development of an operational atmospheric correction algorithm that is widely applicable for a broad range of inland water conditions requires improvements in sensor technology, validation protocols, and algorithm development. Hyperspectral sensors with a high spatial resolution, spectral bands covering the UV-SWIR range at a fine spectral resolution (on the order of 10 nm or less), and a high SNR are required to resolve aerosol properties in an optically complex atmosphere. From a sensor design and mission cost standpoint, there is an inevitable trade-off amongst spatial resolution, spectral resolution, and SNR, necessitating compromises that affect the ability to remotely monitor inland waters. Because of the proximity of inland waters to various terrestrial sources of atmospheric aerosols, *in situ* atmospheric data meant for validation must be collected on the water body as close in time as possible to sensor overpass; using data from sites located even just a few km away might introduce uncertainties. Atmospheric correction algorithms developed for coastal waters or land, which have been applied to inland waters with moderate success, would need to be expanded and tailored for inland waters, which might have a wider variation in biogeo-physical properties than coastal waters. Nevertheless, the algorithms dis-cussed here are promising. The launch of new hyperspectral sensors with high spatial resolution, such as the German Aerospace Center's Environmental Mapping and Analysis Program (EnMAP), the Italian Space Agency's PRecursore IperSpettrale della Missione Applicativa (PRISMA), and NASA's Hyperspectral Infrared Imager (HyspIRI) will pave the way for the development of more robust atmospheric correction algorithms suited for inland waters.

As a preliminary, albeit important, step in the processing of remote sens-ing data for retrieving biophysical products, the need and the desired rigor of atmospheric correction is driven largely by the type of biophysical end-products that are to be retrieved (e.g., Song et al., 2001) and the methods used to retrieve them, whereas the need for adjacency correction is deter-mined by factors such as the local topography, the magnitude of the atmo-spheric AOD, the viewing/illumination geometry, and the spectral contrast between the target of interest and its surroundings. When dealing with just a single image and the desired end-product is a qualitative measure, such as a map showing the qualitative spatial distribution of [chl-a], atmospheric cor-rection is not critical and can be omitted. However, when multitemporal images are involved or when quantitative end-products, such as the concen-trations of chl-a, CDOM, and SPM (discussed in later chapters), are desired, atmospheric correction becomes necessary because differences in the atmo-spheric conditions in a multi-temporal dataset need to be accounted for and quantitative estimation of biophysical products requires the retrieval of the at-surface remote sensing reflectance of water, R_{rs}^w. The needed rigor of atmospheric correction depends on the type of algorithm used to retrieve the biophysical products. Bio-optical algorithms based on ratios of reflectances at closely located spectral bands are more tolerant of errors in atmospheric correction because the atmospheric effects at the spectral bands of interest tend to be similar because of the proximity of the bands and they, to a large extent, cancel out. However, algorithms based on spectral inversions of reflectance in the VNIR region or spectral matching of reflectance with library spectra would be very sensitive to errors in atmospheric correction because the retrieval is strongly dependent on the magnitude of reflectance at all wavelengths across the spectrum and any residual atmospheric effect at any part of the spectrum might be misinterpreted by the algorithm as a spec-tral feature produced by constituents in water, leading to erroneous retrievals. Thus, the choice of atmospheric correction algorithm for a given inland water dataset depends on a number of factors, including how well the atmo-spheric, environmental, and water quality conditions match the assumptions built into the algorithm, the desired end-products, and the bio-optical algorithms that will be used to retrieve the end-products. Nevertheless, reliable and accurate atmospheric correction is necessary for quantitative retrievals of biophysical products such as the concentrations of chl-*a*, CDOM, SPM, and accessory pigments such as phycocyanin, and chl-*a* fluorescence, which are critical indicators of inland water quality and are discussed in the chapters that follow.

ACKNOWLEDGMENTS

We acknowledge Dr. Daniel Odermatt from Odermatt & Brockmann GmbH and staff at the National Research Council of Italy for providing us the *in situ* data that were used in the case study for demonstrating the combined atmospheric and adjacency correction.

REFERENCES

Allan, M.G., Hamilton, D.P., Hicks, B.J., Brabyn, L., 2011. Landsat remote sensing of chlorophyll a concentrations in central North Island lakes of New Zealand. Int. J. Rem. Sens 32, 2037−2055.

Béal, D., Baret, F., Bacour, C., Gu, X.F., 2007. A method for aerosol correction from the spectral variation in the visible and near infrared: Application to the MERIS sensor. Int. J. Rem. Sens. 28, 761−779.

Beltrán-Abaunza, J.M., Kratzer, S., Brockmann, C., 2014. Evaluation of MERIS products from Baltic Sea coastal waters rich in CDOM. Ocean Sci. 10, 377−396.

Berk, A., Anderson, G.P., Bernstein, L.S., Acharya, P.K., Dothe, H., Matthew, M.W., et al., 1999. MODTRAN4 radiative transfer modeling for atmospheric correction, in: Larar, A.M. (Ed.), SPIE Proc., Optical Spectroscopic Techniques and Instrumentation for Atmospheric and Space Research III, Denver, Colorado, July 19−21, 1999. 3756, 348−353.

Berk, A., Anderson, G.P., Acharya, P.K., Bernstein, L.S., Muratov, L., Lee, J., et al., 2006. MODTRANTM 5: 2006 Update, in: Shen, S.S., Lewis, P.E. (Eds.), SPIE Proc., Algorithms and Technologies for Multispectral, Hyperspectral, and Ultraspectral Imagery XII. 6233, 62331F, p. 8, http://dx.doi.org/10.1117/12.665077.

Bernstein, L.S., Adler-Golden, S. M., Sundberg, R.L., Levine, R.Y. Perkins, T.C., Berk, A., et al., 2005a. A new method for atmospheric correction and aerosoloptical property retrieval for VIS-SWIR multi- and hyperspectral imaging sensors: QUAC (QUick atmospheric correction), in: Proc. IEEE Int. Geosci. Rem. Sens. Symp. (IGARSS'05), July 25−29, 2005. 5, 3549-3552, http://dx.doi.org/10.1109/IGARSS.2005.1526613.

Bernstein, L.S., Adler-Golden, S.M., Sundberg, R.L., Levine, R.Y. Perkins, T.C., Berk, A., et al., 2005b. Validation of the QUick Atmospheric Correction (QUAC) algorithm for VNIR-SWIR multi- and hyperspectral imagery, in: Shen, S.S., Lewis, P.E. (Eds.), SPIE Proc., Algorithms and Technologies for Multispectral, Hyperspectral, and Ultraspectral Imagery XI. 5806, 668-678, http://dx.doi.org/10.1117/12.603359.

Borde, R., Ramon, D., Schmechtig, C., Santer, R., 2003. Extension of the DDV concept to retrieve aerosol properties over land from the Modular Optoelectronic Scanner (MOS) sensor. Int. J. Rem. Sens. 24 (7), 1439−1467.

Brando, V.E., Dekker, A.G., 2003. Satellite hyperspectral remote sensing for estimating estuarine and coastal water quality. IEEE Trans. Geosci. Rem. Sens 41, 1378−1387.

Brando, V.E., Anstee, J., Wettle, M., Dekker, A., Phinn, S., Roelfsema, C., 2009. A physics based retrieval and quality assessment of bathymetry from suboptimal hyperspectral data. Rem. Sens. Enviro. 113, 755−770.

Brando, V.E., Braga, F., Zaggia, L., Giardino, C., Bresciani, M., Matta, E., et al., 2015. High-resolution satellite turbidity and sea surface temperature observations of river plume interactions during a significant flood event. Ocean Sci. 11 (6), 909−920.

Campbell, G., Phinn, S.R., Dekker, A.G., Brando, V.E., 2011. Remote sensing of water quality in an Australian tropical freshwater impoundment using matrix inversion and MERIS images. Rem. Sens. Environ. 115 (9), 2402−2414.

Chandrasekhar, S., 1960. Radiative Transfer. Dover, New York.

Chavez, P.S., 1996. Image-based atmospheric corrections − revisited and improved. Photogramm. Eng. Rem. Sens. 62 (9), 1025−1036.

Cox, C., Munk, W., 1954. Statistics of the sea surface derived from sun glitter. J. Mar. Res. 13, 198−227.

De Haan, J.F., Kokke, J.M.M., 1996. Remote sensing algorithm development toolkit I: Operationalization of atmospheric correction methods for tidal and inland waters. Netherlands Remote Sensing Board (BCRS) publication, p. 91. Rijkswaterstaat Survey Dept. Technical Report.

Doerffer, R., Fischer, J., 1994. Concentrations of chlorophyll, suspended matter, and gelbstoff in Case II water derived from satellite coastal zone color scanner data with inverse modeling methods. J. Geophys. Res. 99 (C4), 7457−7466.

Doerffer, R., Schiller, H., 2007. The MERIS Case 2 water algorithm. Int. J. Rem. Sens. 28 (3−4), 517−535.

Doerffer, R., Schiller, H., 2008. MERIS lake water algorithm for BEAM − MERIS algorithm theoretical basis document. Inst. Coastal Res. GKSS Res. Center, Geesthacht, Germany, Rep. GKSS-KOF-MERIS-ATBD01, Vol. 1.

Dogliotti, A.I., Ruddick, K., Nechad, B., Lasta, C., 2011. Improving water reflectance retrieval from MODIS imagery in the highly turbid waters of La Plata River, in: Proceedings of VI International Conference «Current problems in optics of natural waters» (ONW'2011), St. Petersburg, Russia, September 6−9, 2011. pp. 8.

Doron, M., Belanger, S., Doxaran, D., Babin, M., 2011. Spectral variations in the near-infrared ocean reflectance. Rem. Sens. Environ. 115 (7), 1617−1631.

Doxaran, D., Froidefond, J.-M., Castaing, P., 2003. Remote-sensing reflectance of turbid sediment-dominated waters. Reduction of sediment type variations and changing illumination conditions effects by use of reflectance ratios. Appl. Opt. 42 (15), 2623−2634.

Feng, L., Hu, C., Han, X., Chen, X., Qi, L., 2014. Long-term distribution patterns of chlorophyll-a concentration in China's largest freshwater lake: MERIS full-resolution observations with a practical approach. Rem. Sens. 7 (1), 275−299.

Gao, B.-C., Heidebrecht, K.B., Goetz, A.F. H., 1993. Derivation of scaled surface reflectances from AVIRIS data. Rem. Sens. Environ. 44, 165−178.

Gao, B.-C., Montes, M.J., Ahmad, Z., Davis, C.O., 2000. Atmospheric correction algorithm for hyperspectral remote sensing of ocean color from space. Appl. Opt. 39, 887−896.

Gao, B.-C., Montes, M.J., Li, R.-R., Dierssen, H.M., Davis, C.O., 2007. An atmospheric correction algorithm for remote sensing of bright coastal waters using MODIS land and ocean channels in the solar spectral region. IEEE Trans. Geosci. Rem. Sens. 45 (6), 1835−1843.

Gong, S.Q., Huang, J.H., Li, Y.M., Wang, H.J., 2008. Comparison of atmospheric correction algorithms for TM image in inland waters. Int. J. Rem. Sens. 29 (8), 2199−2210.

Gordon, H.R., 1978. Removal of atmospheric effects from satellite imagery of the oceans. Appl. Opt. 17 (10), 1631−1636.

Gordon, H.R., Clark, D.K., 1981. Clear water radiances for atmospheric correction of coastal zone color scanner imagery. Appl. Opt. 20 (24), 4175−4180.

Gordon, H.R., Wang, M., 1994. Retrieval of water-leaving radiance and aerosol optical thickness over the oceans with SeaWiFS: a preliminary algorithm. Appl. Opt. 33 (3), 443−452.

Goyens, C., Jamet, C., Ruddick, K.G., 2013. Spectral relationships for atmospheric correction. I. Validation of red and near infra-red marine reflectance relationships. Opt. Exp. 21 (18), 21162−21175.

Grey, W.M.F., North, P.R.J., Los, S.O., 2006. Computationally efficient method for retrieving aerosol optical depth from ATSR-2 and AATSR data. Appl. Opt. 45, 2786−2795.

Guanter, L., Del Carmen González-Sanpedro, M., Moreno, J., 2007. A method for the atmospheric correction of ENVISAT/MERIS data over land targets. Int. J. Rem. Sens. 28 (3−4), 709−728.

Guanter, L., Ruiz-Verdu, A., Odermat, D., Giardino, C., Simis, S., Estelles, V., et al., 2010. Atmospheric correction of ENVISAT/MERIS data over inland waters: validation for European lakes. Rem. Sens. Environ. 114 (3), 467−480.

He, Q., Chen, C., 2014. A new approach for atmospheric correction of MODIS imagery in turbid coastal waters: a case study for the Pearl River Estuary. Rem. Sens. Lett. 5 (3), 249−257.

He, X.Q., Bai, Y., Pan, D.L., Tang, J.W., Wang, D.F., 2011. Atmospheric correction of satellite ocean color imagery using the ultraviolet wavelength for highly turbid waters. Opt. Exp. 20, 20754−20770.

Heege, T., Fischer, J., 2004. Mapping of water constituents in Lake Constance using multi-spectral airborne scanner data and a physically based processing scheme. Can. J. Rem. Sens. 30 (1), 77−86.

Holben, B.N., Eck, T.F., Slutsker, I., Tanre, D., Buis, J.P., Setzer, A., et al., 1998. AERONET − A federated instrument network and data archive for aerosol characterization. Rem. Sens. Environ. 66 (1), 1−16.

Hovis, W.A., Leung, K.C., 1977. Remote sensing of ocean color. Appl. Opt. 16 (2), 158−166.

Hu, C., Carder, K.L., Muller-Karger, F.E., 2000. Atmospheric correction of SeaWiFS imagery over turbid coastal waters: a practical method. Rem. Sens. Environ. 74, 195−206.

Jaelani, L.M., Matsushita, B., Yang, W., Fukushima, T., 2013. Evaluation of four MERIS atmospheric correction algorithms in Lake Kasumigaura, Japan. Int. J. Rem. Sens. 34 (24), 8967−8985.

Jamet, C., Loisel, H., Kuchinke, C.P., Ruddick, K., Zibordi, G., Feng, H., 2011. Comparison of three SeaWiFS atmospheric correction algorithms for turbid waters using AERONET-OC measurements. Rem. Sens. Environ. 115 (8), 1955−1965.

Jiang, L., Wang, M., 2014. Improved near-infrared ocean reflectance correction algorithm for satellite ocean color data processing. Opt. Exp. 22, 21657−21678.

Kaufman, Y.J., Sendra, C., 1988. Algorithm for automatic atmospheric corrections to visible near-IR satellite imagery. Int. J. Rem. Sens. 9, 1357−1381.

Kaufman, Y.J., Tanre, D., 1992. Atmospherically resistant vegetation index (ARVI) for EOS-MODIS. IEEE Trans. Geosci. Rem. Sens. 30 (2), 261−270.

Kay, S., Hedley, J.D., Lavender, S., 2009. Sun glint correction of high and low spatial resolution images of aquatic scenes: a review of methods for visible and near-infrared wavelengths. Rem. Sens. 1, 697−730.

Kiselev, V., Bulgarelli, B., Heege, T., 2015. Sensor independent adjacency correction algorithm for coastal and inland water systems. Rem. Sens. Environ. 157, 85−95.

Knaeps, E., Dogliotti, A.I., Raymaekers, D., Ruddick, K., Sterckx, S., 2012. *In situ* evidence of non-zero reflectance in the OLCI 1020 nm band for a turbid estuary. Rem. Sens. Environ. 120, 133−144.

Kokhanovsky, A.A., Breon, F.-M., Cacciari, A., Carboni, E., Diner, D., Di Nicolantio, W., et al., 2007. Aerosol remote sensing over land: a comparison of satellite retrievals using different algorithms and instruments. Atmos. Res. 85, 372−394. Available from: http://dx.doi.org/ 10.1016/j.atmosres.2007.02.008.

Kou, L., Labrie, D., Chylek, P., 1993. Refractive indices of water and ice in the 0.65−2.5 m spectral range. Appl. Opt. 32, 3531−3540.

Kratzer, S., Vinterhav, C., 2010. Improvement of MERIS level 2 products in Baltic Sea coastal areas by applying the Improved Contrast between Ocean and Land processor (ICOL) − data analysis and validation. Oceanologia 52 (2), 211−236.

Kuchinke, C.P., Gordon, H.R., Franz, B., 2009. Spectral optimization for constituent retrieval in Case 2 waters I: implementation and performance. Rem. Sens. Environ. 113 (3), 571−587.

Kutser, T., 2004. Quantitative detection of chlorophyll in cyanobacterial blooms by satellite remote sensing. Limnol. Oceanogr. 49 (6), 2179−2189.

Majozi, N.P., Salama, S.M., Bernard, S., Harper, D.M., 2012. Water quality and remote sensing: a case study of Lake Naivasha, Kenya, in: Proc. 16th SANCIAHS National Hydrology Symposium, University of Pretoria, October 1−3, 2012, p. 13.

Mancino, G., Nolè, A., Urbano, V., Amato, M., Ferrara, A., 2009. Assessing water quality by remote sensing in small lakes: The case study of Monticchio Lakes in Southern Italy. iForest 2, 154−161.

Martin, J., Eugenio, F., Marcello, J., Medina, A., 2016. Automatic sun glint removal of multi-spectral high-resolution Worldview-2 imagery for retrieving coastal shallow water parameters. Rem. Sens. 8 (1), 37. Available from: http://dx.doi.org/10.3390/rs8010037.

Matthews, M.W., Bernard, S., Robertson, L., 2012. An algorithm for detecting trophic status (chlorophyll-*a*), cyanobacterial-dominance, surface scums and floating vegetation in inland and coastal waters. Rem. Sens. Environ. 124, 637−652.

Mobley, C.D., 2015. Polarized reflectance and transmittance properties of windblown sea surfaces. Appl. Opt. 54 (15), 4828−4849.

Montes, M.J., Gao, B.-C., Davis, C.O., 2001. A new algorithm for atmospheric correction of hyperspectral remote sensing data, in: Roper, W. E. (Ed.), SPIE Proc., Geo-Spatial Image and Data Exploitation II. 4383, 23-30.

Moses, W.J., Gitelson, A.A., Berdnikov, S., Povazhnyy, V., 2009. Estimation of chlorophyll-*a* concentration in case II waters using MODIS and MERIS data − successes and challenges. Environ. Res. Lett. 4 (045005), 8.

Moses, W.J., Gitelson, A.A., Perk, R.L., Gurlin, D., Rundquist, D.C., Leavitt, B.C., et al., 2012. Estimation of chlorophyll-a concentration in turbid productive waters using airborne hyperspectral data. Wat. Res. 46, 993−1004.

Moses, W.J., Ackleson, S.G., Hair, J.W., Hostetler, C.A., Miller, W.D., 2016. Spatial scales of optical variability in the coastal ocean: implications for remote sensing and *in situ* sampling. J. Geophys. Res.: Oceans.121. Available from: http://dx.doi.org/10.1002/ 2016JC011767.

Noh, Y.M., Muller, D., Lee, H., Choi, T.J., 2013. Influence of biogenic pollen on optical properties of atmospheric aerosols observed by lidar over Gwangju. South Korea. Atmos. Env. 69, 139−147. Available from: http://dx.doi.org/10.1016/j.atmosenv.2012.12.018.

Odermatt, D., Giardino, C., Heege, T., 2010. Chlorophyll retrieval with MERIS Case-2-Regional in perialpine lakes. Rem. Sens. Environ. 114, 607−617.

Oo, M., Vargas, M., Gilerson, A., Gross, B., Moshary, F., Ahmed, S., 2008. Improving atmospheric correction for highly productive coastal waters using the short wave infrared retrieval algorithm with water-leaving reflectance constraints at 412 nm. Appl. Opt. 47, 3846–3859.

Osińska-Skotak K., 2005. Influence of atmospheric correction on determination of lake water quality parameters based on CHRIS/PROBA images, in: Proc. 25th EARSeL Symposium, June 6–11, 2005, Porto, Portugal, pp. 129-135.

PACE SDT, 2012. Pre-Aerosol, Clouds, and ocean Ecosystem (PACE) Mission Science Definition Team Report, Del Castillo, C. E. and Platnick, S. (co-chairs), NASA Goddard Space Flight Center, Greenbelt, Maryland, p. 308, available at http://decadal.gsfc.nasa.gov/PACE/PACE_SDT_Report_final.pdf; accessed on March 10, 2016.

Pahlevan, N., Lee, Z., Wei, J., Schaaf, C.B., Schott, J.R., Berk, A., 2014. On-orbit radiometric characterization of OLI (Landsat-8) for applications in aquatic remote sensing. Rem. Sens. Environ. 154, 272–284.

Palmer, S.C.J., Hunter, P.D., Lankester, T., Hubbard, S., Spyrakos, E., Tyler, A.N., et al., 2015. Validation of Envisat MERIS algorithm for chlorophyll retrieval in a large, turbid and optically-complex shallow lake. Rem. Sens. Environ. 157, 158–169.

Patterson, K., 2014. Reducing ocean surface specular reflection in WorldView-2 images, in: Hou, W., Arnone, R.A. (Eds.), SPIE Proc., Ocean Sensing and Monitoring VI. 91110B. http://dx.doi.org/10.1117/12.2050480.

Pope, R.M., Fry, E.S., 1997. Absorption spectrum (380–700 nm) of pure water. II. Integrating cavity measurements. Appl. Opt. 36, 8710–8723.

Ramon, D., Santer, R., 2005. Aerosol over land with MERIS, present and future, in: Lacoste, H. (Ed.), Proc. MERIS-(A)ATSR Workshop. September 26–30, 2005, Frascati, Italy, p. 8.

Reinersman, P.N., Carder, K.L., 1995. Monte Carlo simulation of the atmospheric point-spread function with an application to correction for the adjacency effect. Appl. Opt. 34 (21), 4453–4471.

Ruddick, K.G., Ovidio, F., Rijkeboer, M., 2000. Atmospheric correction of SeaWiFS imagery for turbid coastal and inland waters. Appl. Opt. 39 (6), 897–912.

Ruddick, K., De Cauwer, V., Park, Y., 2006. Seaborne measurements of near infrared water-leaving reflectance: the similarity spectrum for turbid waters. Limnol. Oceanogr. 51 (2), 1167–1179.

Santer, R., Schmechtig, C., 2000. Adjacency effects of water surfaces: primary scattering approximation and sensitivity study. Appl. Opt. 39 (3), 361–375.

Santer, R., Zagolski, F., 2009. Improved Contrast between Ocean and Land (ICOL) algorithm theoretical basis document, D6. Tech. Rep. Université du Littoral Côte d'Opale, Wimereux, France, p. 15.

Santer, R., Zagolski, F., Gilson, M., 2007. Improved Contrast between Ocean and Land (ICOL) algorithm theoretical basis document. Tech. Rep. Université du Littoral Côte d'Opale, Wimereux, France, Version 0.1.

Schroeder, Th, Schaale, M., Fischer, J., 2007. Retrieval of atmospheric and oceanic properties from MERIS measurements: a new Case-2 water processor for BEAM. Int. J. Rem. Sens. 28 (24), 5627–5632.

Shi, W., Wang, M., 2009. An assessment of the ocean black pixel assumption for the MODIS SWIR bands. Rem. Sens. Environ. 113, 1587–1597.

Siegel, D.A., Wang, M., Maritorena, S., Robinson, W., 2000. Atmospheric correction of satellite ocean color imagery: the black pixel assumption. Appl. Opt. 39 (21), 3582–3591.

Song, C., Woodcock, C.E., Seto, K.C., Lenney, M.P., Macomber, S.A., 2001. Classification and change detection using Landsat TM data: when and how to correct atmospheric effects? 75 (2), 230-244.

Steinmetz, F., Deschamps, P.-Y., Ramon, D., 2011. Atmospheric correction in presence of sun glint: application to MERIS. Opt. Exp. 19, 9783–9800.

Sterckx, S., Knaeps, E., Ruddick, K., 2011. Detection and correction of adjacency effects in hyperspectral airborne data of coastal and inland waters: the use of the near infrared similarity spectrum. Int. J. Rem. Sens. 32 (21), 6479–6505.

Sterckx, S., Knaeps, E., Kratzer, S., Ruddick, K., 2015a. SIMilarity Environment Correction (SIMEC) applied to MERIS data over inland and coastal waters. Rem. Sens. Environ. 157, 96–110.

Sterckx, S., Knaeps, E., Adriaensen, S., Reusen, I., De Keukelaere, L., Hunter, P., et al., 2015b. OPERA: an atmospheric correction for land and water, in: Ouwehand, L. (Ed.), Proc. Sentinel-3 for Science Workshop. 2–5 June 2015, Venice, Italy, p. 4.

Tanre, D., Herman, M., Deschamps, P.Y., 1981. Influence of the background contribution upon space measurements of ground reflectance. Appl. Opt. 20 (20), 3676–3684.

Vanhellemont, Q., Ruddick, K., 2015. Advantages of high quality SWIR bands for ocean colour processing: examples from Landsat-8. Rem. Sens. Environ. 161, 89–106.

Van Mol, B., Ruddick, K., 2005. Total suspended matter maps from CHRIS imagery of a small inland water body in Oostende (Belgium). *Proceedings of the 3rd ESA CHRIS/Proba Workshop*. 21–23 March 2005. ESRIN, Frascati, Italy.

Vermote, E.F., Tanre, D., Deuze, J.L., Herman, M., Morcette, J.J., 1997. Second simulation of the satellite signal in the solar spectrum, 6S: an overview. IEEE Trans. Geosci. Rem. Sens. 35 (3), 675–686.

Vidot, J., Santer, R., 2005. Atmospheric correction over inland waters: applications to SeaWiFS. Int. J. Rem. Sens. 26 (17), 3663–3682.

Wang, M., Liu, X., 2013. MODIS ocean color products using the SWIR method, MODIS-SWIR algorithm theoretical basis document. NOAA Product System Development and Implementation, p. 40.

Wang, M.H., Shi, W., 2005. Estimation of ocean contribution at the MODIS near-infrared wavelengths along the east coast of the US: two case studies. Geophys. Res. Lett. 32 (L13606), 5.

Wang, M., Shi, W., 2012. Sensor noise effects of the SWIR bands on MODIS-derived ocean color products. IEEE Trans. Geosci. Rem. Sens. 50, 3280–3292.

Wang, F., Zhou, B., Xu, J., Song, L., Wang, X., 2009. Application of neural network and MODIS 250 m imagery for estimating suspended sediments concentration in Hangzhou Bay, China. Environ. Geol. 56 (6), 1093–1101.

Wang, M., Shi, W., Tang, J., 2011. Water property monitoring and assessment for China's inland Lake Taihu from MODIS-Aqua measurements. Remote Sens. Environ. 115, 841–845.

Wang, M., Nim, C.J., Son, S., Shi, W., 2012. Characterization of turbidity in Florida's Lake Okeechobee and Caloosahatchee and St. Lucie estuaries using MODIS-Aqua measurements. Wat. Res. 46, 5410–5422.

Wong, S.W., Nichol, J.E., Lee, K.H., 2009. Modeling of aerosol vertical profiles using GIS and remote sensing. Sensors 9, 4380–4389. Available from: http://dx.doi.org/10.3390/s90604380.

Yu, H., Chin, M., Winker, D.M., Omar, A.H., Liu, A., Kittaka, C., et al., 2010. Global view of aerosol vertical distributions from CALIPSO lidar measurements and GOCART simulations: regional and seasonal variations. J. Geophys. Res. 115, D00H30. Available from: http://dx.doi.org/10.1029/2009JD013364.

Zeng, Q., Zhao, Y., Tian, L.-Q., Chen, X.-L., 2013. Evaluation on the atmospheric correction methods for water color remote sensing by using HJ-1A/1B CCD image-taking Poyang Lake in China as a case. Spectroscopy Spectral Anal. 33, 1320–1326.

Zhang, R., Wang, L., Khalizov, A.F., Zhao, J., Zheng, J., McGraw, R.L., et al., 2009. Formation of nanoparticles of blue haze enhanced by anthropogenic pollution. Proc. Natl. Acad. Sci. USA 106 (42), 17650–17654. Available from: http://dx.doi.org/10.1073/pnas.0910125106.

Zhou, W., Wang, S., Zhou, Y., Troy, A., 2006. Mapping the concentrations of total suspended matter in Lake Taihu, China, using Landsat-5 TM data. Int. J. Rem. Sens. 27 (6), 1177–1191.

Chapter 4

Bio-optical Modeling of Colored Dissolved Organic Matter

Tiit Kutser[1], Sampsa Koponen[2], Kari Y. Kallio[2], Tonio Fincke[3], and Birgot Paavel[1]

[1]*University of Tartu, Tallinn, Estonia,* [2]*Finnish Environment Institute, Helsinki, Finland,* [3]*Brockmann Consult, Geesthacht, Germany*

4.1 CARBON IN INLAND WATERS

The necessity to understand and predict ever-changing climate on Earth increased dramatically the need to understand global carbon cycle. The largely unaccounted form of carbon is dissolved organic carbon (DOC) in different water bodies. It was assumed that the DOC pool in oceans is largely old and refractory and thus unavailable for biological consumption (Hansell and Carlson, 2001). It was viewed as unimportant to carbon fluxes. The view has changed during recent years and global estimates of ocean DOC pool have been produced (Siegel et al., 2002). However, the global estimates are based on models developed for optically simple open ocean (Case I) waters. These remote sensing estimates (Siegel et al., 2002) exclude near coastal pixels from image processing. It means that the near shore areas with the highest DOC are excluded from the global analysis and the results for optically complex waters (the whole Baltic Sea, parts of North Sea, Arctic Ocean, etc.) are not reliable as the Case I water remote sensing algorithms don't work there.

The situation is much more complex in the case of inland waters. Relatively recently carbon-climate models, for instance those used by the Intergovernmental Panel for Climate Change (IPCC) or the Integrated Global Observing Strategy (IGOS), ignored inland waters treating them as inert pipes simply transporting terrigenous organic carbon into the oceans (IPCC, 2007). More recent estimates (Cole et al., 2007; Tranvik et al., 2009; Battin et al., 2008, 2009) show that lakes are by no means inert pipes transporting carbon from land to oceans. They are rather carbon hot spots in land system.

Bio-optical Modeling and Remote Sensing of Inland Waters.
DOI: http://dx.doi.org/10.1016/B978-0-12-804644-9.00004-5

101

Tranvik et al. (2009) estimated that land exports of carbon to inland waters are twice as high as land exports of carbon to the ocean. Most of this carbon is either subsequently exported to the oceans ($0.9 \, \mathrm{Pg \, C \, y^{-1}}$), is buried ($0.6 \, \mathrm{Pg \, C \, y^{-1}}$), or is oxidized and is outgassed to atmosphere as carbon dioxide (CO_2) and methane (CH_4) (at least $0.9 \, \mathrm{Pg \, C \, y^{-1}}$). Lake sediments may contain as much as $820 \, \mathrm{Pg \, C}$ (Cole et al., 2007). Emission of methane from lakes is greater than emissions from oceans (Bastviken et al., 2004). Net CO_2 outgassing from inland waters worldwide has most recently been estimated at $2.1 \, \mathrm{Pg \, C \, y^{-1}}$ (Raymond et al., 2013) and global methane emission has been estimated to be $103 \, \mathrm{Tg \, CH_4 \, y^{-1}}$ (Bastviken et al., 2011). However, the implications of inland waters for the terrestrial carbon cycle (e.g., lateral fluxes) and for the marine carbon cycle (e.g., carbon sink) remain elusive. Thus, the need for integrating the inland and terrestrial carbon cycles into a "boundless" cycle is thus evident and becomes increasingly recognized.

The importance of lakes in the global carbon cycle is now recognized also by the IPCC (2013). However, the global estimate of lake carbon content and determining the true role of lakes in the global carbon cycle remains a goal still to be achieved. There is need for improvement in many aspects. For example, both Tranvik et al. (2009) and IPCC (2013) used statistical estimates on the numbers of lakes on Earth and upscaled *in situ* data from a few thousand lakes on the statistical estimate of total lake number and area. Remote sensing has already helped to make significant correction in the number and area of lakes on Earth. Verpoorter et al. (2014) used 14.25 m pan-sharpened Landsat imagery to produce global estimate of lake numbers, the area, and several other parameters of each lake as well as global statistics necessary for different studies. However, determining the global carbon content of inland waters and carbon fluxes is still a target for future remote sensing studies.

More than 90% of carbon in lakes is in the form of DOC (Wetzel, 2001). However, the DOC in inland waters is important not only from global carbon cycle point of view but it also has public health effects. DOC is a substrate for microorganisms, and hence promotes fouling of water, causing problems of taste, odor, and hygiene. Moreover, disinfection of water by chlorination, which is a crucial step to maintain the water quality throughout supply systems, is quenched by DOC, resulting in the need for higher doses of chlorine. Chlorination of water rich in DOC results in the formation of mutagenic chlorinated organic by-products (McDonald and Komulainen, 2005). In a study of more than 600,000 inhabitants in 56 Finnish towns, the Finnish Cancer Registry together with historic information on water quality and chlorination (Koivusalo et al., 1997) revealed enhanced risks of bladder and rectal cancer in connection to chlorination of water rich in DOC. A corresponding study in Norway (Magnus et al., 1999), connecting the Medical Birth Registry to the Norwegian waterworks registry, demonstrated enhanced frequencies of birth defects. Accordingly, epidemiological studies imply that the increase in certain forms of cancer is a plausible side effect of increasing DOC caused by climate change. In addition, DOC has been shown to have hormone-like effects on vertebrates (Steinberg et al., 2004). Brownification of lakes is taking place in many boreal regions where the lake water is an important source of drinking water (Williamson et al., 2015). Variable amount of carbon in the water requires different amounts of coagulants and other chemicals and therefore, routine monitoring of DOC and CDOM is needed in order to provide drinking water with sufficient quality and adjust the treatment processes when needed (Eikebrokk et al., 2004). However, the water treatment plants are not only interested in the amount of carbon in the water, that can be mapped with remote sensing using CDOM as a proxy, but also in the quality of the carbon. The latter is often described through a parameter called SUVA (Specific Ultraviolet Absorption). Thus, there is strong need from public health point of view to monitor changes in lake water DOC content and quality in order to make right management decisions and this cannot be done relaying solely on *in situ* methods.

Increasing amount of CDOM, especially in boreal and Arctic lakes and coastal waters, reduces the amount of light available for photosynthesis (Thrane et al., 2014). The elevated amount of CDOM may change the whole ecosystem from regular food web to microbial loop by reducing dramatically the amount of light available for photosynthesis (Williamson et al., 2015). The increased amount of CDOM increases also water temperature because CDOM absorbs light very strongly in blue part of spectrum where the intensity of the solar radiation is the highest. Highly absorbing waters change the ice regime of inland and coastal waters as the dark waters accelerate melting of ice (Kutser, 2010). On the other hand, absorption of light by CDOM protects aquatic organisms from excessive sunlight. Understanding the changes taking place in aquatic environments due to the variable amount of CDOM is one of the aspects that makes remote sensing mapping of lake CDOM over wide areas very desirable.

4.2 OPTICAL PROPERTIES OF CDOM

Most of the carbon in inland waters is in dissolved form or DOC (Wetzel, 2001). Therefore, it is highly desirable to map inland water DOC content with remote sensing. The only part of electromagnetic radiation that can penetrate the water surface and give us information about the water constituents is visible light. Consequently, CDOM, the optically active part of DOC, is the parameter that can be detected with remote sensing instruments. There is need in estimating CDOM concentration in inland waters, as was mentioned above, but CDOM can also be used as a proxy for DOC. This is possible only if there is reasonable correlation between the total amount of DOC and CDOM. Many studies have shown that this is often the case (Tranvik, 1990; Kallio, 1999; Molot and Dillon, 1997; Erlandsson et al., 2012; Kutser et al., 2016a).

However, there are waterbodies where the CDOM/DOC relationship is weak or has seasonal variations (Hestir et al., 2015; Toming et al., 2016a).

CDOM, called also gelbstoff or yellow substance (Kalle, 1938), gilvin (Kirk, 1976) or aquatic humus is a mixture organic molecules resulting from breakdown of higher plants, algae and bacteria and absorbing light in UV and visible part of spectrum. Therefore, it makes water yellow to brown (see Fig. 4.1) depending on the amount of CDOM. Quantity of the CDOM can be expressed through carbon content or fluorescence (Kowalczuk et al., 2009), but usually it is described as absorption of filtered water at a certain wavelength. Ocean color community uses 0.2 μm filters to separate dissolved fraction from particulates (Pegau et al., 2003), while inland water community

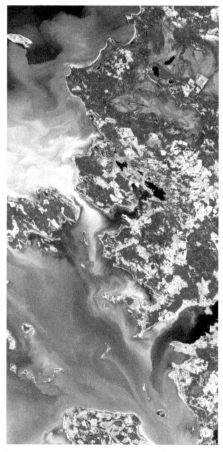

FIGURE 4.1 True color image of ALI satellite sensor (30 m spatial resolution) of the west coast of Estonia acquired on September 1, 2005. CDOM-rich water transported to the Matsalu Bay and Suur Strait by River Kasari is seen in lower right.

uses also larger (0.45 and 0.7 μm) pore size filters (Strömbeck and Pierson, 2001; Sipelgas et al., 2004; Toming et al., 2009). For example, Reinart et al. 2004 reported for Estonian lakes that the correlation between the absorption coefficients measured with 0.2 and 0.7 μm filters was above $R^2 = 0.99$ and the difference in values was less than 4%.

The absorption coefficient of CDOM can be described with the equation (Bricaud et al. 1981):

$$a(\lambda) = a(\lambda_{ref})e^{-S(\lambda - \lambda_{ref})} \qquad (4.1)$$

where a is absorption coefficient (m^{-1}), λ is wavelength, λ_{ref} is a reference wavelength and S is the slope factor. Different authors tend to use different reference wavelengths to express the amount of CDOM. Wavelengths 380, 400, and 420 nm are most commonly used, but satellite remote sensing community uses also central wavelengths of satellite bands like 412 or 443 nm. It has been noted that the slope factor is an indicator of molecular weight, source, and photobleaching of CDOM (Hayase and Tsubota, 1985; Carder et al., 1989; Stedmon and Markager, 2005; Helms et al., 2008). On the other hand, Su et al. (2015) have found that the absorption coefficient behaves differently in alpine lakes above and below the tree line.

Absorption coefficient at 400 nm of the Estonian lakes shown in the Fig. 4.2 varied between 0.64 and 63.05 m^{-1} (Kutser et al., 2016b). The lowest lake CDOM absorption values are probably in Antarctic lakes like Lake Vanda (Vincent et al. 1998) where the CDOM absorption is hard to measure

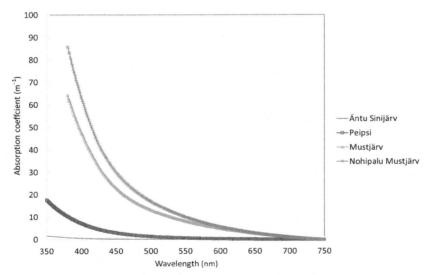

FIGURE 4.2 Absorption coefficient spectra of four Estonian lakes. Measurements of the two most absorbing lakes were carried out diluting the lake water 1:1 with MilliQ water, but still not enough light penetrated the 5 cm cuvette to get noise-free readings below 380 nm.

and consequently fluorometric methods have been used to determine the amount of CDOM. Alpine lakes, like Lake Garda, are also known for their low CDOM content with $a_{CDOM}(400)$ typically about 0.1 m^{-1} (Giardino et al., 2014). Higher end of the reported CDOM values in lakes reaches 70 m^{-1} when recalculated to 400 nm (Brezonik et al., 2015; Su et al., 2015; Kutser et al., 2016b).

The slope factor S in Eq. (4.1) is an indicator of CDOM quality as was mentioned above. The slope factor for lakes varies between 0.006 and 0.036 m^{-1} (Kallio, 1999; Reinart et al., 2004; Müller et al., 2014, Nima et al., 2016) with typical values around $0.017-0.018 \text{ m}^{-1}$. It must be mentioned that different authors use different parts of spectrum to fit the exponential function through it and this changes the value of the slope factor. Remote sensing scientist use mostly shorter wavelengths of visible part of spectrum (400–550 nm), while limnologist and chemists tend to use UV part of spectrum. Loiselle et al. (2009) have demonstrated the variability of the slope factor based on the wavelength range it was calculated for.

One of the main optical parameters used for describing quality of DOC is SUVA. It is a ratio of absorbance at 254 nm to DOC concentration. As the absorbance in UV part of spectrum is determined by CDOM we may say that CDOM is also potential proxy not only for mapping DOC quantity but also quality. Remote sensing mapping of SUVA with MERIS data has been demonstrated (Kutser et al., 2015b) but up to now only in one lake. Further studies on this topic are needed.

There are other constituents in water that make remote sensing estimation of CDOM complicated. For example, iron associated to dissolved organic matter also hampers our capability to estimate lake CDOM and DOC content (Köhler et al., 2013; Kutser et al., 2015a) as both iron and carbon make water brown. Lakes are sometimes also rich in particulate material that absorbs light in a similar way like CDOM—high absorption in UV decreasing exponentially with increasing wavelength (Kirk, 1980). The detrital particular material in lakes consists of dead plankton cells, decaying fractions of aquatic and terrestrial plants. Separating the dissolved and particulate fraction from each other based on water reflectance spectra is very difficult and sometimes nearly impossible (Del Castillo, 2005; Siegel et al., 2013; Wei and Lee, 2015). Therefore, some remote sensing algorithms and inversion methods retrieve concentrations of colored detrital matter (CDM = CDOM + detrital matter) rather than CDOM.

4.3 REMOTE SENSING OF CDOM

It is clear from the summary above that mapping and monitoring of inland (and coastal) waters CDOM content is of critical importance from practical point of view (drinking water) as well as from scientific point of view (inland water ecology, radiative heating, global carbon cycle studies, etc.).

Remote sensing is the only realistic mean that should enable us to get the CDOM information with temporal frequency and spatial coverage needed.

For a quite long time the satellite remote sensing was hampered by lack of suitable sensors (Palmer et al., 2015). Ocean color sensors, like Coastal Zone Color Scanner (CZCS), Sea-viewing Wide Field-of-view Sensor (SeaWiFS), Moderate Resolution Imaging Spectroradiometer (MODIS), Medium Resolution Imaging Spectrometer (MERIS), had coarse (300−1200 m) spatial resolution not suitable for most lakes. Medium resolution sensors, like Landsat series, did not have sufficient radiometric resolution, especially for dark CDOM-rich lakes (Kallio et al., 2008; Kutser, 2012). Airborne and high spatial resolution commercial sensors data is expensive to use at large scale and with high frequency. The launch of Landsat 8 and Sentinel 2 has changed the situation significantly. Data with 10−30 m spatial resolution and high radiometric resolution is freely available with high temporal resolution. Especially if to combine Sentinel 2A and 2B (revisit time at equator 5 days with two sensors) with Landsat 8 (16 day revisit time). This opens great new opportunities for lake studies and near real time monitoring, especially at higher latitudes of the Northern Hemisphere where the majority of lakes are and where the data acquisition will start taking place almost in every second day after Sentinel 2B will be in orbit.

The amount of colored dissolved organic matter in Open Ocean, Antarctic, and alpine lake waters is usually so small that it is actually difficult to measure its absorption. Even in many coastal waters CDOM got attention of remote sensing scientist only as a cause of a small error in chlorophyll-a (Chl-a) estimates (Cipollini and Corsini, 1994). Highly colored coastal waters also exist (Clementson et al., 2004; Kutser et al., 2009a). These can be found near rivers bringing highly colored water to coastal zone (Fig. 4.1). Very high CDOM values and nearly black waters can most often be found in inland waterbodies (Kutser et al., 2016b). Majority of the lakes in the world are located in high latitudes including the boreal zone (Verpoorter et al., 2014). Many of these lakes are CDOM-rich and/or suffer from brownification (Williamson et al., 2015). Therefore, it is not surprising that CDOM remote sensing got more attention in the regions where it dominates in the formation of water color although mapping of phytoplankton biomass (Chl-a) has been the Holy Grail for aquatic remote sensing scientists and CDOM has got much less attention. Optical properties of the Baltic Sea, many boreal lakes around the world, and large rivers (e.g., Amazon, Mississippi, Orinoco, and Arctic rivers) waters are dominated by the absorption of CDOM. Consequently, the development of CDOM retrieval algorithms and methods comes particularly from these regions.

The approaches for retrieving water quality parameters from remote sensing data can be divided into two broad groups—empirical and analytical. Analytical, or physics based, methods are using radiative transfer equation or bio-optical models in retrieving either IOPs, concentrations of optically active substances, or both. The empirical remote sensing algorithms are based on statistical relationship between a color index (single band value, band ratio, spectral difference, or their combination) and a water quality parameter. The empirical remote sensing algorithms have been used since the launch of first satellite sensors. The first satellites were multispectral and the color indices and empirical relationships between them and water quality parameters were basically the only option to use. Using the empirical algorithms is easy and computationally fast, data without atmospheric correction may be used in some cases (Kallio et al., 2008; Kutser, 2012; Toming et al., 2016b). As the algorithms are purely statistical, one may try to estimate water parameters, which do not have direct effect on the water color. For example, DOC and several other carbon components, Secchi depth, iron and total phosphorus concentrations have been estimated in lakes using empirical algorithms (Kutser et al., 1995, 2015a; Kallio et al., 2001). Most of the CDOM retrieval algorithms that have been developed are also empirical (Vertucci and Likens, 1989; Arenz et al., 1996; Kutser et al., 1998, 2005a,b, 2009b; Kallio et al., 2001, 2008; Del Vecchio and Subramanijam, 2004; Brezonik et al., 2005, 2015; Kowalczuk et al., 2010; review paper by Odermatt et al., 2012; Zhu et al., 2014 and many others). Several algorithms use the ratio of blue and red bands as CDOM absorption is the highest in blue and often nearly negligible in red part of spectrum (Fig. 4.2). However, water leaving signal in the blue part of spectrum is often negligible (see Fig. 4.3) in highly absorbing lakes and coastal waters due to high concentration of CDOM. Atmospheric correction

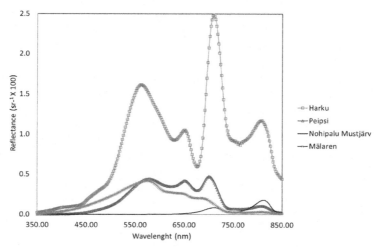

FIGURE 4.3 Remote sensing reflectance spectra of several boreal lakes. Concentrations of optically active substance are given in the Table 4.1. The reflectance spectra are measured without sun and sky glint with the methodology described in Kutser et al. (2013).

TABLE 4.1 Concentrations of Optically Active Substances (Chl-a, Total Suspended Matter, CDOM) in Lakes Reflectance Spectra of Which are Presented in Fig. 4.3

	Chl-a (mg/m^3)	TSM (mg/L)	$a_{CDOM}(400)$ (m^{-1})
Harku	203.3	44.8	6.54
Peipsi	34.34	7.0	6.55
Nohipalu Mustjärv	4.67	12.0	63.05
Mälaren	7.13	21.59	3.31

errors are also more pronounced in blue bands. Therefore, green to red ratios proved to be more useful in lakes (Kutser et al., 2005a,b; Zhu et al., 2014).

It is seen in the Fig. 4.3 that in some lakes the water reflectance is nearly negligible in almost whole visible part of spectrum (Kutser et al., 2016b). These lakes appear visually black. It is obvious that neither blue to red nor green to red band ratios allow to estimate lake CDOM content in such lakes. The figure also illustrates why the blue band is not very useful in many lakes. For example, the sampling station in Lake Mälaren used in Fig. 4.3 is quite typical for boreal lakes—Chl-a concentration is not very high and CDOM values are below average for lakes in boreal zone (Kutser et al., 2009b). Lake Harku water has bright green coloration due to high phytoplankton biomass despite CDOM values above 6 m^{-1}. Lake Peipsi, on the other hand, is brownish in color despite the amount of CDOM that is exactly the same than in Lake Harku. This is because relatively low phytoplankton biomass. Coloration of the lakes is depending on relative contribution of optically active constituents. For example, Duan et al. (2014, 2016) studied "black blooms" in Lake Taihu caused by high amount of CDOM where the "high" amount was $a_{CDOM}(400) = 3.3$ m^{-1}. In boreal lakes, such CDOM concentrations are relatively low and sometimes lake waters with nearly double amount of CDOM (Lake Harku in Fig. 4.3) appear bright green.

4.4 CDOM RETRIEVAL WITH BIO-OPTICAL MODELS

Physics based (analytical) methods using bio-optical models for retrieving water quality parameters (Chl-a, CDOM and suspended matter) water depth and benthic habitat type or all of them at the same time were introduced in the nineties of the last century. Retrieving suspended matter concentrations is discussed in Chapter 5 of this book and Chl-a retrieval in Chapter 6.

There are different approaches to retrieve water quality parameters from reflectance data using the full spectrum rather than just a few bands (Odermatt et al., 2012). For example, spectral matching techniques that either run a forward bio-optical model continuously while processing the data or using modeled in advance spectral libraries (Arst and Kutser, 1992, 1994; Kutser, 1997, 2004; Kutser et al., 2001, 2002, 2006; Pierson and Strombeck, 2001; Mobley et al., 2005; Lesser and Mobley, 2007; Van Der Woerd and Pasterkamp, 2008; Koponen et al., 2015). Different procedures from simple minimizing of the integral between the measured and modeled spectra to Spectral Angle Mapper (Kruse et al., 1993) have been used in spectral matching of the modeled spectra with measured ones. There are methods that invert inherent optical water properties from reflectance using matrix inversion (Hoge and Lyon, 1996; Lee et al., 1999, 2002; Brando and Dekker, 2003; Heege and Fischer, 2004; Giardino et al., 2007, 2012; Cheng et al., 2009; Mouw et al., 2013, Wei and Lee, 2015). The approaches using neural networks (Doerffer and Fischer, 1994; Doerffer and Schiller, 2007; Alikas and Reinart 2008; Odermatt et al., 2008; Duan et al., 2012; Binding et al., 2011; Kallio et al., 2015) also belong to the analytical methods.

Bio-optical models were used in lake remote sensing since the beginning of eighties of the last century (Bukata et al., 1981). However, in this study, the Lake Ontario model was used for testing different band-ratio type algorithms and the full-modeled spectra were not used in retrieving water constituents directly. The first attempts to retrieve Chl-a, CDOM, and TSS concentrations simultaneously by comparing full remote sensing spectra with full modeled spectra were presented in the Second Circumpolar Remote Sensing Conference in 1992 and published in the proceedings of the symposium (Arst and Kutser, 1992). A journal version of the study was published later in a special issue of the Polar Research (Arst and Kutser, 1994). The model was developed for both the Baltic Sea and Estonian lakes and performed well in both types of waters as optically they are quite similar. Unfortunately, there were no actual CDOM measurement results available for the studied waterbodies at that time. Therefore, the accuracy of the model estimate was determined only for Chl-a.

The design of this model was improved during following years. The basis of this model was a semi-analytical model proposed by Gordon et al. (1975, 1988) that related the ratio of total backscattering and absorption coefficients to the reflectance just below the water surface:

$$R(0-,\lambda) = C\frac{b_b(\lambda)}{a(\lambda) + b_b(\lambda)}, \tag{4.2}$$

Where $R(0-,\lambda)$ is irradiance reflectance just below the water surface, $b_b(\lambda)$ is the total backscattering coefficient, $a(\lambda)$ is the total absorption coefficient, and the value of C depends on the solar zenith angle (for sun at the zenith $C = 0.32$, Gordon et al., 1975). Further Monte-Carlo studies (Kirk, 1984) have found the coefficient C to be a function of solar altitude that is reasonably well expressed as a linear function of μ_0, the cosine of the zenith angle of the refracted photons:

$$C(\mu_0) = -0.629\,\mu_0 + 0.975. \tag{4.3}$$

For sun at zenith this relationship predicts $C = 0.346$ (Kirk, 1984). As the light passes through the water-air interface, it undergoes refraction that increases its angle to the vertical. Combining these effects with the effect of internal reflection, Austin (1980) proposed the factor of 0.544 for relating radiance just above the surface with radiance just below the surface. Thus, we can calculate the diffuse component of remote sensing reflectance just above the water surface:

$$r_D(\lambda) = 0.544(-0.629\mu_0 + 0.975)\frac{b_b(\lambda)}{a(\lambda) + b_b(\lambda)}. \tag{4.4}$$

The total absorption and backscattering coefficients are additive over the constituents of the medium by the definition of inherent optical properties (which requires the absence of multiple interactions). It was assumed that there are three optically active components in the water: phytoplankton (expressed as Chl-a), CDOM, and suspended matter. Under these conditions the total spectral absorption coefficient, $a(\lambda)$, is described by:

$$a(\lambda) = a_w(\lambda) + a^*_{Phy}(\lambda)C_{Chl} + a_{CDOM}(\lambda) + a^*_{SM}(\lambda)C_{SM}, \tag{4.5}$$

where a_w is the absorption coefficient of pure water, $a^*_{Phy}(\lambda)$ is the chlorophyll-specific spectral absorption coefficient of phytoplankton, $a_{CDOM}(\lambda)$ is the spectral absorption coefficient of CDOM, and $a^*_{SM}(\lambda)$ is the specific absorption coefficient of suspended matter. C_{Chl} and C_{SM} are concentrations of chlorophyll-a and total suspended matter.

The total spectral backscattering coefficient $b_b(\lambda)$ can be described:

$$b_b(\lambda) = 0.5b_w(\lambda) + b^*_{b,Phy}(\lambda)C_{Chl} + b^*_{b,SM}(\lambda)C_{SM}, \tag{4.6}$$

where b_w is the scattering coefficient of pure water and it is assumed that the backscattering probability is 50% in pure water. $b^*_{b,Ph}$ is chlorophyll-specific backscattering coefficient of phytoplankton and $b^*_{b,SM}$ is suspended sediment specific spectral backscattering coefficient of suspended matter.

The model was written in Turbo Pascal and was continuously modified using different parameters from published literature or derived combining *in situ* measurements and model simulations. For example, the phytoplankton specific absorption coefficient spectra were later calculated using formula by Bricaud et al. (1995) instead of using a constant spectrum, solar zenith angle was taken into account through the formula by Kirk (1984) (Eq. 4.3) not using constant value proposed by Gordon et al. (1975), the equation for calculating dry weight of phytoplankton from Chl-a concentration was taken from Hoogenboom and Dekker (1997), and the specific backscattering spectra of different phytoplankton species were taken from Ahn et al. (1992) instead of using constant values. Earlier versions of the model (Arst and Kutser, 1992, 1994) produced CDOM estimates in mg/L using the carbon specific absorption coefficient of 0.565 (at 380 nm) proposed by Højerslev (1980). Later versions of the model (Kutser, 1997; Kutser et al., 2001) used CDOM absorption coefficient in m^{-1} instead of concentration in mg/L as shown in Eq. (4.5). Many of the characteristics used were from ocean waters and/or for phytoplankton species not present in boreal lakes and the Baltic Sea. Consequently, the model was suboptimal for lakes without all the improvements that were done over the years. Nevertheless, as can be seen in Fig. 4.4 the model was performing well in the case of Estonian and Finnish lakes. Correlation between the measured and estimated CDOM was

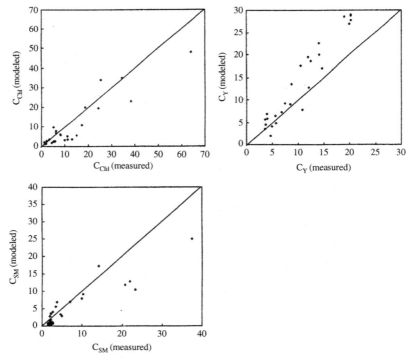

FIGURE 4.4 Correlations between the concentrations of Chl-a (C_{Chl}), CDOM (C_Y) and TSS (C_{SM}) measured *in situ* and estimated from reflectance spectra using the model by Kutser (1997). *In situ* data was collected in 20 Estonian and Finnish lakes (27 sampling stations) in 1994−1995. R was 0.92, for Chl-a, 0.97 for CDOM, and 0.91 for suspended matter. Cy is expressed in mg/L using specific absorption coefficient by Højerslev (1980). Reflectances were measured from boat. *From Kutser (1997)*.

$R = 0.97$. Some overestimation of CDOM values can be seen in the case of very high CDOM concentrations.

The results shown in Fig. 4.4 were obtained for the actually measured (from boat) reflectance spectra. An attempt was made to use also normalized (to value at 520 nm) reflectance spectra. For Chl-*a* and CDOM the results were nearly the same ($R = 0.94$ and 0.95 respectively), but using the normalized spectra did not allow to estimate the concentrations of suspended matter. This is understandable as the phytoplankton and CDOM absorption change the shape or reflectance spectra while variable concentrations of suspended matter mainly change the absolute values of reflectance due to backscattering (that is spectrally relatively flat). Normalizing the spectra removes most of the impact of TSS.

At later stages, some extra modeling exercises were carried out (Kutser et al., 2001) in order to determine specific inherent optical properties (SIOPs) of the optically active substances and some empirical parameters needed for the model. The reason for that was lack of actual *in situ* and laboratory data. Instruments measuring absorption, scattering and backscattering coefficients were not commercially available at that time, IOP and SIOP data available in literature was mainly from clear oceanic waters and not directly suitable for a lake model. The 1997 version of the model slightly overestimated CDOM concentration as can be seen in Fig. 4.4. The modified version (Kutser et al., 2001) predicted CDOM better. Correlation between measured and estimated CDOM was slightly lower ($R^2 = 0.86$), but the estimates were on 1:1 line. The range of CDOM absorption at 400 nm, $a_{CDOM}(400)$, was between 1.2 and 14 m^{-1} in this study whereas the range went up to 20 m^{-1} in the earlier study (Kutser, 1997).

Retrieval of the Chl-*a*, CDOM and TSS with the above described model was carried out in a very simplistic way. It was assumed that the concentrations in lake water match the concentrations used in the model simulation if the measured reflectance spectrum matched with a modeled one. The match was calculated based on the minimum value of the difference between the reflectance spectra—basically as a difference between the integrals of the measured and modeled spectra over the visible part of spectrum. The design of the model and *in situ* data comparison was time consuming in that sense that for each measured reflectance spectrum the model was run again with all possible combinations of concentrations. This was reasonable in the case of point measurements collected from boats, ships and helicopters. Nowadays, when most of the remote sensing data used is imagery, not point measurements, the spectral library (or look-up-table, LUT) is produced with a model in advance and then compared with measured reflectance spectra. The computing time was less than 2−3 minutes for an *in situ* hyperspectral reflectance spectrum with 10 nm resolution. This was not prohibitive on desktop computers of that time. Processing of large hyperspectral image data with this modeling approach would have been too slow, at least on regular desktop computers.

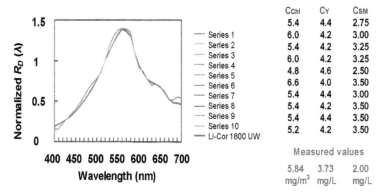

FIGURE 4.5 *In situ* measured reflectance spectrum (using Li-Cor 1800 UW) and the ten best fitting reflectance spectra produced by the bio-optical model. Concentrations used to simulate the ten best fitting spectra are given in the table together with the actual concentrations measured from water samples. C_{Chl} is concentration of Chl-*a*, C_Y is concentration of CDOM (in mg/L using specific absorption coefficient by Højerslev [1980]), and C_{SM} is concentration of TSS. *From Kutser (1997).*

Retrieving water quality parameters from reflectance data is an inverse problem. One issue with solving inverse problems is non-uniqueness of the results. In the case of water remote sensing it means that many different combinations of Chl-*a*, CDOM and TSM may give perfectly identical reflectance spectra. In the models by Kutser (1997) and Kutser et al. (2001) the issue was addressed by selecting not just the set of concentrations that produced the best fitting reflectance spectrum, but also selecting ten best fitting spectra with their respective concentrations (Fig. 4.5). It made possible to use as the solution either the concentrations of the best fitting spectrum or average concentrations of 3, 5, or 10 most similar modeled spectra. The tests revealed that in most cases there was no need to use the average concentrations as the best fitting reflectance spectrum was usually simulated with the concentrations that matched the best with *in situ* values.

It was obvious since the beginning of the model tests that using a single concentration step for the whole range of possible concentrations was not reasonable. For example, changing chlorophyll by 0.1 mg/m^3 when the concentrations are in tens to hundreds or changing CDOM with the step of 0.1 m^{-1} when the values approach tens is not feasible as such small changes in concentrations have negligible effect on reflectance spectra that no remote sensing instrument can detect. Therefore, variable grid steps were used in the model simulations. For example, 5 different chlorophyll concentration steps were used for different concentrations ranges, 5 steps for suspended matter and 3 for CDOM. The 2001 model used $a_{CDOM}(400)$ step of 0.25 m^{-1} for concentrations below 3 m^{-1}; 0.5 m^{-1} for the range between 3 and 6 m^{-1}

and $1 \, \text{m}^{-1}$ for CDOM values above $6 \, \text{m}^{-1}$ (Kutser et al., 2001). These concentration steps were selected based on the analyses of changes they caused in reflectance spectra. The variable concentration steps allowed to reduce modeling time in orders of magnitude compared to using single grid steps for the whole range of concentrations and this approach was probably one of the reasons why there was no problem with the uniqueness of the results.

Pierson and Strömbeck (2001) developed a bio-optical model for Lake Mälaren in Sweden and retrieved concentrations of optically active substances using spectral matching technique similar to that described above. The model was based on the equation by Kirk (1984). The main difference between the Lake Mälaren model and the model for Estonian and Finnish lakes described above was the number of output parameters. The model by Kutser (1997) used an empirical relationship calculating dry weight of organic particles in order to have both organic and inorganic particles taken into account, but the model predicted only total suspended matter besides CDOM and Chl-a. The model by Pierson and Strömbeck estimated organic (SPOM) and inorganic particulate matter (SPIM) concentrations separately although the SPOM estimates were poor. The retrieval of CDOM by this model was good as can be seen in the scatterplot in their paper. Like in the model by Kutser (1997) the CDOM values were overestimated. Unfortunately, Pierson and Strömbeck did not provide correlation coefficients for their scatterplots. The range of $a_{CDOM}(400)$ was between 1.77 and $4.74 \, \text{m}^{-1}$. Kallio et al. (2005) applied spectral matching technique (maximum likelihood) to irradiance reflectance spectra measured in field in Finnish lakes. CDOM ($a_{CDOM}(400)$ range $0.3-18.2 \, \text{m}^{-1}$) was estimated with good accuracy ($R^2 = 0.96$, standard error 29%) without systematic under- or overestimation.

Specific optical properties of phytoplankton are highly variable. For example, the specific absorption coefficient of Chl-a may vary within a single species depending on growth conditions (Fujiki and Taguchi, 2002). Natural phytoplankton assemblages vary in wide range. It is obvious that if the specific absorption and scattering coefficients of a single species are highly variable then the optical properties of phytoplankton assemblages consisting of many species are even more variable. To certain extent, this variability can be taken into account in modeling by using variable phytoplankton absorption coefficient like proposed by Bricaud et al. (1995) or developed for lake waters, e.g., by Ylöstalo et al. (2014) and Paavel et al. (2016). Nevertheless, from optical point of view a correct approach would be retrieving absorption and scattering coefficients not concentrations of optically active substances. The concentrations of Chl-a and TSM for each particular study site can be then calculated multiplying the retrieved absorption or scattering coefficient value with specific absorption or scattering coefficient of local phytoplankton and particles assemblages. The amount of CDOM is usually expressed through absorption coefficient value at a single wavelength. Thus, there is no need to convert the CDOM absorption into other units unless there is need to express CDOM in units of carbon.

Inversion techniques first retrieving inherent optical water properties, IOPs, like the total absorption and backscattering coefficients and then concentrations of optically active substances have been used both in lake and ocean environments (Bukata et al., 1981; Hoge and Lyon, 1996; Lee et al., 2002, Wei and Lee, 2015). For example, Cheng et al. (2009) tuned the quasi-analytical algorithm (QAA) by Lee et al. (2002) for turbid waters of Meiliang Bay of Lake Taihu by modifying some empirical steps in the inversion algorithm. However, the tuned algorithm failed to retrieve total absorption during one of the study years. Unfortunately, this study does not provide information about retrieval of CDOM with the tuned QAA. Mishra et al. (2014) developed a QAA algorithm for highly turbid cyanobacteria-dominated inland waters. They retrieved a_{CDOM} from total absorption using two band ratio algorithms. The results were realistic, but no attempt was made to estimate a_{CDOM}. Mouw et al. (2013) evaluated seven bio-optical inversion algorithms in Lake Superior. Their aim was to improve Chl-a retrieval in waters where absorption of light by CDOM exceeds the absorption by Chl-a. Direct correlations between measured and estimate CDOM was not provided in the paper. The correlations between measured and estimated sum of absorption by CDOM and detrital particles were below 0.6 for different MODIS and MERIS bands. Although the absorption of light by CDOM in the Lake Superior is dominating in the formation of water reflectance spectra and seriously hampering the retrieval of Chl-a, its concentration is relatively low ($a_{CDOM}(400)$ below 0.4 m^{-1}), especially in comparison to boreal lake waters. Campbell et al. (2011) tested matrix inversion method with MERIS data for several Australian reservoirs, but the results for CDOM retrieval were also poor. Wei and Lee (2015) showed that a UV band should improve the CDM retrieval with the QAA, but unfortunately, there are no UV bands available on remote sensing sensors.

One of the bio-optical inversion methods tested by Mouw et al. (2013) was MERIS neural network approach by Doerffer and Schiller (2007). Neural networks have to be trained with data from particular waterbodies in order to provide reliable results. One of the MERIS "standard" processors is "Boreal lakes" (Koponen et al., 2008; Kallio et al., 2015). The performance of the boreal lakes processor in the estimation of CDOM was poor in Finnish lakes and underestimated the measured values significantly (Kallio et al., 2015). Underestimated CDOM was mainly due to overestimation of R_{rs} in blue and problems in partitioning of absorption between phytoplankton and CDOM. Measured $a_{CDOM}(400)$ values in the studied lakes were between 0.9 and 10.1 m^{-1}. Alikas and Reinart (2008) tested MERIS Case 2 processor in Estonian and Swedish large lakes and found no correlation between the measured and MERIS estimated CDOM concentrations ($0.5 < a_{CDOM}(400) < 7$ m^{-1}). Duan et al. (2012) tested several MERIS processors in Lake Taihu,

China, but the performance of the processors in retrieving CDOM was poor $(0.5 < a_{CDOM}(400) < 4 \text{ m}^{-1})$. Philipson et al. (2016) tested MERIS FUB processor (Schroeder et al. 2007) in Lake Vänern, Sweden. Correlation between measured absorbance at 420 nm and the MERIS CDOM product was high $(r^2 = 0.87)$; however, the MERIS product linearly underestimated the water absorbance and a local correction coefficient was needed.

One of the key problems both in the case of QAA type inversions (Lee et al., 2002) and neural network type inversions (Doerffer and Schiller, 2007) seems to be decomposition of the total absorption into components (phytoplankton, CDOM, and detritus). Retrieval of the total absorption coefficient spectra is nearly always better than retrieval of CDOM. The same results were obtained for lake Mälaren in Sweden (Kutser et al., 2015b) where the correlation between measured and estimated CDOM concentration was poor, but that retrieved with CoastColour Baltic Sea processor total absorption coefficient was in good correlation with several carbon fractions (DOC, TOC, and DIC) in the lake. Moreover, calculating the green to red band ratio (similar to that used Kutser et al., 2005a,b) from MERIS reflectance data allowed to predict CDOM concentration in the lake very well $(R^2 = 0.81)$.

Spectral matching techniques, like the SIOCS (Sensor-Independent Ocean Colour Processor) operator have been recently used in processing of lake reflectance data (Koponen et al., 2015). The SIOCS operator encapsulates an inverse model. It is available as plugin for the Sentinel Toolbox. It has two major characteristics: The first is its applicability to different sensors; the second is that it allows for interchanging components. Users may choose between optimization methods (Levenberg Marquardt (Moré, 1977) or Simplex (Nelder and Mead, 1965)), methods for setting the IOP starting values or the forward model. Currently, a LUT-based forward model is included in the operator. Depending on the content of the LUT, the SIOCS operator can be applied to different sensors. Users can extend the operator by creating their own components and share them with others. The SIOCS operator is available for free from ferryscope.org.

The input for the SIOCS processing are the measured spectrum aside and an initial guess of the IOP values (based on local knowledge or band-ratio algorithms estimates). This initial guess is applied to the forward model. The forward model returns a spectrum that is expected for the setting of IOPs. The difference between the modeled spectrum and the actually measured spectrum is computed by the use of a cost function. This cost function will deliver a goodness of fit between the spectra. If this goodness of fit falls below a given threshold, the breaking criterion is met and the IOP guess is accepted as output. If the breaking criterion is not met, the IOP estimate will be altered by an optimization step. This new IOP estimate will again be applied to the forward model. At this point the procedure enters a cycle which is repeated until the breaking criterion is met (this could also mean that a maximum number of iterations has been reached).

An important property of SIOCS is that its core components are exchangeable. A user can choose from a set of predefined components. As the components are designed as plug-ins into a framework, it is also possible to implement new components and use them with the operator (Fig. 4.6).

Exchangeable components are the IOP initialization method, the cost function, the optimizer, and the forward model. The IOP initialization determines the way the initial IOP estimates are derived. Provided are methods which set the same value for each spectrum or which take their guess from an auxiliary product. The cost functions compare the modeled and the measured spectra to each other. An optimizer performs the optimization step to improve the IOP estimate.

As for the forward model, there is currently one provided: A forward model, which derives a spectrum by interpolating values from a LUT. One such LUT holds spectra for different IOPs. For each IOP, sampling points are defined. These sampling points span a value range for each IOP. A spectrum is contained for each permutation of these sampling points. For example, if there are five sampling points for CDOM absorption and five sampling points for total particle scattering, there would be 25 different spectra in the LUT. A LUT may hold highly resolved spectra, though in order to keep the LUT size small and to make the handling feasible, the spectra are usually provided in a form to match the spectral characteristics of a specific sensor. When using the LUT Forward Model, the choice of the LUT determines on which sensor the SIOCS can be applied.

In the LUT forward model, for a given set of input IOP values, the set of enclosing sampling points is determined and the spectra for each of the resulting sampling point combinations are retrieved. Each of these spectra is assigned a weight according to how similar the respective sampling point values are to the input IOP values. By adding the weighted spectra, a modeled spectrum is interpolated and returned by the model.

The development of SIOCS is continued. The SIOCS Operator will, for example, be extended by new components. CDOM retrieval results for Finnish lakes with the current version of SIOCS are shown in Fig. 4.7.

The best results in retrieval of lake CDOM with bio-optical modeling have been achieved with spectral matching techniques (Kutser, 1997; Kutser et al., 2001; Pierson and Strömbeck, 2001; Koponen et al., 2015). However, field reflectance spectra, not satellite imagery, was used in all these studies. This allows the assumption that one of the main problems in using different kinds of physics based approaches in retrieving lake CDOM (and other water constituents) is atmospheric correction. Toming et al. (2016a,b) have shown that Sen2cor atmospheric correction in SNAP produces lake reflectance spectra that are very similar to field reflectances measured in the same lakes before Sentinel-2 image acquisition. Nevertheless, CDOM (and other water quality parameters) retrieval with band ratio algorithms still performed better in the case of top of atmosphere reflectance spectra rather than in the case of

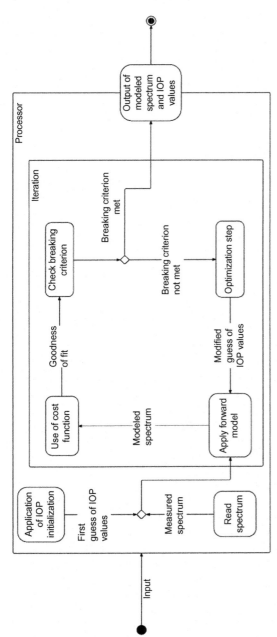

FIGURE 4.6 Description of the SIOCS operator.

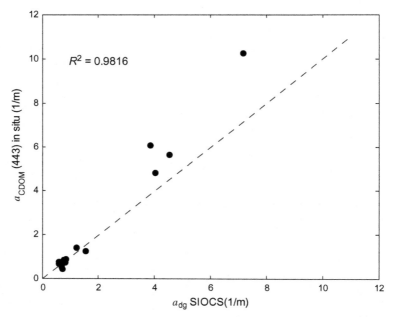

FIGURE 4.7   Correlation between measured CDOM concentrations and predicted with the SIOCS operator. Hyperspectral field reflectance data was recalculated into Sentinel-3 OLCI bands (Koponen et al., 2015).

atmospherically corrected imagery (Toming et al., 2016b). In black, CDOM-rich, lakes the water leaving signal is negligible in most of the visible part of spectrum (Kutser et al., 2016b) and in clear lakes it is also very small. Therefore, it is easy to understand that even a small error in atmospheric correction may be in the same range than the whole water leaving signal. Consequently, these errors have impact on the concentrations of Chl-*a*, CDOM and TSM retrieved from the satellite reflectance spectra.

One of the possible solutions, besides improving the atmospheric correction methods (discussed in the Chapter 3 of this book), could be using of top of atmosphere spectra. For example, Kutser et al. (2002, 2006) showed in the case of coral reef remote sensing that using a top of atmosphere spectral library with a raw image produced much better results than using above water spectral library with atmospherically corrected image. The differences were relatively small in the case of very shallow waters with coral reefs and sand, i.e., in the case of bright targets. However, most of the deeper water areas (deeper than 6–7 m) remained unclassified in the atmospherically corrected image. It means that the shape of atmospherically corrected spectra for the areas with low water leaving signal were not correct as the Spectral Angle Mapper, comparing the shape, not absolute values of spectra, did not manage to classify the areas with low reflectance. Both

CDOM-rich and clear water lakes have low reflectance and their atmospheric correction is more problematic than it is for turbid lakes. Using a forward atmospheric model and producing a top of atmosphere spectral library seems to be less prone to errors than atmospherically correcting satellite imagery and using above water level spectral library.

Different spectral matching and inversion techniques like BOMBER (Bio-Optical Model Based tool for Estimating water quality and bottom properties from Remote sensing images), SIOCS, SAMBUCA (Semi-Analytical Model for Bathymetry, Un-mixing, and Concentration Assessment) have become freely available and will soon be available as plugins in image processing software like SNAP (SentiNel Application Platform) and others. The launch of Landsat 8 and Sentinel 2 opened great new opportunities in lake remote sensing. Majority of lakes on Earth are small (Verpoorter et al., 2014) and the small lakes are more important from carbon processing point of view. Small lakes are also less covered with conventional monitoring programs. The new sensors and the availability of different processors using bio-optical models should improve the accuracy of lake remote sensing products in the near future.

4.5 FINAL CONSIDERATIONS

Mapping of DOC and CDOM content of inland and coastal waters is of critical importance from several points of view as was described above. New satellite sensors provide greater spatial and radiometric resolution improving our ability to study and monitor lakes. Field radiometers on buoys, jetties, and ships are becoming water monitoring devices on their own right rather than just calibration and validation devices of satellite and airborne sensors. Thus, there is large flow of data that can be used in monitoring inland water quality. Analytical methods based on bio-optical models and different model inversion techniques have made good progress during the last three decades. Progress in computer technology together with advancement in methodologies used in the bio-optical models have reduced significantly processing time and made the use of the models feasible even in the case of high spatial and spectral resolution airborne data or large satellite imagery. Empirical algorithms are sensor specific (depending on band configuration), waterbody specific (cannot be used in different waterbodies without tuning) and even seasonal (due to high seasonal variability of SIOPs) in some waterbodies (Ligi et al., 2017; Simis et al., 2017). Validation of the band-ratio type algorithms requires extensive amount of match-up data. Physic based methods relying on bio-optical models are free of all these shortcomings. Therefore, it can be foreseen that the use of bio-optical models increase significantly in interpretation of remote sensing data collected in optically complex inland and coastal waters.

REFERENCES

Ahn, Y.-H., Bricaud, A., Morel, A., 1992. Light backscattering efficiency and related properties of some phytoplankters. Deep Sea Res. 39, 1835–1855.

Alikas, K., Reinart, A., 2008. Validation of the MERIS products on large European lakes – Peipsi, Vänern and Vättern. Hydrobiol 18, 161–168.

Arenz, R., Lewis, W., Saunders, J., 1996. Determination of chlorophyll and dissolved organic carbon from reflectance data for Colorado reservoirs. Int. J. Rem. Sens. 17, 1547–1565.

Arst, H., Kutser, T., 1992. Data processing and interpretation of sea radiance factor measurements. Proc. 2nd Circumpolar Symp. Remote Sens. Arctic Environ. pp. 41-42.

Arst, H., Kutser, T., 1994. Data processing and interpretation of sea radiance factor measurements. Polar. Res. 13, 3–12.

Austin, R.W., 1980. Gulf of Mexico, ocean-colour surface-truth measurements. Boundary-Layer Meteorol. 18, 269–285.

Bastviken, D., Cole, J.J., Pace, M.L., Tranvik, L., 2004. Methane emissions from lakes: dependence of lake characteristics, two regional assessments, and a global estimate. Global. Biogeochem. Cycles18. Available from: http://dx.doi.org/10.1029/2004GB002238.

Bastviken, D., Tranvik, L.J., Downing, J.A., Crill, P.M., Enrich-Prast, A., 2011. Freshwater methane emissions offset the continental carbon sink. Science 331, 50.

Battin, T.J., Kaplan, L.A., Findlay, S., Hopkinson, C.S., Marti, E., Packman, A.I., et al., 2008. Nature Geosci 1, 95–100.

Battin, T.J., Luyssaert, S., Kaplan, L.A., Aufdenkampe, A.K., Richter, A., Tranvik, L.J., 2009. The boundless carbon cycle. Nature Geosci 2, 598–600.

Binding, C.E., Greenberg, T.A., Jerome, J.H., Bukata, R.P., Letourneau, G., 2011. An assessment of MERIS algal products during an intense bloom in Lake of the Woods. J. Plankton Res. 33, 793–806.

Brando, V.E., Dekker, A.G., 2003. Satellite hyperspectral remote sensing for estimating estuarine and coastal water quality IEEE Trans. Geosci. Rem. Sens 41, 1378–1387.

Brezonik, P.L., Menken, K., Bauer, M.E., 2005. Landsat-based remote sensing of lake water quality characteristics, including chlorophyll and colored dissolved organic matter (CDOM). Lake and Reservoir Management 21, 373–382.

Brezonik, P.L., Olmanson, L.G., Finlay, J.C., Bauer, M.E., 2015. Factors affecting the measurements of CDOM by remote sensing of optically complex inland waters. Remote Sens. Environ. 157, 199–215.

Bricaud, A., Morel, A., Prieur, L., 1981. Absorption by dissolved organic matter of the sea (yellow substance) in the UV and visible domains. Limnol. Oceanogr. 26, 43–53.

Bricaud, A., Babin, M., Morel, A., Claustre, H., 1995. Variability in the chlorophyll-specific absorption coefficients of natural phytoplankton: analysis and parameterisation. J. Geophys. Res. 100 (C7), 13321–13332.

Bukata, R.P., Bruton, J.E., Jerome, J.H., Jain, S.C., Zwick, H.H., 1981. Optical water quality model of Lake Ontario. 2: Determination of chlorophyll a and suspended mineral concentrations of natural waters from submersible and low altitude optical sensors. Appl. Optics 20, 1704–1714.

Campbell, G., Phinn, S.R., Dekker, A.G., Brando, V.E., 2011. Remote sensing of water quality in an Australian tropical freshwater impoundment using matrix inversion and MERIS images. Remote Sens. Environ. 115, 2402–2414.

Carder, K.L., Steward, R.G., Harvey, G.R., Ortner, P.B., 1989. Marine humic and fulvic acids: Their effects on remote sensing of ocean chlorophyll. Limnol. Oceanogr. 34, 68–81.

Cheng, F.L., Yun, M.L., Yong, Z., Deyong, S., Bin, Y., 2009. Validation of a quasianalytical algorithm for highly turbid eutrophic water of Meiliang Bay in Taihu Lake, China. IEEE. Trans. Geosci. Remote. Sens. 47, 2492−2500.

Cipollini, P., Corsini, G., 1994. The effect of yellow substance on pigment concentration retrieval using blue to green ratio. Proc. IEEE Oceans'94 1, A772−A777.

Clementson, L.A., Parslow, J.S., Turnbull, A.R., Bonham, P.I., 2004. Properties of light absorption in a highly coloured estuarine system in south-east Australia which is prone to blooms of the toxic dinoflagellate Gymnodinium catenatum. Estuar. Coast. Shelf Sci. 60, 101−112.

Cole, J.J., Prairie, Y.T., Caraco, N.F., McDowell, W.H., Tranvik, L.J., Striegl, R.G., et al., 2007. Plumbing the global carbon cycle: integrating inland waters into the terrestrial carbon budget. Ecosystems 10, 171−184.

Del Castillo, C.E., 2005. Remote sensing of organic matter in coastal waters. In: Miller, R.L., Del Castillo, C.E., McKee, B.A. (Eds.), Remote Sensing of Coastal Aquatic Environments: Technologies, Techniques and Applications. Springer, pp. 157−180.

Del Vecchio, R., Subramanijam, A., 2004. Influence of the Amazon River on the surface optical properties of the western tropical North Atlantic Ocean. J. Geophys. Res. 109, C11001.

Doerffer, R., Fischer, J., 1994. Concentrations of chlorophyll, suspended matter, and gelbstoff in case II waters derived from satellite coastal zone color scanner data with inverse modeling methods. J. Geophys. Res. Oceans 99, 7457−7466.

Doerffer, R., Schiller, H., 2007. The MERIS Case 2 water algorithm. Int. J. Remote. Sens. 28, 517−535.

Duan, H., Ma, R., Simis, S.G.H., Zhang, Y., 2012. Validation of MERIS Case-2 water products in Lake Taihu, China. GIScience & Remote Sens. 49, 873−894.

Duan, H., Ma, R., Loiselle, S.A., Shen, Q., Yin, H., Zhang, Y., 2014. Optical characterization of black water blooms in eutrophic waters. Sci. Total. Environ. 2014, 174−183.

Duan, H., Loiselle, S.A., Li, Z., Shen, Q., Du, Y., Ma, R., 2016. A new insight into black blooms: synergies between optical and chemical factors. Estuar. Coast. Shelf Sci. 2016 (175), 118−125.

Eikebrokk, B., Vogt, R.D., Liltved, H., 2004. NOM increase in Northern European source waters: discussion of possible causes and impacts on coagulation/contact filtration processes. Water Sci. Technol. Water Supply 4, 47−54.

Erlandsson, M., Futter, M.N., Kothawala, D.N., Köhler, S.J., 2012. Variability in spectral absorbance metrics across boreal lake waters. J. Environ. Monitoring 14, 2643−2652.

Fujiki, T., Taguchi, S., 2002. Variability in chlorophyll a specific absorption coefficient in marine phytoplankton as a function of cell size and irradiance. J. Plankton Res. 24, 859−874.

Giardino, C., Brando, V.E., Dekker, A.G., Strömbeck, N., Candiani, G., 2007. Assessment of water quality in Lake Garda (Italy) using Hyperion. Remote Sens. Environ. 109, 183−195.

Giardino, C., Candiani, G., Bresciani, M., Lee, Z.P., Gagliano, S., Pepe, M., 2012. BOMBER: a tool for estimating water quality and bottom properties from remote sensing images. Computers & Geosci. 45, 313−318.

Giardino, C., Bresciani, C., Stroppiana, D., Oggioni, A., Morabito, G., 2014. Optical remote sensing of lakes: an overview on Lake Maggiore. J. Limnol.(73), 201−214.

Gordon, H.R., Brown, O.B., Jacobs, M.M., 1975. Computed relationships between the inherent and apparent optical properties of a flat, homogenous ocean. Appl. Opt. 14, 417−427.

Gordon, H.R., Brown, O.B., Evans, R.H., Brown, J.W., Smith, R.C., Baker, K.S., et al., 1988. A semianalytic radiance model of ocean colour. J. Geophys. Res. 93, 10909−10924.

Hansell, D.A., Carlson, C.A., 2001. Marine dissolved organic matte and the carbon cycle. Oceanography 14, 41−49.

Hayase, K., Tsubota, H., 1985. Sedimentary humic-acid and fulvic-acid as fluorescent organic materials. Geochim. Cosmochim. Acta. 49, 159−163.

Heege, T., Fischer, J., 2004. Mapping of water constituents in Lake Constance using multispectral airborne scanner data and a physically based processing scheme. Can. J. Remote Sens. 30, 77−86.

Helms, J.R., Stubbins, A., Ritchie, J.D., Minor, E.C., Kieber, D.J., Mopper, K., 2008. Absorption spectral slopes and slope ratios as indicators of molecular weight, source, and photobleaching of chromophoric dissolved organic matter. Limnol. Oceanogr. 53, 955−969.

Hestir, E., Brando, V.E., Campbell, G., Dekker, A.G., Malthus, T., 2015. The relationship between dissolved organic matter absorption and dissolved organic carbon in reservoirs along a temperate to tropical gradient. Remote Sens. Environ. 156, 395−402.

Hoge, F.E., Lyon, P.E., 1996. Satellite retrieval of inherent optical properties by linear matrix inversion of oceanic radiance models: an analysis of model and radiance measurement errors. J. Geophys. Res. 101, 16631−16648.

Højerslev, N.K., 1980. On the origin of yellow substance in the marine environment. Univ. Copenhagen Inst. Phys. Oceanogr. Rep. p. 42.

Hoogenboom, J., Dekker, A.G., 1997. Simulation of the medium-resolution imaging spectrometer MERIS performance for detecting chlorophyll-*a* over turbid inland waters. SPIE Proc. 2963, 440−447.

IPCC Fourth Assessment Report. 2007.

IPCC Fifth Assessment Report. Working Group I Report "Climate Change 2013: The Physical Science Basis."

Kalle, K., 1938. Zum Problem des Meercswasserfarbc. Ann. Ilydrol. Mar. Mitt. 66, 1−13.

Kallio, K., 1999. Absorption properties of dissolved organic matter in Finnish lakes. Proc. Estonian Acad. Sci., Biol. Ecol. 48, 75−83.

Kallio, K., Kutser, T., Hannonen, T., Koponen, S., Pulliainen, J., Vepsäläinen, J., et al., 2001. Retrieval of water quality from airborne imaging spectrometry of various lake types in different seasons. Sci. Total. Environ. 268, 59−77.

Kallio, K., Pulliainen, J., Ylöstalo, P., 2005. MERIS, MODIS and ETM channel configurations in the estimation of lake water quality from subsurface reflectance with semi-analytical and empirical algorithms. Geophysica 41, 31−55.

Kallio, K., Attila, J., Härmä, P., Koponen, S., Pulliainen, J., Hyytiäinen, U.-M., et al., 2008. Landsat ETM + images in the estimation of seasonal lake water quality in Boreal River Basins. Environ. Manag. 42, 511−522.

Kallio, K., Koponen, S., Ylöstalo, P., Kervinen, M., Pyhälahti, T., Attila, J., 2015. Validation of MERIS spectral inversion processors using reflectance, IOP and water quality measurements in boreal lakes. Remote Sens. Environ. 157, 147−157.

Kirk, J.T.O., 1976. Yellow substance (Gelbstoff) and its contribution to the attenuation of photosynthetically active radiation in some inland and coastal south-eastern Australian waters. Aust. J. Mar. Freshwater Res. 27, 61−71.

Kirk, J.T.O., 1980. Spectral absorption properties of natural waters: contribution of the soluble and particulate fractions of light absorption in some inland waters of South-eastern Australia. A Aust. J. Mar. Freshwater Res. 31, 287−296.

Kirk, J.T.O., 1984. Dependence of relationship between inherent and apparent optical properties of water on solar altitude. Limnol. Oceanogr. 29, 350−356.

Köhler, S.J., Kothawala, D., Futter, M.N., Liungman, O., Tranvik, L., 2013. In-lake processes offset increased terrestrial inputs of dissolved organic carbon and color to lakes. PLoS. ONE. 8, e70598.

Koivusalo, M., Pukkala, E., Vartiainen, T., Jaakkola, J.J., Hakulinen, T., 1997. Drinking water chlorination and cancer—a historical cohort study in Finland. Cancer. Causes. Control. 8, 192–200.

Koponen, S., Ruiz-Verdu, A., Heege, T., Heblinski, J., Sorensen, K., Kallio, K., et al., 2008. Development of MERIS lake water algorithms. ESA Validation Report.

Koponen et al., 2015. GLaSS Deliverable D5.5, 2015. Global Lakes Sentinel Services, D5.5: Boreal lakes case study results. SYKE, WI, TO, BG. Available via: www.glass-project.eu/downloads.

Kowalczuk, P., Durako, M.J., Young, H., Kahn, A.E., Cooper, W.J., Gonsior, M., 2009. Characterization of dissolved organic matter fluorescence in the South Atlantic Bight with use of PARAFAC model: interannual variability. Marine Chem. 113, 182–196.

Kowalczuk, P., Darecki, M., Zablocka, M., Gorecka, I., 2010. Validation of empirical and semi-analytical remote sensing algorithms for estimating absorption by Coloured Dissolved Organic Matter in the Baltic Sea from SeaWiFS and MODIS imagery. Oceanologia 52, 171–196.

Kruse, F.A., Lefkoff, A.B., Boardman, J.B., Heidebrecht, K.B., Shapiro, A.T., Barloon, P.J., et al., 1993. The Spectral Image Processing System (SIPS)—interactive visualization and analysis of imaging spectrometer data. Remote Sens. Environ. 44, 145–163.

Kutser, T., 1997. Estimation of water quality in turbid inland and coastal waters by passive optical remote sensing. Dissertationes Geophysicales Universitas Tartuensis. Tartu University Press, p. 8.

Kutser, T., 2004. Quantitative detection of chlorophyll in cyanobacterial blooms by satellite remote sensing. Limnol. Oceanogr. 49, 2179–2189.

Kutser, T., 2010. Global change and remote sensing of CDOM in Arctic coastal waters. Proc. IGARSS'10 Conf. (389 – 392). IEEE Geoscience and Remote Sensing Society.

Kutser, T., 2012. The possibility of using the Landsat image archive for monitoring long time trends in coloured dissolved organic matter concentration in lake waters. Remote Sens. Environ. 123, 334–338.

Kutser, T., Arst, H., Miller, T., Käärmann, L., Milius, A., 1995. Telespectrometrical estimation of water transparency, chlorophyll a and total phosphorus concentrations on Lake Peipsi. Int. J. Remote. Sens. 16, 3069–3085.

Kutser, T., Arst, H., Mäekivi, S., Kallaste, K., 1998. Estimation of the water quality of the Baltic Sea and some lakes in Estonia and Finland by passive optical remote sensing measurements on board a vessel. Lakes and Reserv.; Res. Manag. 3, 53–66.

Kutser, T., Herlevi, A., Kallio, K., Arst, H., 2001. A hyperspectral model for interpretation of passive optical remote sensing data from turbid lakes. Sci. Total. Environ. 268, 47–58.

Kutser, T., Miller, I., Jupp, D.L.B., 2002. Mapping coral reef benthic habitat with hyperspectral space borne sensor. Proc. Ocean Optics XVI, Santa Fe (CD-ROM).

Kutser, T., Pierson, D.C., Tranvik, L., Reinart, A., Sobek, S., Kallio, K., 2005a. Using satellite remote sensing to estimate the coloured dissolved organic matter absorption coefficient in lakes. Ecosystems 8, 709–720.

Kutser, T., Pierson, D.C., Kallio, K.Y., Reinart, A., Sobek, S., 2005b. Mapping lake CDOM by satellite remote sensing. Remote Sens. Environ. 94, 535–540.

Kutser, T., Miller, I., Jupp, D.L.B., 2006. Mapping coral reef benthic substrates using hyperspectral space borne images and spectral libraries. Estuar. Coast. Shelf Sci. 70, 449–460.

Kutser, T., Paavel, B., Metsamaa, L., Vahtmäe, E., 2009a. Mapping coloured dissolved organic matter concentration in coastal waters. Int. J. Remote. Sens. 30, 5843–5849.

Kutser, T., Tranvik, L., Pierson, D.C., 2009b. Variations in colored dissolved organic matter between boreal lakes studied by satellite remote sensing. J. Appl. Remote Sens. 3, 033538.

Kutser, T., Vahtmäe, E., Paavel, B., Kauer, T., 2013. Removing glint effects from field radiometry data measured in optically complex coastal and inland waters. Remote Sens. Environ. 133, 85–89.

Kutser, T., Alikas, K., Kothawala, D.N., Köhler, S.J., 2015a. Impact of iron associated to organic matter on remote sensing estimates of lake carbon content. Remote Sens. Environ. 156, 109–116.

Kutser, T., Verpoorter, C., Paavel, B., Tranvik, L.J., 2015b. Estimating lake carbon fractions from remote sensing data. Remote Sens. Environ. 157, 138–146.

Kutser, T., Casal Pascual, G., Barbosa, C., Paavel, B., Ferreira, R., Carvalho, L., et al., 2016a. Mapping inland water carbon content with Landsat 8 data. Int. J. Remote. Sens. 37, 2950–2961.

Kutser, T., Paavel, P., Verpoorter, C., Ligi, M., Soomets, T., Toming, K., et al., 2016b. Remote Sensing of black lakes and using 810 nm reflectance peak for retrieving water quality parameters of optically complex waters. Remote Sens 8, 497.

Lee, Z.P., Carder, K.L., Mobley, C.D., Steward, R.G., Patch, J.S., 1999. Hyperspectral remote sensing for shallow waters. 2. Deriving bottom depths and water properties by optimization. Appl. Opt. 38, 3831–3843.

Lee, Z.P., Carder, K.L., Arnone, R.A., 2002. Deriving inherent optical properties from water color: a multiband quasi-analytical algorithm for optically deep waters. Appl. Opt. 41, 5755–5772.

Lesser, M.P., Mobley, C.D., 2007. Bathymetry, water optical properties, and benthic classification of coral reefs using hyperspectral remote sensing imagery. Coral Reefs 26, 819–829.

Ligi, M., Kutser, T., Kallio, K., Attila, J., Koponen, S., Paavel, B., et al., 2017. Testing the performance of empirical remote sensing algorithms in the Baltic Sea waters with modelled and in situ reflectance data. Oceanologia 57, 57–68.

Loiselle, S.A., Bracchini, L., Dattilo, A.M., Ricci, M., Tognazzi, A., Cózar, A., et al., 2009. Optical characterization of chromophoric dissolved organic matter using wavelength distribution of absorption spectral slopes. Limnol. Oceanogr. 54, 590–597.

Magnus, P., Jaakola, J.J., Skrondal, A., Alexander, J., Becher, G., Krog, T., et al., 1999. Water chlorination and birth defects. Epidemiol. 10, 513–520.

McDonald, T.A., Komulainen, H., 2005. Carcinogenicity of the chlorination disinfection by-product MX. J. Environ. Sci. Health. C. Environ. Carcinog. Ecotoxicol. Rev. 23, 163–214.

Mishra, S., Mishra, D.R., Lee, Z.P., 2014. Bio-optical inversion in highly turbid and cyanobacteria-dominated waters. IEEE. Trans. Geosci. Remote. Sens. 52, 375–388.

Mobley, C.D., Sundman, L.K., Davis, C.O., Bowles, J.H., Downes, T.V., Leathers, R.A., et al., 2005. Interpretation of hyperspectral remote-sensing imagery by spectrum matching and look-up tables. Appl. Opt. 44, 3576–3592.

Molot, L.A., Dillon, P.J., 1997. Colour – mass balances and colour – dissolved organic carbon relationships in lakes and streams in central Ontario. Cananadian J Fish. Aquatic Sci. 54, 2789–2795.

Moré, J.J., 1977. The Levenberg-Marquardt algorithm. Implementation and theory. In G.A. Watson (ed.): Numerical Analysis. Lecture Notes in Mathematics. Springer, Berlin, Heidelberg, 630.

Mouw, C.B., Chen, H., McKinley, G.A., Effler, S., O'Donnell, D., Perkins, M.G., et al., 2013. Evaluation and optimization of bio-optical inversion algorithms for remote sensing of Lake Superior's optical properties. J. Geophys. Res.: Oceans 118, 1696–1714.

Müller, R.A., Kothawala, D.N., Podgrajsek, E., Sahlée, E., Koehler, B., Tranvik, L.J., et al., 2014. Hourly, daily, and seasonal variability in the absorption spectra of chromophoric dissolved organic matter in a eutrophic, humic lake. J. Geophys. Res. Biogeosci. 119, 1985–1998.

Nelder, J.A., Mead, A., 1965. A simplex method for function minimization. Computer J. 7, 308–313.

Nima, C., Hamre, B., Frette, Ø., Erga, S.R., Chen, Y.-C., Zhao, L., et al., 2016. Impact of particulate and dissolved material on light absorption properties in a High-Altitude Lake in Tibet. China Hydrobiol. 768, 63–79.

Odermatt, D., Heege, T., Nieke, J., Kneubühler, M., Itten, K., 2008. Water quality monitoring for Lake Constance with a physically based algorithm for MERIS data. Sensors 8, 4582–4599.

Odermatt, D., Gitelson, A., Brando, V.E., Schaepman, M., 2012. Review of constituent retrieval in optically deep and complex waters from satellite imagery. Remote Sens. Environ. 118, 116–126.

Paavel, B., Kangro, K., Arst, H., Reinart, A., Kutser, T., Nõges, T., 2016. Parameterization of chlorophyll-specific phytoplankton absorption coefficients for productive lake waters. J. Limnol. Available from: http://dx.doi.org/10.4081/jlimnol.2016.1426.

Palmer, S.C.J., Kutser, T., Hunter, P.D., 2015. Remote sensing of inland waters: challenges, progress and future directions. Remote Sens. Environ. 157, 1–8.

Pegau, S., Zaneveld, J.R.V., Mitchell, B.G., Mueller, J.L., Kahru, M., Wieland, J., et al., 2003. Ocean Optics Protocols For Satellite Ocean Color Sensor Validation, Revision 4, Volume IV: Inherent Optical Properties: Instruments, Characterizations, Field Measurements and Data Analysis Protocols. James L. Mueller, Giulietta S. Fargion and Charles R. McClain, Editors NASA/TM-2003-211621/Rev4-Vol. IV.

Philipson, P., Kratzer, S., Mustapha, S.B., Strömbeck, N., Stelzer, K., 2016. Satellite-based water quality monitoring in Lake Vänern, Sweden. Int. J. Remote. Sens. 37, 3939–3960.

Pierson, D.C., Strombeck, N., 2001. Estimation of radiance reflectance and the concentrations of optically active substances in Lake Malaren, Sweden, based on direct and inverse solutions of a simple model. Sci. Total. Environ. 268, 171–188.

Raymond, P.A., Hartmann, J., Lauerwald, R., Sobek, S., McDonald, C., Hoover, M., et al., 2013. Global carbon dioxide emissions from inland waters. Nature. 503, 355–359.

Reinart, A., Paavel, B., Tuvike, L., 2004. Effect of coloured dissolved organic matter on the attenuation of photosynthetically active radiation in Lake Peipsi. Proc. Estonian Acad. Sci. Biol. Ecol. 53, 88–105.

Schroeder, T., Schaale, M., Fischer, J., 2007. Retrieval of atmospheric and oceanic properties from MERIS measurements: a new Case 2 water processor for BEAM. Int. J. Remote. Sens. 28, 5627–5632.

Siegel, D.A., Maritorena, S., Nelson, N.B., Hansell, D.A., Lorenzi-Kayser, M., 2002. Global distribution and dynamics of colored dissolved and detrital organic materials. J. Geophys. Res. C12, 1–14.

Siegel, D.A., Behrenfeld, M.J., Maritorena, S., McClain, C.R., Antoine, D., Bailey, S.W., et al., 2013. Regional to global assessments of phytoplankton dynamics from the SeaWiFS mission. Remote Sens. Environ. 135, 77–91.

Simis, S.G.H., Ylöstalo, P., Spilling, K., Kutser, T., 2017. Contrasting seasonality in optical-waters of northwestern Estonia: in situ measurements. Boreal. Env. Res. 9, 447–456. Biogeochemical properties of the Baltic Sea. PLoS One (in press).

Sipelgas, L., Arst, H., Raudsepp, U., Kouts, T., Lindfors, A., 2004. Optical properties of coastal waters of northwestern Estonia: in situ measurements. Boreal. Env. Res. 9, 447–456.

Stedmon, C.A., Markager, S., 2005. Tracing the production and degradation of autochthonous fractions of dissolved organic matter by fluorescence analysis. Limnol. Oceanogr. 50, 1415–1426.

Steinberg, C.E.W., Höss, S., Kloas, W., Lutz, I., Meinelt, T., Pflugmacher, S., et al., 2004. Hormonelike effects of humic substances on fish, amphibians, and invertebrates. Envir. Toxicol 19, 409–411.

Strömbeck, N., Pierson, D.C., 2001. The effects of variability in the inherent optical properties on estimations of chlorophylla by remote sensing in Swedish freshwaters. Sci. Total Env. 268, 123–137.

Su, Y., Chen, F., Liu, Z., 2015. Comparison of optical properties of chromophoric dissolved organic matter (CDOM) in alpine lakes above or below the tree line: insights into sources of CDOM. Photochem. Photobiol. Sci. 14, 1047–1062.

Thrane, J.-E., Hessen, D.O., Andersen, T., 2014. The absorption of light in lakes: negative impact of dissolved organic carbon on primary productivity. Ecosystems 17, 1040–1052.

Toming, K., Arst, H., Paavel, B., Laas, A., Nõges, T., 2009. Optical properties of coastal waters of northwestern Estonia: in situ measurements. Boreal. Env. Res. 14, 959–970.

Toming, K., Kutser, T., Tuvikene, L., Viik, M., Nõges, T., 2016a. Dissolved organic carbon and its potential predictors in eutrophic lakes. Water. Res. 102, 32–40.

Toming, K., Kutser, T., Laas, A., Sepp, M., 2016b. First experiences in mapping lake water quality parameters with Sentinel-2 MSI imagery. Remote Sensing 8, 640.

Tranvik, L.J., 1990. Bacterioplankton growth on fractions of dissolved organic carbon of different molecular weights from humic and clear waters. Appl. Environ. Microbiol. 56, 1672–1677.

Tranvik, L.J., Downing, J.A., Cotner, J.B., Loiselle, S., Striegl, R.G., Ballatore, T.J., et al., 2009. Lakes and impoundments as regulators of carbon cycling and climate. Limnol. Oceanogr. 54, 2298–2314.

Van Der Woerd, H.J., Pasterkamp, R., 2008. HYDROPT: a fast and flexible method to retrieve chlorophyll-a from multispectral satellite observations of optically complex coastal waters. Remote Sens. Environ. 112, 1795–1807.

Verpoorter, C., Kutser, T., Tranvik, L.J., Seekell, D., 2014. A global inventory of lakes Based on high-resolution satellite imagery. Geophys. Res. Letters 41, 6396–6402.

Vertucci, A., Likens, G.E., 1989. Spectral reflectance and water quality of Adirondack mountain region lakes. Limnol. Oceanogr. 34, 1656–1672.

Vincent, W.F., Rae, R., Laurion, I., Howard-Williams, C., Priscu, J.C., 1998. Transparency of Antarctic ice-covered lakes to solar UV radiation. Limnol. Oceanogr. 43, 618–624.

Wei, J., Lee, Z., 2015. Retrieval of phytoplankton and colored detrital matter absorption coefficients with remote sensing reflectance in an ultraviolet band. Appl. Optics 54, 636–649.

Wetzel, R.G., 2001. Limnology; lake and river ecosystems, third ed. Elsevier.

Williamson, C.E., Overholt, E.P., Pilla, R.M., Leach, T., Brentrup, J.A., Knoll, L.B., et al., 2015. Ecological consequences of long-term browning in lakes. Nature Sci. Rep. 5, 18666.

Ylöstalo, P., Kallio, K., Seppälä, J., 2014. Absorption properties of in-water constituents and their variation among various lake types in the boreal region. Remote Sens. Environ. 148, 190–205.

Zhu, W., Yu, Q., Tian, Y., Becker, B.L., Zheng, T., Carrick, H.J., 2014. An assessment of remote sensing algorithms for colored dissolved organic matter in complex freshwater environments. Remote Sens. Environ. 140, 766–778.

Chapter 5

Bio-optical Modeling of Total Suspended Solids

Claudia Giardino[1], Mariano Bresciani[1], Federica Braga[2], Ilaria Cazzaniga[1], Liesbeth De Keukelaere[3], Els Knaeps[3], and Vittorio E. Brando[4]

[1]National Research Council of Italy, Institute for Electromagnetic Sensing of the Environment (CNR-IREA), Milan, Italy, [2]National Research Council of Italy, Institute of Marine Sciences (CNR-ISMAR), Venice, Italy, [3]Flemish Institute for Technological Research (VITO), Mol, Belgium, [4]National Research Council of Italy, Institute of Atmospheric Sciences and Climate (CNR-ISAC), Rome, Italy

5.1 INTRODUCTION

Suspended solids play a fundamental role in the aquatic ecosystem as they regulate two major transport routes of materials and contaminants: the dissolved transport in the pelagic water and the particulate benthic sedimentation route (Wetzel, 1983; Håkanson, 2006). The presence of total suspended solids (TSS) in water has an impact on primary producers (Zhang et al., 2008), through affecting the amount of light penetrating through the water column that restricts the rate at which benthic algae, phytoplankton, and macrophytes can assimilate energy through photosynthesis.

The TSS in inland waters includes material supplied by tributaries (allochthonous), material produced within the water column (autochthonous), and resuspended material (Håkanson and Peters, 1995). TSS is a metabolically active component: the carbon content inside TSS is at lower trophic levels a fundamental source of energy for bacteria and plankton (Wetzel, 2001; Kalff, 2002). These suspended particles can originate from soil erosion, runoff, discharges, stirred bottom sediments, or algal blooms. Suspended solids can be comprised of organic and inorganic materials. The chemical composition of TSS varies significantly among and within inland waters. Many types of substances may be present, including clay minerals, humic substances, bacterial colonies, living and dead plankton, and detritus (Gustafsson and Gschwend, 1997).

TSS and turbidity can be used as an indicator for water clarity (e.g., Secchi disk depth, color and depth of the photic zone), and as a macro-descriptor for

Bio-optical Modeling and Remote Sensing of Inland Waters.
DOI: http://dx.doi.org/10.1016/B978-0-12-804644-9.00005-7

129

water quality since it directly relates to many variables of general use in lake management (Baban, 1999). Monitoring of TSS is important because high concentrations of particulate matter can impede light penetration into lower water layers, cause shallow lakes and bays to silt, and smother benthic habitats—impacting both living organisms and eggs (e.g., Shaw and Richardson, 2001). As particles of silt, clay, and other organic materials settle to the bottom, they can suffocate newly hatched larvae and impede the survival of zoobenthos (Håkanson and Boulion, 2002). Fine particulate material can also clog or damage sensitive gill structures of fish, decrease their resistance to diseases and potentially interfere with their life. If light penetration in lower water layers is reduced significantly, macrophyte growth may decrease, which will in turn impact the organisms depending on them for food and shelter (Lloyd et al., 1987; Petr, 2000; Havens, 2003) and deteriorate the oxygenation state of the water (Ryan, 1991). However, it is worth noting that this mechanism is not so important for the planktonic species including surface phytoplankton, and floating-leaved or free-floating macrophytes (Bilotta and Brazier, 2008). In addition to the just mentioned impacts, high concentrations of TSS cause undesirable aesthetic effects (Lloyd et al., 1987), higher costs of water treatment (Ryan, 1991), reduced navigability of channels and decreased longevity of dams and reservoirs (Butcher et al., 1993; Verstraeten and Poesen, 2000). The effect of TSS on aquatic biota depends on the concentration, the duration of exposure, the chemical composition, and the particle-size distribution (Bilotta and Brazier, 2008).

TSS measurements are typically based on field data gathering that although accurate and useful might be limited in both space and time leading to potentially biased results. In the last decades, satellite remote sensing has been recognized as an effective tool to provide a synoptic high frequency description of suspended sediments in inland waters, to address water quality monitoring including establishment of early warning systems (Imen et al., 2015) or historic assessment, and to investigate a variety of topics including sedimentation processes, river plume discharge or the ecosystems dynamics. The aim of this chapter is to present how remote sensing technology and bio-optical modeling support TSS measurements in inland waters. Few case studies with respect to remotely sensed sensors are then used as examples to support the introduction discussion points in Section 5.2.

5.2 OPTICAL PROPERTIES OF PARTICLES

The interaction of light with suspended particles produces a signal on water reflectance that can be detected from a remote distance (Bukata et al., 1995). Therefore, the signal sensed by remote detectors can be used to estimate the presence and abundance of TSS (e.g., Sathyendranath et al., 1989). For linking reflectance measurements to the concentration of water constituents, including suspended solids, the inherent optical properties (IOPs) describing the spectral light absorption and backscattering are key parameters.

In the optically complex waters typical of inland, estuarine and coastal systems, the absorption coefficients of particles matter $a_p(\lambda)$ (m^{-1}) is usually decoupled into a component caused by phytoplankton [$a_{phy}(\lambda)$, in m^{-1}] and one caused by nonalgal particles [$a_{NAP}(\lambda)$, in m^{-1}]. The first, $a_{phy}(\lambda)$, might be considered as a proxy for chlorophyll-a concentration [chl-a], while the latter, $a_{NAP}(\lambda)$, which can be also partitioned into contributions by minerogenic particles and organic detritus (Peng and Effler, 2013), can be related to both TSS (Babin et al., 2003a) and its fraction, defined either as tripton (Brando and Dekker, 2003), which is the nonalgal component of TSS, or as bleached particles (Strömbeck and Pierson, 2001).

In inland waters, the composition of these particles might vary significantly since it is governed by the characteristics of the drainage basin and hydrology, resuspension of bottom deposits driven by water motion, or by autochthonous production of primary producers (Wetzel, 1983). The spectral variation in magnitude and shape of absorption coefficients of Nonalgal particles (NAP) has been documented in several studies. For American and Australasian inland waters these characteristics were described by Davies-Colley and Vant (1987), Gallie and Murtha (1992), Kirk (1994), Bukata et al. (2001), Dall'Olmo and Gitelson (2006), Binding et al. (2008), Campbell et al. (2011) and Peng and Effler (2013). Ma et al. (2006) determined the absorption coefficients of different water types (e.g., dominated by detritus or phytoplankton) of Lake Taihu in China. In Europe, the absorption properties have been presented for different freshwater systems: Spanish reservoirs (Simis et al., 2007); Boreal lakes (Strömbeck and Pierson, 2001; Kutser et al., 2005; Paavel et al., 2007; Ylöstalo et al., 2014); perialpine lakes (Keller, 2001; Albert and Peter, 2006; Giardino et al., 2007); small lakes in England (George and Malthus, 2001); and Dutch lakes and reservoirs (Dekker, 1993). At continental scale, the variations in the light absorption coefficients were accurately documented by Babin et al. (2003a). Their study largely increased the knowledge of variations in the shape and magnitude of absorption for about 350 stations in various European coastal waters. Bio-optical models approximate the spectral absorption of NAP with the formula:

$$a_{NAP}(\lambda) = a_{NAP}(\lambda_{ref})e^{-S_{NAP}(\lambda - \lambda_{ref})} \qquad (5.1)$$

where λ_{ref} is a reference wavelength in the blue and S_{NAP} is the shape factor of the exponential curve which depends on the organic or mineral composition. S_{NAP} can be calculated for each measured spectra using a linear regression between the wavelength and the natural logarithm of $a_{NAP}(\lambda)$. The fitting is typically computed between 380 and 730 nm, but excluding the 400−480 and 620−710 nm ranges to avoid any residual pigment absorption (Babin et al., 2003a).

The spectral backscattering coefficient of particles $b_{b,p}(\lambda)$ is often considered to be a proxy for TSS concentration, the size and refractive index of a particle assemblage, while its slopes are known to be an indicator of the composition of the particles (Tzortziou et al., 2006; D'Sa et al., 2007). Similar as for $a_p(\lambda)$, the $b_{b,p}(\lambda)$ might be decoupled into contributions due to phytoplankton and its nonalgal component or tripton. Bernard and Matthews (2013) provided an accurate description of light scattering by planktonic algae and cyanobacteria, while Babin et al. (2003b) deeply investigated light scattering properties of particles in coastal and open ocean waters as related to the particle mass concentration. In bio-optical modeling, the spectral backscattering due to particles $b_{b,p}(\lambda)$ it can be approximated with the formula:

$$b_{b,p}(\lambda) = b_{b,p}(\lambda_{\text{ref}})\left(\frac{\lambda_{\text{ref}}}{\lambda}\right)^{-\gamma_{bb}} \tag{5.2}$$

where λ_{ref} is a reference wavelength and γ_{bb} is the power law exponent of particulate backscattering coefficient. Mobley (1994) related the magnitude of power law exponent of particulate backscattering coefficient to the size of particles, meaning that low values (~ 0.3, dimensionless) correspond to larger particles, whilst high values (~ 1.7, dimensionless) are linked to smaller particles. Similarly, the power law exponent of backscattering γ_{bb} is an indicator of the composition of particles (Twardowski et al., 2007). Blondeau-Patissier et al. (2009) found higher values ($\gamma_{bb} = 1.70$) for the reef waters in the Great Barrier Reef (Australia) where the average concentration of TSS was $1.5\,\text{g m}^{-3}$; lower values were found in the Fitzroy River estuary (0.71) where the average TSS was $5.95\,\text{g m}^{-3}$. For a series of European lakes distributed from boreal to subalpine ecoregions the average spectral slope of scattering coefficient was $\gamma_b = 0.76$.

5.2.1 Relationship between IOPs and TSS

Bio-optical modeling provides the link between IOPs described in the previous section (absorption and backscattering coefficients) and both TSS concentrations and apparent optical properties, such irradiance reflectance and remote sensing reflectance. To show the relations between TSS concentration, the absorption $a_{NAP}(\lambda)$ and backscattering properties $b_{b,p}(\lambda)$ of particles and irradiance reflectance, this section uses *in situ* data and a case-2 water bio-optical model, where the irradiance reflectance R (dimensionless, below the water surface) is calculated as a function of absorption and backscattering coefficients due to pure water, particles, phytoplankton, and colored dissolved organic matter (CDOM).

Field data where gathered in a series of sampling campaigns accomplished in a series of Italian lakes belonging to the drainage basin of Po River, where more than 80% of the total Italian lake volume is stored.

The sampled lakes (cf. Fig. 5.4) cover the variety of lake types present in the region: the clear deep and largest lakes (i.e., Garda and Maggiore) and a system of fluvial lakes (Mantua lakes) formed by Mincio River, and estuary of Garda Lake and a tributary of Po River. Some smaller and more turbid basins (further referred to as 'Po Valley lakes') were also included in the analysis to expand the dataset.

For computing absorption features, water samples were collected as integrated water column samples at different depths depending on Secchi disk depth, e.g., 0.5 m in the turbid Mantua lakes, up to 10 m in the clear Lake Garda. A sufficient amount of water (depending on particle load, e.g., 0.3 l in Mantua lakes up to 3 l in Lake Garda) was filtered in the field under low pressure, using Whatman GF/C filters. The absorption of particles retained onto the filters was measured with the filter pad technique (Trüper and Yentsch, 1967) in a dual-beam spectrophotometer. Subsequently, the filters were submerged for at least 1 h into Petri dishes, filled with an acetone 90% solution for bleaching and subsequently in distilled water for still at least 1 h. The absorption spectra of NAP were then measured onto the depigmented filters and the phytoplankton absorption coefficient was obtained by subtracting the absorption coefficient of NAP from the absorption coefficient of particles. Concentrations of TSS were measured gravimetrically. The inorganic fraction was determined by combusting the filters previously used at 550°C, to assess TSS (Strömbeck and Pierson, 2001) and the organic fraction as difference between TSS and the inorganic fraction. A total of 152 points were finally available for showing variability of absorption associated to suspended solids. To estimate the particle backscattering coefficient $b_{b,p}(\lambda)$ a HOBILabs HydroScat-6 sensor was used in a subset of the aforementioned lakes. A total of 52 profiles were gathered in situ in Lake Garda, Lake Maggiore and in Mantua lakes.

Fig. 5.1A illustrates the relationship between $a_{NAP}(440)$ and TSS concentration. Within the studied lakes (cf. Fig. 5.4), $a_{NAP}(440)$ ranged between 0.008 to 0.838 m^{-1}, whereas S_{NAP} varied from 0.02 to 0.005 nm^{-1}. Productive turbid waters of Mantua lakes are characterized by higher absorption than Lake Garda which identify a cluster corresponding to lowest values of TSS and $a_{NAP}(440)$. Lake Maggiore and the Po valley lakes stay in between Mantua and Garda lakes. With respect to Po valley lakes, Lake Maggiore shows overall a lower values of $a_{NAP}(440)$ for similar TSS concentration. When applying a linear regression with an intercept of zero, the steepness of the slope (0.0265 nm^{-1}) is comparable with those of coastal waters (0.031 nm^{-1}) obtained by Babin et al. (2003a) but is much lower than those of boreal humic lakes (0.079) obtained by Ylöstalo et al. (2014) and higher than those reported by Campbell et al. (2011) for tropical impoundments (0.0085). Fig. 5.1B indicates a lack of clear relation between S_{NAP} and $a_{NAP}(440)$, confirming a similar behavior as observed in humic lakes (Ylöstalo et al., 2014), estuaries and coastal waters (Blondeau-Patissier et al.,

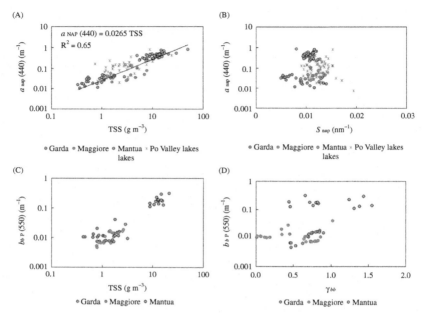

FIGURE 5.1   Absorption and backscattering properties of the particles and relation to TSS; (A) linear regression of $a_{NAP}(440)$ versus TSS concentrations (both in log scale); (B) scatterplot of S_{NAP} versus $a_{NAP}(440)$ (log scale); (C) relation of $b_{b,p}(550)$ versus TSS concentrations (both in log scale); and (D) scatterplot of γ_{bb} versus $b_{b,p}(550)$ (log scale).

2009). Fig. 5.1C illustrates the relationship between $b_{b,p}(550)$ and TSS concentration. This data is rather scattered, partly due to lake differences, natural variability in backscattering of particles and experimental random error (Babin et al., 2003b). In correspondence with Blondeau-Patissier et al. (2014), a relation between $b_{b,p}(550)$ and γ_{bb} (whose average value was 0.68), used to model the backscattering coefficients of particles across the wavelength, was not found (Fig. 5.1D).

To show how variations in optical properties (cf. Fig. 5.1) might affect the remote sensed signal, forward runs of a Case-2 water bio-optical model (Giardino et al., 2012) were performed. The bio-optical model is essentially based on the works from Lee et al. (1998, 1999), where the irradiance reflectance R (dimensionless) is calculated as a function of absorption and backscattering coefficients (due to pure water, particles, phytoplankton and CDOM) and of the ratio of the average cosine of the downwelling light to that of the upwelling light. The bio-optical model was run by keeping the contribution of both phytoplankton and CDOM constant and by changing the slopes S_{NAP} and γ_{bb} used by the bio-optical model to compute $a_{NAP}(\lambda)$ and $b_{b,p}(\lambda)$, respectively. In particular, the average values of the slopes S_{NAP} and γ_{bb} as well as the range between the double standard deviations of the

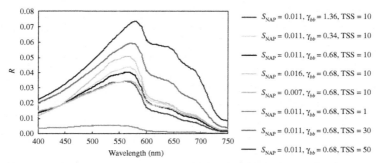

FIGURE 5.2 Irradiance reflectance R as a function of different values (mean values $+/-$ two times the standard deviations) of slopes S_{NAP} and γ_{bb} used to model $a_{NAP}(\lambda)$ and $b_{b,p}(\lambda)$, respectively. For the mean values of S_{NAP} and γ_{bb} (0.011 and 0.68), runs for R spectra for variable TSS concentrations are also plotted. All spectra were generated by keeping the contribution of both phytoplankton and yellow substances fixed; in particular the chlorophyll-a concentration and of CDOM were 2 mg m^{-3} and 0.1 m^{-1}, respectively.

averages were used in the simulation. Moreover, further runs for TSS concentrations of 1, 10, 30, and 50 gm^{-3} were performed for average values of S_{NAP} and γ_{bb}. Fig. 5.2 shows the variation of irradiance reflectance R depending on bio-optics parameters related to TSS as well as depending on TSS concentrations typical of lakes investigated in this study. Overall, the variation of R due to variations of S_{NAP} and γ_{bb}, but with fixed TSS is mostly appreciable in the green, while the variation in spectral shape is barely visible. More pronounced variations of R values and shape are encountered for different TSS concentrations.

5.2.2 Remote Sensing Algorithms for TSS

As discussed in the previous sections the impact of TSS concentrations on reflectance is rather clear and dramatic, therefore suspended solids are one of the most successful parameters that can be measured from remote sensing technology. One of the first mappings of inland water from satellite images was performed in Moon Lake (Mississippi) (Ritchie and Cooper, 1988), with data acquired from the Landsat multispectral scanner. With the advent of Landsat-4 and Thematic Mapper (TM) sensor, the inland water applications had the opportunity to growth as there are examples which range from the eighties (Lathrop and Lillesand, 1986) to nowadays (Alcântara et al., 2016). A prime example of the exploitation of Landsat provides regional maps of water clarity for 10,000 lakes over the 1985−2005 period (Olmanson et al., 2008). Then, the Medium Resolution Imaging Spectrometer (MERIS) had also offered great possibilities for inland water remote sensing, whilst new studies were already accomplished with the recently launched Sentinel-2 (e.g., Toming et al., 2016). A variety of algorithms have been consequently developed: an overview of more recent (2006−2011) publications for estimating TSS has been discussed in Odermatt et al. (2012). In their work, they also made a distinction between semi-empirical methods and spectral inversion procedures of semi-analytical and quasi-analytical models.

Spectral inversion procedures are typically built on matching spectral measurements with bio-optical forward models derived signatures by means of different inversion techniques. For example, Mobley (2004) used a look-up-table approach where least-squares minimization is used to found the closest match between the image spectrum and a database of remote sensing reflectance constructed with the radiative transfer numerical model Hydrolight (Mobley, 1994). In case of bio-optical models are written as a linear equation systems, the matrix inversion method provide a very fast and accurate inversion (Keller, 2001; Giardino et al., 2007). Other inversion techniques such as iterative non linear optimization (or curve-fitting) methods were used (e.g., Lee et al., 1999; Zhang et al., 2008) to provide an estimation of the parameters iteratively, when the gradient of a figure merit (e.g., the root mean square error) reached the minimum. The inversion can also be based on neural-network inversion scheme that was for instance used by Doerffer and Schiller (2007) and Vilas et al. (2011) for retrieval of water quality parameters in optically complex waters. All these inversion procedures are commonly used to retrieve multiple water quality properties at once from a single optical reflectance spectrum (Keller, 2001) and are based on the inversion of a physically based model describing the relationship between the reflectance, optical properties of water components and constituent concentrations (e.g., Brando and Dekker, 2003).

Semi-empirical methods are instead generally used to estimate a single water constituent, based on the relation between the signals captured by optical remote sensing instruments and parameter variations. These relations are included in the statistical analysis by focusing on well-chosen wavebands, which correlate with TSS. A variety of semi-empirical methods have been developed based on the relation between TSS and R (cf. Fig. 5.2) from visible to near infra-red (NIR) wavelengths. The choice of the spectral bands used in regression algorithms depends on the corresponding concentrations ranges, as appropriate bands typically shift toward longer wavelengths for increasing TSS concentrations. As discussed in Ruddick et al. (2006), the retrieval of TSS concentrations at longer wavelengths (700 nm and beyond) is more suitable for high TSS concentrations, while less absorbing portions of the spectrum (e.g., green waveband) are more suitable for low concentrations.

Within this chapter, a sensor generic multiwavelength-switching algorithm based on the Nechad et al. (2010) single band semi-analytical TSS algorithm is shown. It can be applied to any sensor with bands in the Red and NIR and that it is also suitable for extremely high TSS concentrations when the sensor has a spectral band in the short-wave infrared region (1000−1200 nm, Knaeps et al., 2015). Fig. 5.3 shows the relationship

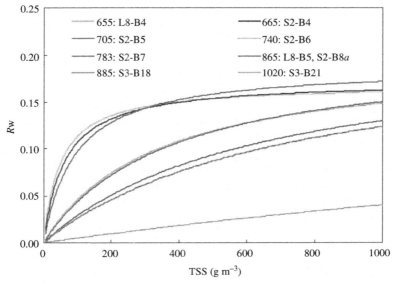

FIGURE 5.3 Semi-analytical algorithms for band settings of sensors onboard of Landsat-8 (L8), Sentinel-2 (S2), and Sentinel-3 (S3). Each line shows the relation between R_w and TSS for different band setting (nm units).

between the Water reflectance (R_w, dimensionless, equal to the remote sensing reflectance R_{rs} times π) and the TSS concentration as described by this semi-analytical algorithm. The relationship is shown for wavelengths available in the Operational Land Imager (OLI), Multispectral Imager's (MSI), and OLCI sensors onboard of Landsat-8 (L8), Sentinel-2 (S2), and Sentinel-3 (S3), respectively.

For low TSS concentrations a linear relation between TSS and water reflectance at shorter wavelengths exists, but saturates at higher TSS concentrations. Longer wavelengths are less sensitive to low TSS concentrations but saturation occurs at much higher TSS concentrations. Hence, a switching between shorter and longer wavelengths is appropriate to cover a wide range of TSS concentrations. Different "families" of spectral bands with similar characteristics (i.e., similar behavior of R to changes in TSS) are defined depending on pure water absorption coefficient and used in a switching approach. The spectral band families are identified as follows:

- family 1: spectral bands in the range $655-708$ nm, with absorption coefficient of pure water (a_w) in the order of $0.37-0.77$ m^{-1});
- family 2: spectral bands in the range $740-865$ nm, with a_w ranging from 2.8 to 4.6 m^{-1}. Above 865 nm, there are strong atmospheric absorption features which limit the use of these spectral bands in the estimation of the TSS concentration; and

- family 3: spectral bands in the range 1010−1100 nm, where although the a_w is ranging from 14 to 35 m^{-1}, a signal is still expected in highly turbid waters. Beyond 1100 nm atmospheric transmittance decreases and pure water absorption increases.

To define the TSS switching rules for the different families, the sensitivity of R_w to changes in TSS was evaluated. This resulted in two TSS threshold ranges: 50−150 and 500−750 gm^{-3}. When the concentration is lower than 50 gm^{-3}, a spectral band of family 1 is preferred. Exceeding 150 gm^{-3}, a spectral band of family 2 will perform better. Between these values, a merging of both families will be applied. Similarly above 750 gm^{-3} a spectral band in the short-wave infrared region is selected and between 500 and 750 gm^{-3} a merging of families 2 and 3 will take place. This merging is performed based on a linear weighting, where the weight is calculated based on the TSS concentration.

Independent of the selected approach, the flexibility to derive suitable algorithms makes every optical sensor a candidate for TSS mapping. In general, as TSS is a parameter which is subjected to a rapid degree of changes, high revisiting time sensors, as the ocean color sensors with a spatial resolution ranging from 300 m to 1 km, have been mostly used for capturing TSS variability in large-medium size lakes and large rivers (e.g. Giardino et al., 2010; Eleveld, 2012; Park and Latrubesse, 2014). The rapid degree of change in TSS patterns can also be observed from geostationary platforms, for example the Spinning Enhanced Visible and InfraRed Imager−Meteosat Second Generation (SEVIRI-MSG) meteorological sensor (Neukermans et al., 2009, 2012) or Geostationary Ocean Color Imager (GOCI) (Lyu et al., 2015). In addition, sensors developed for land applications, such as Landsat with a pixel size of 30 m and with an historical archive of almost 40 years, provide useful imagery to investigate TSS patterns at the finer spatial scale, suitable for inland water bodies (Schiebe et al., 1992; Baban, 1993; Fraser, 1998; Dekker et al., 2001; Brando et al., 2015). In the next section, a selection of these sensors is presented to provide TSS mapping of a couple of inland waters present in the drainage basin of the Po River.

5.3 CASE STUDIES

The most important place where Italian inland waters are located is the drainage basin of the Po River, with its length of 652 km is considered as the longest river of the country, situated in Northern Italy. The Po River basin extends from the Southern Alps in the West to the Adriatic Sea in the East and covers an area of 74,000 km^2. The basin plays a significant role in Italian economy producing 40% of the national Gross domestic products, consuming 48% of national produced energy, and approximately 25% of the

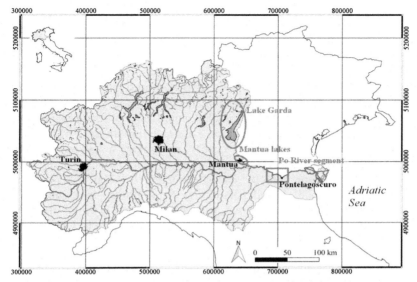

FIGURE 5.4 The Po River basin with indication of target areas investigated in this study.

Italian population is living in. Large population and socioeconomic activities lead to water quality degradation (Laini et al., 2011); challenging management of water and land resources, increase in hydraulic and hydrogeological risks and loss of biodiversity. Consequently, the region is highly vulnerable to droughts and floods and to further degradation of environmental quality.

The Po River has 141 main tributaries and the related river network has a total length of about 6750 and 31,000 km for natural and artificial channels, respectively. About 450 lakes are located in the Po River basin. The Po River basin includes the deep subalpine lakes and some small-medium lakes (Fig. 5.4). Together these represent more than 80% of the total Italian lacustrine volume (Premazzi et al., 2003). The water level of the larger south-alpine lakes of glacial origin is regulated according to given management policies, therefore obtaining a regulation volume of approximately 1.3 km^3.

The next sections show some results for water bodies in the Po River valley. In particular, the subcatchment of the Eastern part of the Po River basin receives additional attention. The state-of-the art knowledge on remote sensing for TSS in Lake Garda, the fluvial Mantua lakes and a segment of Po River is presented in the following sections.

5.3.1 MERIS Time-Series—Lake Garda

Located in the Po river basin (cf. Fig. 5.4) Lake Garda is the largest Italian lake. The most important tributary is Sarca River, which flows in the

Northern part of the lake. In the northern part also, a channel from Lake Ledro flows into the lake for hydropower exploitation. Lake Garda represents an essential strategic resource for agriculture, industry, fishing, and drinking. Moreover, the lake is an important resource for recreation and tourism with their attractions of landscape, mild climate, and water quality. Monitoring of water quality is therefore important to support developing lake management strategies both to preserve a good quality status and to take actions in case of water deterioration, that would make water unavailable for those purposes (Bresciani et al., 2011a,b). As both turbidity and TSS are one of most visible indicators of water quality, we provide a 10-year long time series of TSS based on MERIS imagery.

The MERIS sensor with its spatial resolution of about 300 m at Full Resolution (FR) mode offered possibilities for improved lake and coastal remote sensing for the whole period of its lifetime from 2002 to 2012 (e.g., Harvey et al., 2015; Kiefer et al., 2015). A selection of 140 clear-sky MERIS (top-of-atmosphere radiance level-1) images between 2003 and 2012 were used. Level-1 data were corrected for smile and adjacency effects, the first which accounts for the difference between actual and nominal wavelengths of the solar irradiance, the second caused by the reflection of neighboring pixels, which are atmospherically scattering into the reflectance path of the target pixel. Different correction algorithms exist for adjacency correction (e.g., c-Wombat-c in Brando and Dekker, 2003; SIMEC in Moses et al. in Chapter 3 of this book), ICOL (Improved Contrast between Ocean and Land described in Santer and Schmechtig, 2000) is one of them and used in this case. The smile and ICOL corrected data were then processed with Case-2 Regional (C2R) atmospheric algorithm. C2R (Doerffer and Schiller, 2007), which adopts a neural network inversion approach to both perform the atmospheric correction and retrieve IOPs (Doerffer and Schiller, 2007). The derived IOPs are then converted to constituent concentrations based on measured mass-specific IOP coefficients. IOP and in-water constituent concentration data from cruises in optically complex coastal waters were used to establish the bio-optical model followed by Hydrolight (Mobley, 1994) to train the coupled inverse and forward neural networks. The option of specifying optical properties of Lake Garda was used to suit C2R outputs to the particular case of Lake Garda according to Giardino et al. (2007) and data showed in the previous section.

Fig. 5.5 shows the frequency of TSS concentrations as derived from MERIS imagery and from *in situ* measurements. Overall satellite and field observations are in good agreement as they both indicate higher frequency of TSS concentration for values between 1 and 2.5 gm^{-3}.

A Kendall trend test (Hirsch and Slack, 1984) and the Sen slope estimator were used as in Boyer et al. (1999) and Räike et al. (2003) to analyze tendency of MERIS-derived TSS concentrations from 2003 to 2012. The trend was meaningful (p value < 0.05) and showed a slight increase of TSS

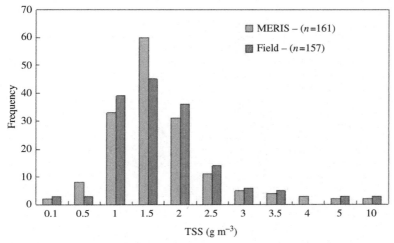

FIGURE 5.5 Frequency of TSS concentrations as derived from MERIS imagery and from *in situ* measurements. The number of samples (*n*) is given between brackets.

(median annual Sen slope 0.014). As the TSS concentrations of Lake Garda for the temporal range investigated in this study were almost stable an average map of the 161 C2R products was produced (Fig. 5.6). The map confirms the limited level of variation of TSS occurring in Lake Garda. The spatial distribution is also rather stable with a slight heterogeneity observed in the Southeastern basin. The basin has more gentle slopes than the other coastal zones of the lake and higher TSS concentrations are due to wind-induced suspended sediment resuspension. Moreover, this area corresponds to the lake emissary whose artificial regulation of levels might affect the water circulation with consequence on TSS accumulation.

5.3.2 Airborne Imaging Spectrometry—Mantua Lakes

The Mantua lake system is composed by three (Superior, Middle, and Lower) small and shallow basins surrounding the city of Mantua (cf. Fig. 5.4). Mantua lakes were artificially formed during the 12th century by damming the Mincio River, a tributary of the Po River, fed by Lake Garda. The level of the three lakes is controlled by an artificial dam to guarantee predefined water levels in each lake. Mantua lakes are mainly important for protecting water birds and plants. Water quality is a major concern as the lakes are characterized by eutrophic levels, problematic organic matter sedimentation, and excess growth of aquatic vegetation. Such conditions become critical especially in summer, when the average flow of the Mincio usually decreases (up to −90%) due to the combination of upstream water diversion for irrigation and little wet depositions, which induce water stagnation

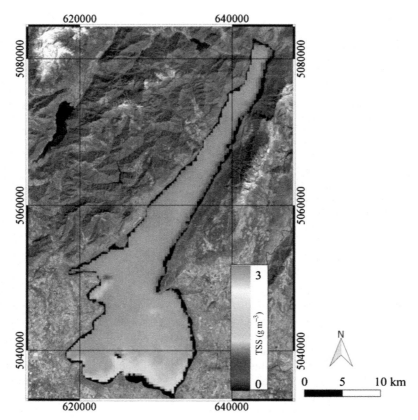

FIGURE 5.6 TSS concentration 10-years mean computed by averaging 161 C2R products from 2003 to 2012. Pixels with bottom effect and water-land mixture are masked.

(Pinardi et al., 2011). Water stagnation favors phytoplankton blooms and making the lakes more turbid than an average situation, with a Secchi depth below the meter. Suspended sediments typically assume a patchy spatial distribution consequent to water circulation mainly depending on Mincio River discharge and winds (Pinardi et al., 2015).

To capture the patchy distribution of TSS in Mantua lakes we used airborne hyperspectral remote sensing data gathering measurements in a contiguous spectrum from visible to NIR with a spatial resolution of about 4 m. The airborne campaigns were performed with two different hyperspectral sensors: the Multispectral Infrared Visible Imaging Spectrometer (MIVIS) and the Airborne Prism Experiment (APEX). MIVIS is a Daedalus modular whiskbroom scanner composed of four spectrometers, which simultaneously record radiation in 102 spectral channels from the visible to thermal infrared red spectral range, with wavelengths ranging between 430 and 12,700 nm

(Bianchi et al., 1994). APEX is a more recent pushbroom sensor recording high spectral resolution data in approximately 100 spectral bands in the wavelength range 426–910 nm (Itten et al., 2008). Both MIVIS and APEX have been demonstrated to be a powerful instrument to assess TSS in Lake Trasimeno (Giardino et al., 2015) and Scheldt River (Knaeps et al., 2010), respectively.

The airborne campaign with MIVIS was performed on July 26, 2007, while APEX flew on September 21, 2011 and September 27, 2014. The sensor radiances were converted into water reflectance values by removing the atmospheric effects; in case of MIVIS the conversion was merely based on match ups with in situ water reflectance data (Bresciani et al., 2009), while for APEX the atmospheric and air-interface correction was relying on physic-based radiative transfer modeling (see Moses et al. in Chapter 3 of this book).

A spectral inversion procedure was then applied to the hyperspectral dataset which typically provide high quality radiometric data to quantitatively retrieve water constituents with a physics-based modeling approach (e.g., Devred et al., 2013; Hestir et al., 2015). In particular, the retrieval of TSS concentration was achieved with the BOMBER, software package (Giardino et al., 2012) which makes use of a spectral inversion of bio-optical models for optically deep and optically shallow waters and of optimization techniques based on the works from Lee et al. (1998, 1999). The inversion applied to imaging spectrometry data simultaneously produced the maps with concentrations of TSS, CDOM, and chl-a (the results for CDOM and chl-a will be not discussed in this chapter; for bio-optical modeling of CDOM see Kutser in Chapter 4 and for bio-optical modeling of chl-a see Matthews in Chapter 6 of this book).

As an example, Fig. 5.7 shows the spatial distribution of TSS retrieved from APEX data acquired in September 2014. TSS concentrations range from 0 to 30 gm^{-3} with an average value of 10 gm^{-3}. The three lakes exhibit homogenous patterns except for two areas in the first largest lake (namely Superior) fed by Mincio River. TSS concentrations were higher in the downstream part of the lake due to the river discharge; lower concentrations were instead observed in the same lake northern the big island (i.e., a Lotus flower island masked as land in the map) due to the role of aquatic vegetation in keeping clear water conditions (Bolpagni et al., 2014; Pinardi et al., 2015). Another area of TSS accumulation is visible where water flows from the middle lake to the eastern lake.

In situ data gathered during the airborne campaigns were subsequently used to evaluate the accuracies of TSS concentrations obtained from MIVIS and APEX imagery. A total of 21 samples were available of which seven synchronous to MIVIS, six collected during the 2011 APEX flight and the remaining eight matching the 2014 APEX acquisition. Fig. 5.8 shows the scatter plot depicting the estimations of TSS derived from airborne

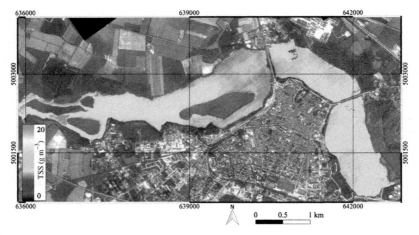

FIGURE 5.7 Map of TSS concentration in the Mantua lake system fed by Mincio River. From west to east they are called Superior, Middle and Lower Lake.

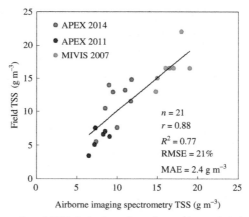

FIGURE 5.8 Scatter plot of TSS derived products from airborne imaging spectrometry and in situ data. The statistics of fitting are given as correlation coefficient (r), coefficient of determination (R^2), relative root means square error (RMSE) and mean absolute error (MAE), number of samples (n) and the 1:1 line.

hyperspectral data versus the in situ measurements sampled at the time of the MIVIS and APEX flights. The airborne data are those obtained by averaging a 3 by 3 pixels area centered on the location of the sampling stations. The TSS concentrations derived from imaging spectrometry and spectral inversion procedure were consistent with the in situ data, with a correlation coefficient r of 0.88, a coefficient of determination R^2 of 0.77, a mean absolute error (MAE) of 2.4 $\mathrm{gm^{-3}}$, a relative root mean square error (RMSE) of 21%, and close to the 1:1 line with a slope of 1.007.

5.3.3   Multitemporal OLI Data—Po River

The investigated segment of Po River (cf. Fig. 5.4) is located nearby its delta (the hinge point), where the river leaves the confines of its valley and divides into distributary channels, reaching the sea across very low slopes of the delta plain. The Po basin encompasses a variety of Alpine and Apennine streams with different hydrology and the climatic conditions along the catchment induce a fluctuating flow. A third of the annual flow is regulated by reservoir management for hydropower and irrigation purposes (Syvitski and Kettner, 2007). Despite this, the river still experiences short-lived event in response to precipitations (Palinkas and Nittrouer, 2007; Tesi et al., 2013). The Po River discharge typically exhibits large interannual variability, always characterized by two seasonal floods: one in the autumn driven by intense rainfall with the onset of Polar front and the other in the spring due to both snow melting and frontal rainfall (Boldrin et al., 2005; Syvitski and Kettner, 2007). At the closure river cross section, which is conventionally located at Pontelagoscuro (90 km upstream the river mouth), the flow of the Po River and its tributaries conveys and it is rarely affected by tides and seawater intrusion. At Pontelagoscuro gauging station, hourly discharge is measured since more than 100 years.

The hydrological behaviors of the Po River have been extensively studied, especially for what refers to the flood regime (e.g., Zanchettini et al., 2008; Montanari, 2012). Although the river discharge exerts a direct control on the transport of organic and inorganic matter, both qualitative and quantitative estimations of the river material are poorly constrained. This information is very important for many scientific applications: i.e., for assessing the input of the land-derived material entering the Adriatic Sea, for estimating the sediment load, for modeling the sediment transport and partitioning through the distributaries and for understanding the importance of event-dominated transport.

In order to fill this knowledge gap, 30-m resolution turbidity maps derived from OLI images were used to investigate temporal variations of suspended matter in the period from October 2014 to November 2015, in period time when also field measurements of TSS were available. OLI is in fact providing interesting data for studying inland waters owing to the improved radiometry (Irons et al., 2012; Pahlevan et al., 2014) and applications to operative monitoring of suspended sediment in rivers are realistic (Lobo et al., 2015).

In total, ten OLI images were selected and processed according to two different operational image-processing chains. This first image-processing chain (IPC#1) includes an adjusted radiometric calibration for water application (Pahlevan et al., 2014) and a subsequent correction for atmospheric effects based on ACOLITE (Atmospheric Correction for OLI "light"), an automatic method for atmospheric correction of L8 and S2 imagery of coastal and inland waters (Vanhellemont and Ruddick, 2014, 2015). In this study, ACOLITE was run selecting the SWIR atmospheric correction mode and a per-pixel variable aerosol type was determined from the ratio between the Rayleigh corrected reflectances in the aerosol correction bands, for pixels where the water reflectance can be assumed as negligible; i.e., in all pixels for the SWIR atmospheric correction. The ACOLITE-derived Water reflectance $[R_w(\lambda)]$ were finally converted in turbidity (T, expressed in formazin nephelometric unit, FNU), according to Dogliotti et al. (2015). Considering the wider range of turbidity in coastal and inland waters, this algorithm uses either the red (655 nm) or NIR (865 nm) OLI bands, depending on the $R_w(\lambda)$ at 655 nm. In case $R_w(655) < 0.05$ (corresponding to T < 15 FNU, so low turbidity), $R_w(\lambda)$ at 655 nm band is used in equation 1. In case of $R_w(655) > 0.07$ (corresponding to T > 45 FNU, so from medium to high turbidity), $R_w(\lambda)$ at 865 nm band is used to avoid the signal saturation at lower wavelengths (Bowers et al., 1998). In between these thresholds, when $0.05 > R_w(655) > 0.07$, the two algorithms are linearly blended to ensure a smooth transition.

The second image-processing-chain (IPC#2), uses the OPERational Atmospheric correction algorithm for land and water (OPERA). Shortly, in the first step the same adjusted radiometric calibration was performed as in IPC#1. Next, adjacency effects were removed using the SIMilarity Environment Correction (SIMEC) integrated in OPERA (Sterckx et al., 2015, and Chapter 3 of this book). The TSS concentrations are extracted from the corrected R_w data by applying the multi-waveband switching algorithm, which is explained in previous section.

Fig. 5.9 shows the TSS concentrations measured at Pontelagoscuro and the values of T and TSS as derived in correspondence of two years of field data from OLI by using the image-processing chains #1 and #2, respectively.

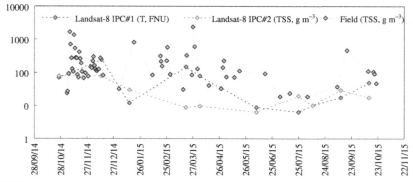

FIGURE 5.9 Temporal variation of TSS gathered from field measurements and OLI imagery, processed according to two operational image-processing chains.

Overall satellite and field observations are in good agreement as they both have the same trend and are able to capture either low or high concentrations of suspended sediments. Field measurements are more frequent and clearly show the dynamics of sediments in rivers. The frequency of cloud-free OLI overpasses of course produce a lower temporal resolution profile but has the advantage of being less time-consuming and of covering a longer river portion and a longer temporal range. More pixels were used to compare the two image-processing chains. Fig. 5.10 shows an overall agreement between methods even if data appear a little more scattered for TSS concentrations around 50 gm^{-3}.

In the Fig. 5.11, turbidity maps produced with IPC#1 (Fig. 5.11A and 5.11B) and TSS maps produced with IPC#2 (Fig. 5.11C and 5.11D) of the Po River segment are shown. They correspond to the minimum and maximum discharge (450 and 2900 m^3 s^{-1}, respectively) recorded synchronously to the OLI time-series acquisitions. When the discharge is low (<500 m^3 s^{-1}), the Po River presents low value of turbidity and TSS (Fig. 5.10A and 5.10C). Moreover, because of low hydrometric level, shallow areas of the riverbed are emerged (i.e., riverbanks, backwaters, and islands). With a moderate flood (2000−4000 m^3 s^{-1}), turbidity and TSS values are generally higher, depending on the characteristics of river regime. In Fig. 5.10B and 5.10D, a turbidity and TSS decreasing trend from East to West (from the sea to the inland) is evident: in fact, the corresponding OLI image was acquired after a flood, when discharge was decreasing. The turbidity and TSS maps are then able to capture the distribution of suspended sediment along the river.

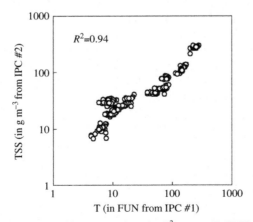

FIGURE 5.10 Scatter plot of TSS concentrations (gm^{-3}) versus T (FNU) derived from OLI imagery of a segment of Po River (cf. Fig. 5.4) processed according to two operational image-processing chains.

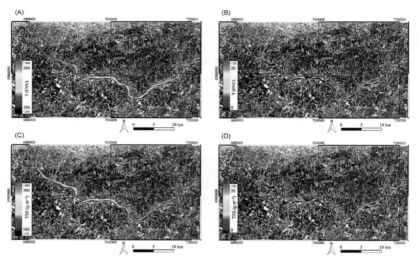

FIGURE 5.11 T (FNU) and TSS (gm^{-3}) maps derived from OLI imagery through two image-processing-chains: (A) IPC#1 map of turbidity on December 12, 2014; (B) IPC#1 map of turbidity on July 24, 2015; (C) IPC#2 map of TSS concentrations on December 12, 2014; and (D) IPC#2 map of TSS concentrations on July 24, 2015.

5.4 CONCLUSIONS

Suspended solids are a key parameter to assess the quality of inland waters. They derive from soil erosion, runoff, land-discharges, algal blooms, or bottom sediment resuspension by wind-waves. Once they enter the fluvial and lacustrine systems, suspended solids and their dispersion pathways become tracers of sediment dynamics. Understanding the processes of sediment suspension, transport and deposition is therefore relevant to monitor hydrologic and morphodynamic processes in floodplains. Many studies have been carried out to monitor TSS temporal and spatial variability using remote sensing. By observing the changes in light attenuation, the use of satellite remote sensing data for assessing TSS in inland waters has been recognized by the scientific community. Sensor technology includes almost any optical sensor operating in visible and NIR bands while a variety of methods has been implemented with a different degree of complexity: going from empirical methods established by direct relationship between remote sensing data and TSS concentrations (e.g., Zhou et al., 2006) to physical models based on radiative transfer theories. In between there are the so-called semi-analytic models, which are based on the relation between the IOPs and water components (e.g., Wang et al. 2007).

Within this chapter, an overview of bio-optical modeling and remote sensing techniques for mapping TSS in inland waters was presented. The bio-optical modeling was relying on a suite of IOPs [$a_{NAP}(\lambda)$ and $b_{b,\mathrm{p}}(\lambda)$, with their shape parameters S_{NAP} and γ_{bb}], which are commonly adopted for relating TSS concentrations to water reflectance. These data were gathered in a series of lakes distributed in the Po River catchment (northern Italy), where TSS concentrations were ranging from 0.3 to 50 gm^{-3}. Changing of irradiance reflectance spectra depending on variability of these IOPs was showed by adopting a forward run of a typical case-2 bio-optical model, for TSS concentrations ranging from 1 to 30 gm^{-3}. A sensor generic multiwavelength switching algorithm sensitive to TSS concentrations ranging from values lower than 50 up to 750 gm^{-3} was also presented for band setting of both L8 and S2.

Three case studies were presented for showing some applications of bio-optical modeling and remote sensing techniques. The case studies illustrated TSS mapping based on characteristics of different optical sensors and addressing the investigation of different inland water types: (i) time series from MERIS revealed stable conditions of low TSS concentration a clear deep lake; (ii) airborne imaging spectrometry allowed investigating the patchy distribution of TSS in three turbid fluvial lakes; and (iii) OLI imagery provided Turbidity and TSS maps for a segment in a river. In all cases, the retrieval of suspended sediment from remotely sensed data was comparable to field measurements.

Main conclusions from three case studies indicate how a multisensor approach is needed to investigate the spatial distribution and frequency of TSS concentrations in inland waters. The variations in optical properties, TSS concentrations (from less than 1 gm^{-3} to up to 1000 gm^{-3}) and morphometric parameters (e.g., from narrow rivers to big lakes) typically met in inland waters might force the choice of sensors and algorithms. With respect to this, operational image processing for TSS mapping is challenging, although practical tools as the C2R neural networks (developed for MERIS) and ACOLITE (for L8 OLI and S2 MSI) might be adopted for a systematic monitoring of TSS in inland waters, presuming they have been successfully compared to in situ data to ensure that the appropriate bio-optical parameterization has been chosen. Due to the high dynamics which usually characterize suspended sediment pattern in inland waters (e.g., river discharge and wind-induced sediment resuspension in shallow lakes), the validation of TSS products derived from remotely sensed data with respect to field data is still challenging due to their mismatching in both temporal and scale observations.

Satellite remote sensing finally offers an alternative option for tracking spatial and temporal variations in TSS. It is especially useful in large, remote, or complex hydrologic environments where also in situ monitoring is insufficient or impractical. Satellite sensors such as MERIS (2002−2013) and OLI have acquired excellent data for assessing TSS concentrations and to investigate fine-scale phenomena typical of inland waters, such wind-induced re-suspension in shallow waters (Eleveld, 2012) or river plume dynamics (Brando et al., 2015). Higher spatial resolution data have been used to assess TSS concentration in the inaccessible region of Himalayan lakes for assessing turbid moraine-dammed glacial lakes with a risk for outburst flood (Giardino et al., 2010). As both S2 and S3 are now placed in orbit, more data will become available to capture the wide range of spatial and temporal variability typical of suspended solids, to develop operational inland water management and for investigating further topics such as to develop methods for TSS products assimilation within ecological and hydrodynamic models (Palmer et al., 2015). In particular, the 10 and 20 m spatial resolution bands of S2 are innovative as spatial resolution is still one of the primary limiting factors in the application of satellite remote sensing to freshwater ecosystems (Özesmi and Bauer, 2002; Hestir et al. 2015). While MERIS and the previous Landsat allow retrospective studies to be performed, OLI, OLCI, and MSI sensors can now be merged to support inland water management or to perform an overall assessment of inland water status. The improved spectral, spatial, and temporal resolutions of these sensors are also allowing for better characterization on other inland water components strongly related to TSS, such as primary production of phytoplankton and macrophytes distribution and growing. The next four book chapters will hence discuss the bio-optical modeling of these primary producers.

ACKNOWLEDGMENTS

MERIS imagery MERIS data were made available through the ESA project MELINOS AO-553. Landsat 8 data were downloaded from USGS, http://glovis.usgs.gov. APEX data were acquired thanks to EUFAR (European Facility for Airborne Research) and particularly to the TA projects HabLakes and HYPPOS. MIVIS data were acquired by Compagnia Generale Riprese Aeree CGR Parma. This study was partly supported by the European Union; thanks to FP7 projects INFORM (GA no. 606865) and GLaSS (GA no. 313256). We are very grateful to Erica Matta and Sara Scuero for supporting image processing and mapping.

REFERENCES

Albert, A., Peter, G., 2006. Inversion of irradiance and remote sensing reflectance in shallow water between 400 and 800 nm for calculations of water and bottom properties. Appl. Optics 45, 2331–2343.

Alcântara, E., Curtarelli, M., Stech, J., 2016. Estimating total suspended matter using the particle backscattering coefficient: results from the Itumbiara hydroelectric reservoir (Goiás State, Brazil). Remote Sens. Lett. 7 (4)), 397–406.

Baban, S.M.J., 1993. Detecting water quality parameters in the Norfolk Broads, U.K., using Landsat imagery. Int. J. Remote. Sens. 14 (7), 1247–1267.

Baban, S.M.J., 1999. Use of remote sensing and geographical information systems in developing lake management strategies. Hydrobiologia. 395/396, 211–226.

Babin, M., Stramski, D., Ferrari, G.M., Claustre, H., Bricaud, A., Obolensky, G., et al., 2003a. Variations in the light absorption coefficients of phytoplankton, nonalgal particles, and dissolved organic matter in coastal waters around Europe. J. Geophys. Res. 108, 3211, 4.1-4.20.

Babin, M., Morel, A., Fournier-Sicre, V., Fell, F., Stramski, D., 2003b. Light scattering properties of marine particles in coastal and open ocean waters as related to the particle mass concentration. Limnol. Oceanogr. 48 (2), 843−859.

Bernard, S., Matthews, M.W., 2013. Using a two-layered sphere model to investigate the impact of gas vacuoles on the inherent optical properties of Microcystis aeruginosa. Biogeosciences 10 (1), 8139−8157.

Bianchi, R., Marino, C.M., Pignatti, S., 1994. Airborne hyperspectral remote sensing in Italy. In: Proceeding SPIE 2318, Recent advances in remote sensing and hyperspectral remote sensing, p. 29.

Bilotta, G.S., Brazier, R.E., 2008. Understanding the influence of suspended solids on water quality and aquatic biota. Water. Res. 42, 2849−2861.

Binding, C.E., Jerome, J.H., Bukata, R.P., Booty, W.G., 2008. Spectral absorption properties of dissolved and particulate matter in Lake Erie. Remote Sensing Environ 112, 1702−1711.

Blondeau-Patissier, D., Brando, V.E., Oubelkheir, K., Dekker, A.G., Clementson, L.A., Daniel, P., 2009. Bio-optical variability of the absorption and scattering properties of the Queensland inshore and reef waters, Australia. J. Geophys. Res.: Oceans 114 (C5), C05003-1−C05003-24.

Blondeau-Patissier, D., Gower, J.F., Dekker, A.G., Phinn, S.R., Brando, V.E., 2014. A review of ocean color remote sensing methods and statistical techniques for the detection, mapping and analysis of phytoplankton blooms in coastal and open oceans. Prog. Oceanogr. 123, 123−144.

Boldrin, A., Langone, L., Miserocchi, S., Turchetto, M., Acri, F., 2005. Po River plume on the Adriatic continental shelf: dispersion and sedimentation of dissolved and suspended matter during different river discharge rates. Mar. Geo. 222, 135−158.

Bolpagni, R., Bresciani, M., Laini, A., Pinardi, M., Matta, E., Ampe, E.M., et al., 2014. Remote sensing of phytoplankton-macrophyte coexistence in shallow hypereutrophic fluvial lakes. Hydrobiologia. 737 (1), 67−76.

Boyer, J.N., Fourqurean, J.W., Jones, R.D., 1999. Seasonal and long-term trends in the water quality of Florida Bay. Estuaries 22 (2B), 417−430.

Brando, V.E., Dekker, A.G., 2003. Satellite hyperspectral remote sensing for estimating estuarine and coastal water quality. IEEE T. Geosci. Remote 41, 1378−1387.

Brando, V.E., Braga, F., Zaggia, L., Giardino, C., Bresciani, M., Matta, E., et al., 2015. High-resolution satellite turbidity and sea surface temperature observations of river plume interactions during a significant flood event. Ocean Sci. 11 (6), 909−920.

Bresciani, M., Giardino, C., Longhi, D., Pinardi, M., Bartoli, M., Vascellari, M., 2009. Imaging spectrometry of productive inland waters. Application to the lakes of Mantua. Ital. J. Remote Sens 41, 147−156.

Bresciani, M., Giardino, C., Bartoli, M., Tavernini, S., Bolpagni, R., Nizzoli, D., 2011a. Recognizing harmful algal bloom based on remote sensing reflectance band ratio. J. Appl. Remote Sens. 5 (1), 053556-1−053556-9.

Bresciani, M., Stroppiana, D., Odermatt, D., Morabito, G., Giardino, C., 2011b. Assessing remotely sensed chlorophyll-a for the implementation of the Water Framework Directive in European perialpine lakes. Sci. Total. Environ. 409 (17), 3083−3091.

Bukata, R.P., Jerome, J.H., Kondratyev, K., Ya, Pozdnyakov, D.V., 1995. Optical Properties and Remote Sensing of Inland Coastal Waters. CRC Press Inc., Boca Raton, Florida.

Bukata, R.P., Pozdnyakov, D.V., Jerome, J.H., Tanis, F.J., 2001. Validation of a radiometric color model applicable to optically complex water bodies. Remote Sens. Environ. 77, 165−172.

Butcher, D.P., Labadz, J.C., Potter, A.W.R., White, P., 1993. Reservoir sedimentation rates in the Southern Pennine region, UK. In: McManus, J., Duck, R.W. (Eds.), Geomorphology and Sedimentology of Lakes and Reservoirs. Wiley, Chichester, pp. 73−93.

Campbell, G., Phinn, S.R., Daniel, P., 2011. The specific inherent optical properties of three sub-tropical and tropical water reservoirs in Queensland, Australia. Hydrobiologia. 658 (1), 233−252.

Dall'Olmo, G., Gitelson, A.A., 2006. Effect of bio-optical parameter variability and uncertainties in reflectance measurements on the remote estimation of chlorophyll-a concentration in turbid productive waters: modeling results. Appl. Optics 45 (15), 3577−3592.

Davies-Colley, R.J., Vant, W.N., 1987. Absorption of light by yellow substance in freshwater lakes. Limnol. Oceanogr. 32 (2), 416−425.

Dekker, A.G., 1993. Detection of optical water quality parameters for eutrophic waters by high resolution remote sensing. Phd Thesis. Vrije Universiteit.

Dekker, A.G., Brando, V.E., Anstee, J.M., Pinnel, N., Kutser, T., Hoogenboom, H.J., et al., 2001. Imaging spectrometry of water. Imaging Spectrometry: Basic principles and prospective applications, vol. IV, Remote Sensing and Digital Image Processing. Kluwer Academic Publishers, Dordrecht, pp. 307−359.

Devred, E., Turpie, K., Moses, W., Klemas, V., Moisan, T., Babin, M., et al., 2013. Future retrievals of water column bio-optical properties using the hyperspectral infrared imager (HyspIRI). Remote Sens. 5, 6812−6837.

Doerffer, R., Schiller, H., 2007. The MERIS Case 2 water algorithm. Int. J. Remote. Sens. 28, 517−535.

Dogliotti, A.I., Ruddick, K.G., Nechad, B., Doxaran, D., Knaeps, E., 2015. A single algorithm to retrieve turbidity from remotely-sensed data in all coastal and estuarine waters. Remote Sens. Environ. 156, 157−168.

D'Sa, E.J., Miller, R.L., McKee, B.A., 2007. Suspended particulate matter dynamics in coastal waters from ocean color: application to the northern Gulf of Mexico. Geophys. Res. Lett. 34, L23611.

Eleveld, M.A., 2012. Wind-induced resuspension in a shallow lake from Medium Resolution Imaging Spectrometer (MERIS) full-resolution reflectances. Water. Resour. Res. 48 (4), W04508-1−W04508-13.

Fraser, R.S., 1998. Multispectral remote sensing of turbidity among Nebraska Sand Hills lakes. Int. J. Remote. Sens. 19, 3011−3016.

Gallie, E.A., Murtha, P.A., 1992. Specific absorption and backscattering spectra for suspended mineral and chlorophyll-a concentrations in Chilko Lake. British Columbia. Remote Sens. Environ. 39, 103−118.

George, D.G., Malthus, T.J., 2001. Using a compact airborne spectrographic imager to monitor phytoplankton biomass in a series of lakes in north Wales. Sci. Total. Environ. 268, 215−226.

Giardino, C., Brando, V.E., Dekker, A.G., Strömbeck, N., Candiani, G., 2007. Assessment of water quality in Lake Garda (Italy) using Hyperion. Remote Sens. Environ. 109, 183−195.

Giardino, C., Bresciani, M., Villa, P., Martinelli, A., 2010. Application of remote sensing in water resource management: the case study of Lake Trasimeno, Italy. Water Resour. Manag. 24 (14), 3885−3899.

Giardino, C., Candiani, G., Bresciani, M., Lee, Z., Gagliano, S., Pepe, M., 2012. BOMBER: a tool for estimating water quality and bottom properties from remote sensing images. Comput. Geosci. 45, 313−318.

Giardino, C., Bresciani, M., Valentini, E., Gasperini, L., Bolpagni, R., Brando, V.E., 2015. Airborne hyperspectral data to assess suspended particulate matter and aquatic vegetation in a shallow and turbid lake. Remote Sens. Environ. 157, 48–57.

Gustafsson, Ö., Gschwend, P.M., 1997. Aquatic colloids: concepts, definitions, and current challenges. Limnol. Oceanogr. 42 (3), 519–528.

Håkanson, L., 2006. Suspended Particulate Matter in Lakes, Rivers, and Marine Systems. The Blackburn Press, Caldwell, New Jersey.

Håkanson, L., Boulion, V., 2002. The Lake Foodweb: Modeling Predation and Abiotic/biotic Interactions. Backhuys Publishers, Leiden.

Håkanson, L., Peters, R.H., 1995. Predictive Limnology-Methods for Predictive Modeling. SPC Academic Publishing, Amsterdam, p. 464.

Harvey, E.T., Kratzer, S., Philipson, P., 2015. Satellite-based water quality monitoring for improved spatial and temporal retrieval of chlorophyll-a in coastal waters. Remote Sens. Environ. 158, 417–430.

Havens, K.E., 2003. Submerged aquatic vegetation correlations with depth and light attenuating materials in a shallow subtropical lake. Hydrobiologia. 493, 173–186.

Hestir, E.L., Brando, V.E., Bresciani, M., Giardino, C., Matta, E., Villa, P., et al., 2015. Measuring freshwater aquatic ecosystems: the need for a hyperspectral global mapping satellite mission. Remote Sens. Environ. 167, 181–195.

Hirsch, R.M., Slack, J.R., 1984. A nonparametric trend test for seasonal data with serial dependence. Water. Resour. Res. 20 (6), 727–732.

Imen, S., Chang, N.B., Yang, Y.J., 2015. Developing the remote sensing-based early warning system for monitoring TSS concentrations in Lake Mead. J. Environ. Manage. 160, 73–89.

Irons, J.R., Dwyer, J.L., Barsi, J.A., 2012. The next Landsat satellite: the Landsat data continuity mission. Remote Sens. Environ. 122, 11–21.

Itten, K.I., Dell'Endice, F., Hueni, A., Kneubühler, M., Schläpfer, D., et al., 2008. APEX – the hyperspectral ESA Airborne Prism Experiment. Sensors 8, 6235–6259.

Kalff, J., 2002. Limnology: Inland Water Ecosystems. Prentice Hall, Upper Saddle River, New Jersey.

Keller, P.A., 2001. Comparison of two inversion techniques of a semi-analytical model for the determination of lake water constituents using imaging spectrometry data. Sci. Total. Environ. 268, 189–196.

Kiefer, I., Odermatt, D., Anneville, O., Wüest, A., Bouffard, D., 2015. Application of remote sensing for the optimization of in-situ sampling for monitoring of phytoplankton abundance in a large lake. Sci. Total Environ. 527, 493–506.

Kirk, J.T.O., 1994. Light & Photosynthesis in Aquatic Ecosystems, second ed. Cambridge University Press, New York.

Knaeps, E., Sterckx, S., Raymaekers, D., 2010. A seasonally robust empirical algorithm to retrieve suspended sediment concentrations in the Scheldt River. Remote Sens 2 (9), 2040–2059.

Knaeps, E., Ruddick, K.G., Doxaran, D., Dogliotti, A.I., Nechad, B., Raymaekers, D., et al., 2015. A SWIR based algorithm to retrieve total suspended matter in extremely turbid waters. Remote Sens. Environ. 168, 66–79.

Kutser, T., Pierson, D.C., Kallio, K.Y., Reinart, A., Sobek, S., 2005. Mapping lake CDOM by satellite remote sensing. Remote Sens. Environ. 94, 535–540.

Laini, A., Bartoli, M., Castaldi, S., Viaroli, P., Capri, E., Trevisan, M., 2011. Greenhouse gases (CO_2, CH_4 and N_2O) in lowland springs within an agricultural impacted watershed (Po River Plain, northern Italy). Chem. Ecol. 27 (2), 177–187.

Lathrop, R.G., Lillesand, T.M., 1986. Use of Thematic Mapper data to assess water quality in Green Bay and Central Lake Michigan, Photogramm. Eng.& Remote Sens. 52, 671–680.

Lee, Z., Carder, K.L., Mobley, C.D., Steward, R.G., Patch, J.S., 1998. Hypespectral remote sensing for shallow waters: 1. A semi-analytical model. Appl. Optics 37, 6329–6338.

Lee, Z., Carder, K.L., Mobley, C.D., Steward, R.G., Patch, J.S., 1999. Hypespectral remote sensing for shallow waters: 2. Deriving bottom depths and water properties by optimization. Appl. Optics 38, 3831–3843.

Lloyd, D.S., Koenings, J.P., LaPerriere, J.D., 1987. Effects of turbidity in fresh waters of Alaska. North Am. J. Fish. Manage. 7, 18–33.

Lobo, F.L., Costa, M.P., Novo, E.M., 2015. Time-series analysis of Landsat-MSS/TM/OLI images over Amazonian waters impacted by gold mining activities. Remote Sens. Environ. 157, 170–184.

Lyu, H., Zhang, J., Zha, G., Wang, Q., Li, Y., 2015. Developing a two-step retrieval method for estimating total suspended solid concentration in Chinese turbid inland lakes using Geostationary Ocean Colour Imager (GOCI) imagery. Int. J. Remote. Sens. 36 (5), 1385–1405.

Ma, R., Tang, J., Dai, J., Zhang, Y., Song, Q., 2006. Absorption and scattering properties of water body in Taihu Lake, China: absorption. Int. J. Remote Sens. 27 (19), 4277–4304.

Mobley, C.D., 1994. Light and Water–Radiative Transfer in Natural Waters. Academic Press, San Diego.

Mobley, C.D., 2004. A Spectrum-Matching and Look-Up-Table Approach to Interpretation of Hyperspectral Remote-Sensing Data. Sequoia Scientific Inc., Bellevue WA, 2004.

Montanari, A., 2012. Hydrology of the Po River: looking for changing patterns in river discharge. Hydrol. Earth Syst. Sci. 16 (10), 3739–3747.

Nechad, B., Ruddick, K.G., Park, Y., 2010. Calibration and validation of a generic multisensor algorithm for mapping of total suspended matter in turbid waters. Remote Sens. Environ. 114 (4), 854–866.

Neukermans, G., Ruddick, K., Bernard, E., Ramon, D., Nechad, B., Deschamps, P.-Y., et al., 2009. Mapping total suspended matter from geostationary satellites: a feasibility study with SEVIRI in the Southern North Sea. Opt. Express. 17, 14029–14052.

Neukermans, G., Ruddick, K., Greenwood, N., 2012. Diurnal variability of turbidity and light attenuation in the southern North Sea from the SEVIRI geostationary sensor. Remote Sens. Environ. 124, 564–580.

Odermatt, D., Gitelson, A., Brando, V.E., Schaepman, M., 2012. Review of constituent retrieval in optically-deep and complex waters from satellite imagery. Remote Sens. Environ. 118, 116–126.

Olmanson, L.G., Bauer, M.E., Brezonik, P.L., 2008. A 20-year Landsat water clarity census of Minnesota's 10,000 lakes. Remote Sens. Environ 112 (11), 4086–4097.

Özesmi, S.L., Bauer, M.E., 2002. Satellite remote sensing of wetlands. Wetl. Ecol. Manag. 10, 381–402.

Paavel, B., Arst, H., Herlevi, A., 2007. Dependence of spectral distribution of inherent optical properties of lake waters on the concentrations of different water constituents. Nord. Hydrol. 38, 265–285.

Pahlevan, N., Lee, Z., Wei, J., Schaaf, C.B., Schott, J.R., Berk, A., 2014. On-orbit radiometric characterization of OLI (Landsat-8) for applications in aquatic remote sensing. Remote Sens. Environ. 154, 272–284.

Palinkas, C.M., Nittrouer, C.A., 2007. Modern sediment accumulation on the Po shelf, Adriatic Sea. Cont. Shelf Res. 27, 489–505.

Palmer, S.C., Kutser, T., Hunter, P.D., 2015. Remote sensing of inland waters: challenges, progress and future directions. Remote Sens. Environ. 157, 1−8.

Park, E., Latrubesse, E.M., 2014. Modeling suspended sediment distribution patterns of the Amazon River using MODIS data. Remote Sens. Environ. 147, 232−242.

Peng, F., Effler, S.W., 2013. Spectral absorption properties of mineral particles in western Lake Erie: insights from individual particle analysis. Limnol. Oceanogr. 58 (5), 1775−1789.

Petr, T., 2000. Interactions between fish and aquatic macrophytes in inland waters. A review. FAO Fisheries Technical Paper. No 396. FAO, Rome, p. 185.

Pinardi, M., Bartoli, M., Longhi, D., Viaroli, P., 2011. Net autotrophy in a fluvial lake: the relative role of phytoplankton and floating-leaved macrophytes. Aquat. Sci. 73, 389−403.

Pinardi, M., Fenocchi, A., Giardino, C., Sibilla, S., Bartoli, M., Bresciani, M., 2015. Assessing potential algal blooms in a shallow fluvial lake by combining hydrodynamic modeling and remote-sensed images. Water 7 (5), 1921−1942.

Premazzi, G., Dal Miglio, A., Cardoso, A.C., Chiaudani, G., 2003. Lake management in Italy: the implications of the Water Framework Directive. Lakes Reserv. Manage. 8, 41−59.

Räike, A., Pietiläinen, O.P., Rekolainen, S., Kauppila, P., Pitkänen, H., Niemi, J., et al., 2003. Trends of phosphorus, nitrogen and chlorophyll a concentrations in Finnish rivers and lakes in 1975−2000. Sci. Total. Environ. 310 (1), 47−59.

Ritchie, J.C., Cooper, C.M., 1988. Comparison of measured suspended sediment concentrations with suspended sediment concentrations estimated from Landsat MSS data. Int. J. Remote Sens. 9, 379−387.

Ruddick, K., De Cauwer, V., Park, Y.J., Moore, G., 2006. Seaborne measurements of near infrared water-leaving reflectance: the similarity spectrum for turbid waters. Limnol. Oceanogr. 51 (2), 1167−1179.

Ryan, P.A., 1991. Environmental effects of sediment on New Zealand streams: a review. New Zealand J. Mar. Freshwater Res. 25, 207−221.

Santer, R., Schmechtig, C., 2000. Adjacency effects on water surfaces: primary scattering approximation and sensitivity study. App. Opt. 39, 361−375.

Sathyendranath, S., Prieur, L., Morel, A., 1989. A three-component model of ocean colour and its application to remote sensing of phytoplankton pigments in coastal waters. Int. J. Remote. Sens. 10 (8), 1373−1394.

Schiebe, F.R., Harrington, J.A., Ritchie, J.C., 1992. Remote sensing of suspended sediments: the Lake Chicot, Arkansas project. Int. J. Remote. Sens. 13, 1487−1509.

Shaw, E.A., Richardson, J.S., 2001. Direct and indirect effects of sediment pulse duration on stream invertebrate assemblages and rainbow trout (Oncorhynchus mykiss) growth and survival. Can. J. Fish. Aquat. Sci. 58 (11), 2213−2221.

Simis, S.G.H., Ruiz-Verdu, A., Domingues-Gomez, J.A., Pena-Martinez, R., Peters, S.W.M., Gons, H.J., 2007. Influence of phytoplankton pigment composition on remote sensing of cyanobacterial biomass. Remote Sens. Environ. 106, 414−427.

Sterckx, S., Knaeps, E., Adriaensen, S., Reusen, I., De Keukelaere, L., Hunter, P., et al. 2015. OPERA: an atmospheric correction for land and water. In Proc. Sentinel-3 for Science Workshop, Ouwehand, L. (Ed.), June 2−5, 2015, Venice, Italy, p. 4.

Strömbeck, N., Pierson, E., 2001. The effects of variability in the inherent optical properties on estimations of chlorophyll a by remote sensing in Swedish freshwater. Sci. Total. Environ. 268, 123−137.

Syvitski, J.P.M., Kettner, A.J., 2007. On the flux of water and sediment into the Northern Adriatic Sea. Cont. Shelf Res. 27, 296−308.

Tesi, T., Miserocchi, S., Acri, F., Langone, L., Boldrin, A., Hatten, J.A., et al., 2013. Flood-driven transport of sediment, particulate organic matter, and nutrients from the Po River watershed to the Mediterranean Sea. J. Hydrol. 498, 144−152.

Toming, K., Kutser, T., Laas, A., Sepp, M., Paavel, B., Nõges, T., 2016. First experiences in mapping lake water quality parameters with Sentinel-2 MSI imagery. Remote Sens. 8 (8), 640.

Trüper, H.G., Yentsch, C.S., 1967. Use of glass fiber filters for the rapid preparation of in vivo absorption spectra of photosynthetic bacteria. J. Bacteriol. 94, 1255−1256.

Twardowski, M.S., Claustre, H., Freeman, S.A., Stramski, D., Huot, Y., 2007. Optical backscattering properties of the "clearest" natural waters. Biogeosciences Discussions 4 (4), 2441−2491.

Tzortziou, M., Herman, J.R., Gallegos, C.L., Neale, P.J., Subramaniam, A., Harding, L.W., et al., 2006. Bio-optics of the Chesapeake Bay from measurements and radiative transfer closure. Estuar. Coast Shelf S. 68 (1), 348−362.

Vanhellemont, Q., Ruddick, K., 2014. Turbid wakes associated with offshore wind turbines observed with Landsat 8. Remote Sens. Environ. 145, 105−115.

Vanhellemont, Q., Ruddick, K., 2015. Advantages of high quality SWIR bands for ocean colour processing: Examples from Landsat-8. Remote Sens. Environ. 161, 89−106.

Verstraeten, G., Poesen, J., 2000. Estimating trap efficiency of small reservoirs and ponds: methods and implications for the assessment of sediment yield. Prog. Phys. Geogr. 24 (2), 219−251.

Vilas, L.G., Spyrakos, E., Palenzuela, J.M.T., 2011. Neural network estimation of chlorophyll a from MERIS full resolution data for the coastal waters of Galician rias (NW Spain). Remote Sens. Environ. 115 (2), 524−535.

Wang, Y.H., Deng, Z.D., Ma, R.H., 2007. Suspended solids concentration estimation in Lake Taihu using field spectra and MODIS data. Acta Scien. Circum. 27 (3), 509−515.

Wetzel, R.G., 1983. Limnology, second ed. Saunders College Publishing, Philadelphia, PA.

Wetzel, R.G., 2001. Limnology. Academic Press, London.

Ylöstalo, P., Kallio, K., Seppälä, J., 2014. Absorption properties of in-water constituents and their variation among various lake types in the boreal region. Remote Sens. Environ. 148, 190−205.

Zanchettini, D., Traverso, P., Tomasino, M., 2008. Po River dis- charge: a preliminary analysis of a 200-year time series. Clim. Chang. 88, 411−433.

Zhang, B., Li, J., Shen, Q., Chen, D., 2008. A bio-optical model based method of estimating total suspended matter of Lake Taihu from near-infrared remote sensing reflectance. Environ. Monit. Assess. 145, 339−347.

Zhou, W., Wang, S., Zhou, Y., Troy, A., 2006. Mapping the concentrations of total suspended matter in Lake Taihu, China, using Landsat-5 TM data. Int. J. Remote. Sens. 27 (6), 1177−1191.

FURTHER READING

Giardino, C., Bresciani, M., Cazzaniga, I., Schenk, K., Rieger, P., Braga, F., et al., 2014. Evaluation of multi-resolution satellite sensors for assessing water quality and bottom depth of Lake Garda. Sensors 14 (12), 24116−24131.

Chapter 6

Bio-optical Modeling of Phytoplankton Chlorophyll-*a*

Mark W. Matthews[1,2]
[1]*CyanoLakes (Pty) Ltd., Cape Town, South Africa,* [2]*University of Cape Town, Cape Town, South Africa*

6.1 INTRODUCTION

Phytoplankton are the earth's most prolific primary producers occurring in the world's oceans and freshwaters. Although microscopic in size, they have global significance as they produce about half of the oxygen that makes our atmosphere suitable for mammals to breathe (Field et al., 1998). They therefore play a critical role in sustaining life on the planet and are of considerable interest to biologists and scientists who study the earth's biogeochemical cycles. Also, considering the vast expanse of the oceans and inland waters, assessing their occurrence and ecology becomes a very challenging task. In inland waters, the need for monitoring is based less on biogeochemical applications and more on management intervention. Increasingly, the world's freshwaters are affected by phytoplankton blooms driven by eutrophication that pose a serious nuisance for water managers and the public. Cyanobacteria (and some species of algae) produce deadly toxins which pose a considerable health risk for recreational water users, and for potable water supplies that may be affected (e.g., Codd, 2000). The negative impacts of eutrophication are manifest by dominant algal and cyanobacteria species which reduce biodiversity and degrade water quality. There is therefore a great need for widespread monitoring and surveillance of inland waters focused on cyanobacteria and algal blooms, and remote sensing methods are proving to be indispensable for this purpose (e.g., Matthews, 2014; Wynne and Stumpf, 2015; Bresciani et al., 2016; Sayers et al., 2016).

Algae and cyanobacteria change the observed light field through absorption, scattering, and inelastic processes, such as fluorescence. Bio-optical models are used to describe this interaction, and are used to obtain information at a global scale by taking advantage of space-borne remote sensing instruments that measure the water surface. These instruments measure the

Bio-optical Modeling and Remote Sensing of Inland Waters.
DOI: http://dx.doi.org/10.1016/B978-0-12-804644-9.00006-9

light field emerging from oceans and inland waters at the top of the atmosphere. In simple terms, the biological characteristics of phytoplankton cause them to interact with the light field and cause changes in the observed color of the water. Hence the term "ocean color radiometry" became the name of the discipline related to measuring the color of the ocean's surface from space, with the primary aim of measuring chlorophyll-a (chl-a) from phytoplankton. However, it became possible to observe even small inland water bodies given advances in the ground sampling distance of modern remote sensing instruments (see Chapter 1 for review). Hence the name "water color radiometry" seems like a more appropriate name for the discipline for the 21st century.

Given that phytoplankton are microscopic, single-celled organisms suspended in water, they cannot be directly observed in their natural environment. Rather, the bulk effect of thousands or millions of cells on the water color is what is observed by the human eye or other sensor through the air-water interface. Therefore, their interaction with light might best be described by the theory of single particle optics, in a manner similar to dust particles in the atmosphere. Using single particle optics as a framework for understanding, the primary biological characteristics affecting the way that light interacts with phytoplankton cells are (1) their cell size and shape, (2) their pigmentation, that is, their assembly of colorful pigments that absorb light, and (3) their cellular structure or chemical composition.

From the perspective of Mie theory (often used to describe the interaction of light with single particles in the size range of phytoplankton),[1] these characteristics relate to: (1) the relative size or the area of the particle intercepted by the light field which affects both absorption and scattering; (2) the imaginary refractive index or 'softness' of the cell which is controlled by absorption; and (3) the real refractive index or 'hardness' of the cell which affects scattering. Thus, the biological characteristics of a phytoplankton cell displayed through its size, shape, pigmentation and intracellular structure governs how light is absorbed and scattered by the cell.

The principles that apply to a single cell can be extended to a population of polydisperse cells suspended in the three-dimensional water column which give rise to the bulk optical properties (absorption and scattering) of a water body and thus its color (Morel and Bricaud, 1986). However, the color observed by satellite remote sensing, or the estimated bulk absorption and scattering coefficients, can be traced back to the interaction of light at the scale of a single cell. Understanding the microscopic scale of light interaction with a cell provides the theoretical foundation into the subject of bio-optical modeling of chl-a. Chl-a pigment is of primary interest because it

---

1. Mie theory describes the interaction of light with a homogeneous spherical particle − for a detailed treatment of Mie theory for modeling how phytoplankton interact with light see Morel and Bricaud (1986).

the most suitable proxy for phytoplankton abundance (Huot et al., 2007). It is thus useful to have a sound understanding of phytoplankton and their characteristics in order to understand how bio-optical modeling might be used to estimate chl-*a* that may be used in models for primary production in the world's oceans and freshwaters.

Phytoplankton cells vary considerably in size and shape between species, and in intracellular structure (intracellular occlusions). Cells vary in size from less than 1 μm (picoplankton) to up to 10 mm (macrophytoplankton), although we are mostly concerned here with nanoplankton and microplankton with sizes less than 200 μm. The variation in size over 9 orders of magnitude is comparable to the variation of "forest trees to the herbs that grow at their bases" as described by C. S. Reynolds (2006). Their forms vary from single cells (unicells) to grouped cells (coenobia), filaments, and large colonies. Their structure differs considerably between genera with species-specific structure, e.g., silica plates in diatoms, or gas vacuoles in cyanobacteria. Many species of algae are motile and able to migrate vertically in the water column. These differences in size, structure and behavior impact on their bio-optical properties. Additional complexity is introduced when considering that tens to hundreds of species may be present in a water column at various abundance levels at any one time. The characteristics of the bulk optical signal from phytoplankton are mostly determined by the dominant species or groups present.

With this in mind, this chapter will investigate bio-optical modeling of chl-*a*, the primary indicator of phytoplankton production, and its estimation using various techniques of detection through different "optical pathways." An "optical pathway" as defined here, refers to the path followed by light through a series of interactions with phytoplankton that enables the inference of the concentration of chl-*a* from the measured signal. These pathways are (1) absorption, (2) fluorescence, and (3) phytoplankton backscattering. Measurements directed at different parts of the electromagnetic spectrum focused on these effects act as pathways for estimating the concentration of chl-*a*. The chapter will introduce physical concepts without delving into unnecessary detail, in order to provide a thorough understanding of the biophysical principles used in the estimation of chl-*a* from remote sensing instruments.

The chapter will pay some attention to the differences between prokaryotic and eukaryotic phytoplankton, and the consequences for estimating chl-*a*. Prokaryotic cyanobacteria are abundant in eutrophic freshwaters and in many coastal seas (e.g., the Baltic Sea); thus, they are of particular importance in inland and optically complex waters. Given their unique bio-optical properties and behavior, methods for estimating their chl-*a* concentration are sometimes distinct from that of algae (eukaryotes). These differences will be discussed and outlined with reference to each of the optical pathways.

Rather than acting as a review paper,[2] the aims of the chapter are firstly, to present an introductory text of the theoretical basis or first principles for quantitatively deriving chl-*a* from remote sensing measurements via the so-called "optical pathways"; secondly, to identify the challenges associated with its estimation, focusing on the complexities introduced by water consti-tuents other than phytoplankton found in optically complex waters; and thirdly, to present methods that have been used to solve some of these chal-lenges, and discuss the limitations thereof. Thus, the chapter does not attempt to review all available algorithms, but rather to highlight the theory behind the estimation of chl-*a* while using several examples of algorithms to illus-trate the principles at work. This should point to reader towards creative thought regarding improvements in techniques and algorithms, and towards better understanding of bio-optical modeling of chl-*a* concentrations in water. Lastly, the estimated sensitivity of the approaches will be briefly con-sidered in view of the available light signal, along with the constraints imposed by current remote sensing instruments in space.

6.2 CHLOROPHYLL-*a*: THE FUNDAMENTAL MEASURE OF PHYTOPLANKTON BIOMASS AND PRODUCTION

Chl-*a* is the photosynthetic green pigment present in all photoautotrophic plants and bacteria, including cyanobacteria. It is most often the primary light-harvesting pigment, responsible for capturing most of the energy from sunlight used in photosynthesis. Chl-*a* absorbs light strongly at wavelengths centered in the blue (near 440 nm) and the red (near 670 nm) parts of the light spectrum. This imparts the typical green coloration to natural waters rich in phytoplankton, but the combination of accessory pigments (e.g., caro-tenoids and phycobilins) alongside chl-*a* at high cell concentrations can pro-duce colors ranging from brown to red (Dierssen et al., 2006). Chl-*a* is one of a group of chlorophyll molecules that have similar absorption characteris-tics that include chlorophyll-*b* and chlorophyll-*c*.

In eukaryotes, that is algae, the chl-*a* molecules are contained and pro-duced within the chloroplast structure. The chl-*a* molecules are embedded in the thylakoid membranes within the chloroplast (Fig. 6.1). Algae typically have one chloroplast, although some species may have several. In prokar-yotes, including planktonic cyanobacteria, the thylakoids containing the chl-*a* molecules are contained within the chromatoplasm, the intracytoplasmic membrane towards the periphery of the cell. Prokaryotes lack the typical chloroplast structure found in eukaryotes, as well as other membrane bound organelles.

---

2. For reviews, see Ruiz-Verdú et al. (2008), Kutser (2009), Matthews (2011), Aurin and Dierssen (2012), Odermatt et al. (2012a,b), and Blondeau-Patissier et al. (2014), Mouw et al. (2015), and Palmer et al. (2015).

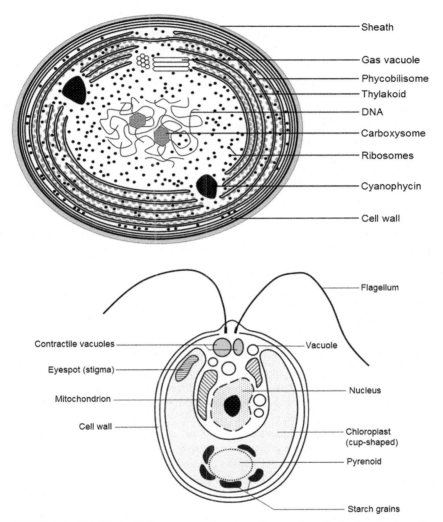

FIGURE 6.1 Simplified cellular structure of prokaryotic cyanobacteria (top) and eukaryotic algae (Chlamydomonas) (bottom) showing differences in cellular structure. *Used with permission from www.cronodon.com.*

The packaging of chl-*a* within the cell membrane, and within the thylakoid and/or chloroplast structure has significant optical consequences. Chl-*a* molecules are bound to proteins to form light-harvesting complexes that have somewhat different optical properties (Johnsen and Sakshaug, 2007). Larger cells have higher pigment packaging than smaller cells, and thus reduced light absorption per unit chl-*a* (Bricaud et al., 2004; Zhang et al., 2012).

The package affect causes the absorption of living cells to be reduced relative to the same pigment concentration in solution (in vitro).

The intracellular chl-a concentration, denoted c_i, depends on many factors. These include inherent variations between species, the history of light exposure of the cell and nutrient availability. Thus the value of c_i is neither constant between species, nor in time and space. The average value of c_i for phytoplankton in natural waters is around $5 \, \mathrm{kg \, m}^{-3}$, with typical values varying between 3 and $7 \, \mathrm{kg \, m}^{-3}$ (Reynolds, 2006).

Light availability is typically the greatest controlling factor for c_i. In low-light conditions, cells increase their production of light-harvesting pigments including chl-a, leading to larger c_i values. In contrast lower c_i values are associated with highlight conditions. This is easily observed when culturing phytoplankton in the laboratory, where low-light cultures appear a darker shade of green than high-light cultures. This process is called photoacclimation, regulation of the pigment content to maximize growth rate in a variable light climate (Johnsen and Sakshaug, 2007). The natural variability of c_i poses a challenge for bio-optical models since this factor has a large controlling effect on both absorption and scattering by the cells (Morel and Bricaud, 1986).

Given the ubiquitous nature of chl-a in phytoplankton, including algae and bacteria, and its significant positive relationship to phytoplankton biomass (Desortová, 1981), chl-a is used globally as a simple measure for phytoplankton in natural waters. Given its relative ease of measurement in the laboratory after extraction in an organic solvent, or direct measurement based on fluorescence in vivo or in vitro, it is routinely measured by government agencies around the world in water quality monitoring programs (Gregor & Marsalek, 2004). In particular, chl-a is a vital indicator in programs monitoring the impacts of eutrophication, ecological health status and health risks from harmful cyanobacteria and algal blooms. Therefore, its estimation using remote sensing is of primary interest to water quality management agencies and water utilities. For scientists, when compared to various alternative optical measurements, chl-a is the best indicator of the maximum photosynthetic rate (Huot et al., 2007).

6.3 OPTICAL PATHWAYS TO ESTIMATE PHYTOPLANKTON CHLOROPHYLL-a

By accurately measuring the changes in the light spectrum using sensitive detectors, the concentration of chl-a pigment might be deduced. The following sections address the following with respect to each of the optical pathways: the bio-optical properties of phytoplankton; how it can be related to the concentration of chl-a; the theoretical basis of measurement; how it is observed in reflectance; the primary methods; complications presented by optically complex waters; approaches for dealing with optically complex waters; and sources of error and likely sensitivity.

6.3.1   Phytoplankton Absorption

The spectral absorption coefficient of phytoplankton (a_{phy}) measured in a spectrophotometer is a combination of absorption from chl-*a* and all associated accessory pigments, and is highly variable in shape (e.g., Fig. 6.2). The concentration of chl-*a* is related to a_{phy} at the chl-*a* absorption maxima at 675 and 440 nm by a power law function, that has been observed in ocean and inland waters with chl-*a* concentrations over 6 orders of magnitude ($0.01-10000$ mg m^{-3}) (e.g., Bricaud et al., 2004; Gilerson et al., 2010; Matthews and Bernard, 2013; Riddick et al., 2015) (see Fig. 6.3). There is

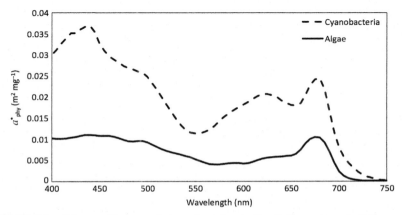

FIGURE 6.2 Chl-*a* specific spectral absorption of a small-celled cyanobacteria (*Microcystis Aeruginosa*) and a large-celled algae (*Ceratium hirundinella*).

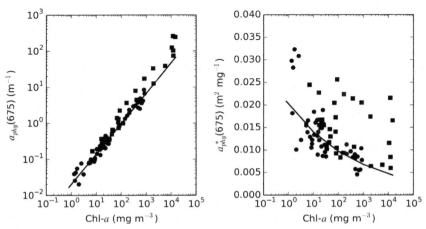

FIGURE 6.3 $a_{phy}(675)$ vs. chl-*a* (left) and corresponding $a^*_{phy}(675)$ versus chl-*a* (right) for three inland waters in South Africa. Circles=algae-dominant, squares=cyanobacteria-dominant using data from Matthews and Bernard (2013). Solid line shows fit for marine data from Bricaud et al. (1995).

some evidence that the relationship for inland waters differs slightly from the well-established relationship in the ocean (see Riddick et al., 2015). Phytoplankton absorption normalized to chl-a (a_{phy}^*) exhibits variability controlled by phytoplankton population size (pigment packaging) and to a lesser degree accessory pigments overlapping the chl-a absorption band, particularly affecting the 440 nm band (Bricaud et al., 1995). An inverse relationship is observed between chl-a and a_{phy}^* caused by the tendency of larger cells to dominate the phytoplankton population size structure at higher chl-a concentrations (Fig. 6.3, Bricaud, et al., 2004).

However, in inland waters this relationship may be violated by high abundance of small-celled cyanobacteria (e.g., Zhang et al., 2012; Matthews and Bernard, 2013). Thus, the conversion from a_{phy} to chl-a appears to be somewhat more variable in inland waters dependent upon the phytoplankton size structure and accessory pigment components (which is affected by physiology). Furthermore, this variability in a_{phy}^* (broadly ranging between 0.005 and 0.08 m^2 mg^{-1} at 440 nm) must be accounted for in bio-optical models utilizing the relationship between absorption and chl-a concentration (or inherent optical property inversions). One method to do this is to utilize modeled size-specific absorption coefficients (Bernard et al., 2007; Robertson Lain et al., 2014), or to use regional, seasonal, or type-specific parameterizations of the chl-a to a_{phy}^* relationship (see Gilerson et al., 2010). It is likely that established relationships may be satisfactory for characterizing the general variability in a_{phy}^*, especially at 675 nm, keeping in mind the theoretical maximum unpackaged chl-a absorption value is 0.027 m^2 mg^{-1} (Johnsen et al., 1994).

Measuring the intensity of absorption of light by phytoplankton is at first glance the most obvious method for estimating the concentration of chl-a. In the laboratory, the pigment is extracted from the sample using various organic solvents, and the subsequent intensity of absorbance centered at the absorption peak and various baselines are measured to calculate the concentration of chl-a (e.g., Ritchie, 2006). This absorption-based approach however poses several challenges for passive spectroradiometers measuring the reflected light spectrum, as against spectrophotometers measuring the loss by absorption and scattering from a beam of light. There are several reasons for this. Firstly, wavelengths coinciding with pigment absorption maxima have fewer photons available for detection by passive sensors. This lowers the signal-to-noise ratio (SNR) in the chl-a absorption wavelengths (at 440 and 665 nm). Particularly at high biomass the very small signal increases the error of measurement beyond reasonable limits. In fact, the 665 nm reflectance band is insensitive to larger chl-a concentrations (e.g., Gitelson et al., 1999). For this reason, algorithms utilizing blue wavelengths less than ± 440 nm are useful only for very low chl-a concentrations. For example, Carder et al. (1999) restricted the use of 412 and 440 nm bands to chl-a concentrations less than 1.5 mg m^{-3} or 0.025 m^{-1} phytoplankton absorption units. For this reason, at moderate to high chl-a concentrations, it is more robust from an engineering perspective to measure longer wavelengths or signal from a scattering or inelastic process (discussed below).

A further reason why the absorption pathway is challenging relates to the natural variability in c_i and a^*_{phy} (discussed above). Thus, the conversion from loss of reflectance (caused by absorption) to chl-*a* for cells in vivo is not constant, and results in increased uncertainty. This also applies to inversions retrieving the phytoplankton absorption coefficient, which is used to estimate chl-*a* concentration using the conversion:

$$\text{Chl-}a = \frac{a_{phy}}{a^*_{phy}} \qquad (6.1)$$

Despite the practical drawbacks of detecting chl-*a* based on absorption-related spectral reflectance features, several highly effective algorithms have been designed on this basis. These approaches most often target effects related to the chl-*a* absorption peak in the blue (near 440 nm), in order to avoid water absorption that affects red wavelengths. The absorption of blue light by increasing concentrations of phytoplankton has been observed in the open ocean as a gradual shift in water color from dark blue to green (e.g., Morel and Prieur, 1977). Thus, an empirical algorithm relating the ratio of blue to green light (488−551 nm) to chl-*a* concentration is the simplest method of estimating chl-*a* concentration in clear ocean and lake waters (e.g., Carder et al., 1999, Fig. 6.4).

The most noteworthy examples of this kind of approach are probably the SeaWiFS and MODIS Ocean Color algorithms that utilize several blue/green ratios to maximize the signal-to-noise at selected wavelengths (O'Reilly et al., 1998). This maximum band ratio algorithm is presented here in its configuration for MERIS (OC4Me) from Antoine (2010):

$$\text{Chl-}a = 10^{\hat{}}(a_0 + a_1 x + a_2 x^2 + a_3 x^3 + a_4 x^4) \qquad (6.2)$$

where

$$x = \log_{10}\left(\frac{R_{rs1}}{R_{rs2}}\right) \qquad (6.3)$$

$$R_{rs1} = \max[R_{rs}(443), R_{rs}(489), R_{rs}(510)] \qquad (6.4)$$

and

$$R_{rs2} = R_{rs}(560)$$

$$a_0 = 0.450, a_1 = -3.259, a_2 = 3.523, a_3 = -3.359, a_4 = 0.950$$

The algorithm has proved to be so effective it remains the core algorithm for chl-*a* used for ocean waters for MERIS and OLCI, albeit its derivation is not purely empirical (Antoine, 2010). These algorithms have been applied globally and remain the benchmark for ocean color and sometimes inland

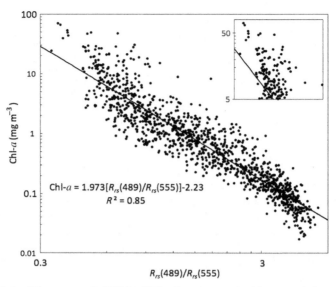

FIGURE 6.4   Chl-*a* versus $R_{rs}(489)/R_{rs}(555)$. Data are used with acknowledgment of the NASA bio-Optical Marine Algorithm Dataset (NOMAD). *As cited in Werdell and Bailey (2005).*

water studies today (see Odermatt et al., 2012a,b; Sayers et al., 2015). Many inland water studies also utilize green/blue ratios in clear lake waters with a good degree of correlation between chl-*a* (e.g., Odermatt et al., 2012a,b; Matsushita et al., 2015).

In optically complex waters that have significant absorption of either Chromophoric Dissolved Organic Matter (CDOM) or particulate matter, the absorption signal in the blue will be the sum of these components (see Chapters 1 and 2 for detailed overview of theory of optical properties of water). In particular, the absorption by humic and fulvic acids is potent across the a_{phy} maximum near 440 nm, decreasing exponentially towards the red. The absorption properties of tripton (nonliving matter) are similar to the exponential function of CDOM but have less steep slopes (see Babin et al., 2003; Binding et al., 2008; Matthews and Bernard, 2013). This causes much uncertainty when solving for the additive total absorption coefficient, and significant ambiguity in solutions given the nonuniqueness (or bijection) of the solution (see Defoin-Platel and Chami (2007) for discussion of ambiguity). Thus, the measurement of the effect of phytoplankton absorption on reflectance will have to account for other absorbing substances through the use of a multicomponent (additive) bio-optical model. Since the number of unknowns makes the system undetermined, there is no rigorous solution, and iterative processes with empirical laws must be employed (see discussion in Morel and Prieur, 1977).

Because of the influence of these other components, a simple empirical relationship between blue/green wavelengths and chl-*a* becomes severely degraded in waters with high detrital and CDOM absorption (in cases of very dark waters, the signal may be so small as to be unusable). As a result, retrieval of the underlying a_{phy} coefficient is necessary prior to estimation of chl-*a* via inversion of a bio-optical physical model. Studies have therefore focused on the retrieval of a_{phy} (and other constituent absorption coefficients) from reflectance measurements in order to indirectly estimate chl-*a*. These inherent optical properties (IOP) inversion algorithms are focused on the retrieval of the absorption and backscattering coefficients and their parts; the discussion here is limited to those aimed at the estimation of chl-*a* through retrieval of a_{phy}. Estimation of chl-*a* is dependent upon accurate retrieval of a_{phy}, but even if this is achieved the subsequent conversion to chl-*a* will be subject to error (from sources discussed above).

A great number and variety of IOP inversion methods have been suggested such that review here would be protracted and detailed investigations of each technique is beyond the scope of this chapter.[3] The discussion here only highlights some approaches that are useful for further understanding of the complexities of a_{phy} retrieval in inland waters, and for illustrating some of the most promising results for applications.

The reflectance approximation provides a link between reflectance and the IOPs that is commonly used in semi-analytical or algebraic inversion procedures (see Chapter 1 for types of bio-optical models). The observed reflectance is a function of backscattering and absorption $R = f/Q(b_b/(a + b_b))$ (Gordon et al., 1975). It is worth noting that the f/Q factor is subject to uncertainties in highly turbid waters and assumptions from the single scattering approximation (Aurin and Dierssen, 2012; Lee et al., 2011; Piskozub and McKee, 2011). The reflectance approximation is very commonly used as the basis for solutions solving sets of equations from multiple wavelengths for three to four unknowns using linear matrix inversion or nonlinear-least squares optimization. The reflectance can also be expressed in simpler terms ($R \propto b_b/a$) or as a function of K_d (Sathyendranath and Platt, 1997). These semi-analytical approaches typically include various empirical parameterizations or approximations for the IOPs, and sometimes include empirical relationships between reflectance and the IOPs being solved for.

These models by necessity make several assumptions regarding the shape and magnitude of a_{phy}, absorption by CDOM and detritus (often combined) and particulate backscattering. Variations is these parameterizations can result

---

3. For reviews of IOP inversion algorithms, see Gordon, 2002; IOCCG, 2006; Odermatt et al., 2012a,b; Mouw et al., 2015; Li et al., 2013.

is significant differences in algorithm estimates. Some algorithms apply parameterizations in order to account for variability particularly in the shape of a_{phy} based on biomass or regional relationships (e.g., Carder et al., 2004).

It is encouraging that, despite the more extreme constituent concentration ranges and greater variability in IOPs, several IOP inversions have been demonstrated in turbid inland and coastal waters (e.g., Salama et al., 2009; Santini et al., 2010; Aurin and Dierssen, 2012). The quasi-analytical algorithm in particular has been parameterized for inland waters (Yang et al., 2013; Mishra et al., 2014; Pan et al., 2015); however, there has been little examination of the accuracy of chl-a estimates from a_{phy}. The simpler inversion model of Li et al. (2013) based on earlier semi-analytical derivations of Gons et al. (2008) and Simis et al. (2007) retrieved chl-a from a_t rather than a_{phy}, and shows promising application particularly in high biomass scenarios.

Indirect methods of retrieving the a_{phy} or chl-a include alternative nonlinear optimization approaches, such as spectral decomposition using Gaussian oscillators in phytoplankton rich waters (e.g., Wang et al., 2016), or neural networks (Doerffer and Schiller, 2007), whose estimates of a_{phy} can be correlated to chl-a using local parameterizations (e.g., Odermatt et al., 2012a,b). The former is distinct from algebraic semi-analytical inversions in that they do not depend on any knowledge of the IOPs of the water under investigation, but are rather mathematical solutions utilizing synthetic training datasets and statistical methods. The use of the entire spectrum of information by these techniques may lead to advantages over semi-analytical inversions utilizing fewer wavebands.

Having discussed firstly, empirical relationships between reflectance and chl-a, and then bio-optical inversions to derive a_{phy} as required in optically complex waters, the reader has a good overview of the main challenges associated with estimating chl-a based on absorption-related features of reflectance in both clear and complex waters. The reader should keep in mind that the blue/cyan/green spectral region is subject to very low signal in highly absorbing waters, meaning that there are limits as far as the sensitivity of remote sensors is concerned. This poses a constraint on the range over which absorption-based approaches can be expected to perform, from an engineering perspective.

A further challenge arises for absorption-based approaches from atmospheric effects in blue wavelengths (see Chapter 3 for detailed overview of atmospheric correction). Absorbing and scattering aerosols are known to have the greatest effect on blue wavelengths which makes extrapolation of optical thickness estimates from red to blue wavelengths subject to errors, particularly in waters where the signal from near infra-red (NIR) wavelengths cannot be assumed to be black as a result of particulate backscattering. Further, spectrally variant bands more than 200 nm apart have significantly differing aerosol attenuation. Thus, absorption-related approaches are almost entirely reliant on accurate retrieval of R_{rs} (or the accuracy of coupled air-water models). Further, scattering by stray light into the sensor field of view is significant over small inland targets, requiring special corrections in spatially constrained waters (this issue also applies to other pathways reviewed here, not only absorption).

From the perspective of sensor design, accurate absorption-based chl-*a* measurements from space demand very high SNRs in precisely placed bands at multiple wavelengths (IOCCG, 1998, 2012). Given its importance for the approach, there is also a need for bands dedicated to atmospheric correction. Since absorption-based approaches have defined historical ocean color missions, these optimal requirements are present in instruments such as SeaWiFS, MODIS, MERIS, VIIRS, OCM, POLDER, OLCI, etc. (see Chapter 1 for instruments). However, most of these instruments do not have sufficiently high spatial resolution for inland waters less than approximately 5 km^2. Further, full resolution bands do not have the same sensitivity as reduced resolution bands. Thus, absorption-based approaches using lower sensitivity full resolution ocean color datasets on smaller targets will not prove as useful or accurate as those used for open ocean waters.

Provided the challenges from atmospheric effects can be sufficiently addressed to derive accurate estimates of R_{rs}, there are clear advantages of utilizing the absorption pathway in clear waters since they allow accurate chl-*a* estimates at very low concentrations less than 0.1 mg m^{-3}. Empirical and semi-analytical blue/green algorithms are likely to provide the highest accuracies for clear oligotrophic waters with chl-*a* less than 10 mg m^{-3}, outperforming the other pathways. The approach is likely to provide the highest accuracy for most of the area of the world's largest inland water bodies (e.g., Great Lakes, African Rift Lakes) since these are typically oligotrophic (e.g., Sayers et al., 2015). Making some assumptions, the absorption-based approach is likely to be applicable for a large portion of the world's inland water surface area (more than 40%[4] larger than 10 km^2).

There are also considerable benefits to be had with coupling absorption approaches with the other pathways in combined or hybrid algorithms that deal with distinct low and high chl-*a* concentration ranges (see Matsushita et al., 2015). In complex and turbid waters, IOP-based inversions for the a_{phy} coefficient have been performed with considerable success; however, the conversion from a_{phy} to chl-*a* introduces additional uncertainty. IOP inversions incorporating red wavelength bands provide most promise for higher chl-*a* concentration ranges.

6.3.2 Phytoplankton Fluorescence

Part of the light energy that is absorbed by phytoplankton pigments is reemitted as fluorescence[5] (see Chapter 7 for detailed treatment of

---

4. Based on rough calculation using data in Verpoorter et al. (2014)
5. For detailed review on algal fluorescence, see Huot and Babin (2010) and references therein. For a detailed review on cyanobacteria fluorescence, see Campbell et al. (1998)

fluorescence). The balance of the energy is used in photosynthetic reactions (charge separation) or dissipated as heat through molecular vibrations, known as nonphotochemical quenching (NPQ). The processes of fluorescence and NPQ are used to discard excess energy not used in photosynthesis. Pigment-laden light-harvesting complexes (LHCs) and chl-*a* molecules contained in photosystems 1 (PSI) and 2 (PSII) absorb light for photosynthesis, a part of which is reemitted as photons causing fluorescence. The fluorescence is emitted at a longer wavelength than it is absorbed due to a loss of energy in the process. The chl-*a* molecules contained in the photosystems have different absorption and fluorescence emission characteristics, as these are bound to different protein complexes (Johnsen and Sakshaug, 2007). The great degree of variability in pigment-protein complexes results in significant variability in fluorescence as for absorption (*ibid.*). PSI fluoresces light at wavelengths larger than 700 nm, while PSII fluorescence band is centered around 685 nm and at longer wavelengths around 730–740 nm.

In eukaryotic algae, the majority of fluorescence (±95%) arises from photosystem II since this contains the majority of chl-*a* molecules (up to 80%) and the LHCs transfer most of their energy to this photosystem. Thus, studies have focused on chl-*a* fluorescence detection in the band centered near 685 nm corresponding to PSII emissions (Fig. 6.5). However, in cyanobacteria, the majority of chl-*a* (>70%) is contained in photosystem I which fluoresces at longer wavelengths (Simis et al., 2012). In addition, the main light-harvesting antennae in cyanobacteria are phycobilisomes that contain the pigments phycoerythrin, phycocyanin, and allophycocyanin which transfer electrons to either PSI or PSII. State transitions occur in cyanobacteria causing energy to flow to PSI, thus reducing fluorescence

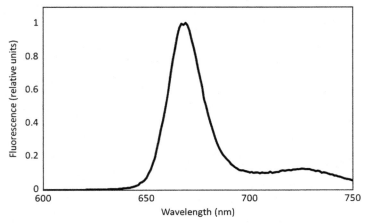

FIGURE 6.5 Chl-*a* fluorescence emission spectrum (in vitro). *Used with permission from Huot and Babin (2010).*

emissions from PSII (Campbell et al., 1998). There is evidence that the phycobilisomes also possess variable fluorescence that overlaps with the chl-*a* fluorescence emission of PSII (Simis et al., 2012). As a consequence, the natural fluorescence signal from cyanobacteria around 685 nm is much weaker than that in algae.

Unlike methods aimed at detecting chl-*a* absorption, fluorescence is an "active" process whereby a detectable signal is emitted from phytoplankton to be measured by a sensor. Estimating chl-*a* from fluorescence signals from PSII is based on the assumption that it is proportional to the concentration of chl-*a* in the cell. In algae, because most of the chl-*a* is contained in photosystem II, this assumption is generally valid. However, this is typically not the case for cyanobacteria as already explained, especially at high concentrations when significant reabsorption may occur. This conclusion has been supported by observations using in vivo fluorometers and satellite spectroradiometers where limited fluorescence from cyanobacteria has been detected (e.g., Seppälä et al., 2007; Matthews et al., 2012). In fact, a considerable absorption trough exists near 685 nm in the case of cyanobacteria-dominant waters (observed by Wynne et al., 2008). Thus, the discussion here is generally limited to estimating chl-*a* from algal fluorescence, given that different approaches are required for cyanobacteria (e.g., utilizing PSI emissions, and suitably designed fluorometers, c.f. Simis et al., 2012).

Estimating the concentration of chl-*a* using in vivo fluorescence is routinely used on underway measurement systems (Lorenzen, 1966). These instruments measure the PSII emission in a band near 680 nm under a controlled illumination source as a proxy for chl-*a* concentrations. The theoretical principle, illustrated by Huot and Babin (2010), is that chl-*a* is proportional to fluorescence emitted from PSII provided that the light intensity, chl-*a* absorption, fluorescence quantum yield (proportion of light absorbed to that emitted as fluorescence, typically around 1% (e.g., Gilerson et al., 2007)), and the intracellular fluorescence reabsorption factor are constant (see equation 6 in Huot and Babin, 2010). Given that this is generally not the case with natural samples, in vivo fluorescence only provides a rough estimate of chl-*a* concentration. In vitro estimates of chl-*a* through fluorescence in the laboratory are more accurate as these assumptions are generally valid once chl-*a* is dissolved in an organic solvent such as acetone.

In natural waters phytoplankton also fluoresce strongly in the presence of sunlight, which is known as sun-induced chl-*a* fluorescence or SICF. Studies began investigating whether the SICF signal could be observed in the red wavelengths of upwelling radiance and reflectance measurements. Scientists discovered that the SICF signal was readily observed in reflectance measurements, even at high altitudes using passive airborne spectroradiometers (e.g., Neville and Gower, 1977). SICF is evident in reflectance measurements by a distinct peak near 680 nm, and moving toward longer wavelengths as the phytoplankton concentration increases. The fluorescence emission per unit chl-*a* varies depending on the physiological state of the phytoplankton. The fluorescence peak near 680 nm should not be confused with the red-peak from scattering around 700 nm, which is discussed in detail below.

Using these observations as a starting point, algorithms based on reflectance were developed that targeted the fluorescence line height (FLH) using the same principle as the fluorescence emission peak measurement used in in vivo fluorometry (e.g., Letelier and Abbott, 1999; Gower et al., 1999). Ocean color satellite sensors were configured to detect fluorescence by adding narrow red wavelength bands specifically aimed at the detection of SICF in the oceans (e.g., MODIS, MERIS, and OLCI). The FLH measures the height of the reflectance peak caused by SICF and typically has a baseline on either side of the peak centered at the chl-*a* absorption maximum near 665 nm, and the red edge caused by water absorption near 700 or 750 nm:

$$FLH = L_1 - L_{baseline}$$

where

$$L_{baseline} = L_0 + (L_2 - L_0) \times \frac{\lambda_1 - \lambda_0}{\lambda_2 - \lambda_0}$$

where L is radiance, λ is wavelength, and $I=1$ is centered on the fluorescence peak.

The FLH itself is of much use for understanding phytoplankton behavior, particularly if an independent estimate of chl-*a* can be simultaneously obtained.

The chl-*a* concentration can be correlated to the FLH from reflectance in natural waters (e.g., Fell et al., 2003; Gons et al., 2008; Matthews et al., 2012; Moreno-Madriñán and Fischer, 2013). Obtaining a consistent relationship between chl-*a* and FLH is based on the same assumptions used in vivo, with the additional constraint dependent on the downward attenuation coefficient, K_d. Since a^*_{phy}, the fluorescence reabsorption factor and the quantum yield vary widely in nature (Fell et al., 2003), FLH provides only a gross index for chl-*a* concentration. The greatest source of uncertainly arises from variability in the chl-*a*:fluorescence ratio (quantum yield) which can readily vary between 0.08 and 0.01 $W\,m^{-2}\,sr^{-1}\,um^{-1}$ per mg chl-*a* (Abbott and Letelier, 1999), enacting variability in the limit of detection of chl-*a* from radiometry. There is some evidence however, that in coastal waters the quantum yield is relatively stable near 1% (Gilerson et al., 2007; Zhou et al., 2008).

In addition, water turbidity can reduce the sensitivity of FLH algorithms. Variability in K_d in turbid waters not associated with phytoplankton introduces additional uncertainty since it controls the availability of PAR (excitation wavelengths in the blue) and attenuates the upwelling radiance derived from fluorescence. Mckee et al. (2007) showed that whilst CDOM absorptions of less than 1 m^{-1} at 440 nm have little effect on FLH, concentrations of mineral particles larger than 5 $g\,m^{-3}$ cause significant interruption due to enhanced background radiance resulting from particulate scattering. The studies of Gilerson et al. (2007, 2008) confirmed these results showing that attenuation was a more important determinant of fluorescence than the quantum yield factor, and that MODIS and MERIS FLH estimates are significantly degraded by higher concentrations of mineral particulate matter. Thus FLH is most useful in less turbid coastal and inland waters, with lower chl-*a* concentration, or in waters where phytoplankton are optically dominant. The equations relating FLH to the concentrations of chl-*a*, NAP, and CDOM absorption developed by Gilerson et al. (2007) may be useful for improving FLH determinations in complex waters.

It is also important to note that while FLH signals can be hampered by mineral or particulate backscattering, it is also overwhelmed by phytoplankton backscattering at chl-*a* biomass larger than approx. 20 mg m^{-3}; this is the optical pathway discussed in the next section. Thus from the perspective of inland waters, FLH is appropriate for oligotrophic to mesotrophic waters, but typically not eutrophic waters with chl-*a* larger than ± 20 mg m^{-3} where the scattering signal becomes dominant (for preliminary analysis on the contribution of fluorescence versus scattering, see Tao et al., 2013; Gilerson et al., 2008).

FLH algorithms place high demands on instrument spectral sensitivity (SNR) and the precise placement of narrow bandwidths. This means that current FLH estimates from satellite are limited to a few ocean color instruments, such as MODIS, MERIS, and OLCI (the FLEX mission will add to this capability along with hyperspectral missions). Encouragingly, application with these instruments has proved to be much use in inland and coastal waters (e.g., Gons et al., 2008; Moreno-Madriñán and Fischer, 2013; Matthews and Odermatt, 2015). In addition to passive detection, active fluorescence sensing (light detection and ranging—LiDAR) also has promise for chl-*a* retrievals in inland complex waters (e.g., Palmer et al., 2013). There is also evidence that SICF signal in water can be observed in the oxygen absorption band around 760 nm (Lu et al., 2016), as commonly performed in plants (Meroni et al., 2009). Development of methods using this alternative fluorescence signal could improve the use for cyanobacteria populations containing more chl-*a* in PS I and in turbid waters with higher concentrations of suspended sediment.

Atmospheric turbidity also plays a significant role in the sensitivity of FLH algorithms (Abbott and Letelier, 1999). However, given the spectral proximity of the bands used (typically <100 nm apart), and the baseline peak measurement method, these serve to most often effectively normalize the signal for atmospheric effects, albeit the attenuation from higher concentrations of absorbing aerosols may dampen the FLH signal (*ibid.*). Studies have shown that partial atmospheric corrections accounting only for Rayleigh scattering and absorption, which is more easily performed than aerosol corrections, demonstrate fluorescence signals (Matthews et al., 2012;

Matthews and Odermatt, 2015). These recent results accord with the first studies showing that the fluorescence signal was clearly detectable at high altitudes from airborne sensors (Neville and Gower, 1977). Thus, FLH methods do not depend on a full aerosol correction to derive the R_{rs}, which makes application more feasible in complex/bright waters with non-negligible NIR reflectance.[6]

In summary, FLH provides a gross estimate of chl-a in waters with chl-a concentrations between approximately 1 and 20 mg m^{-3}, from sensitive ocean color sensors with high SNR, although the lower limit of sensitivity may be as poor as 8 mg m^{-3} depending on the fluorescence quantum yield and atmospheric and water turbidity. Attenuation in turbid waters reduces the available FLH signal, but preliminary methods already exist to account for the effects of other absorption and scattering components in complex waters (e.g., Gilerson et al., 2008; Lu et al., 2016). The FLH provides a reasonable indication of chl-a concentration, with much promise for applications in inland and complex waters.

6.3.3   Phytoplankton Scattering

Having explored two direct ways of estimating chl-a using absorption and fluorescence, a third indirect method is now considered, that of particulate or phytoplankton backscattering ($b_{b,phy}$). Light scattering by phytoplankton may at its simplest be described by a population of homogeneous spheres using Lorenz-Mie modeling (e.g., Bricaud and Morel, 1986; Stramski et al., 2001). This however is often criticized as being too simplistic, and in particular, of not adequately representing experimentally observed backscattering from phytoplankton (e.g., Volten et al., 1998; Zhou et al., 2012, Whitmire et al., 2010). Alternative two or three layered models providing for additional structures representing a chloroplast or cell wall structure more adequately reproduce observed scattering by phytoplankton (e.g., Quinby-Hunt et al., 1989; Kitchen and Zaneveld, 1992; Quirantes and Bernard, 2006; Matthews and Bernard, 2013). These models may more adequately be used to produce synthetic backscattering coefficients that may be used in modeling studies.

Backscattering spectra of phytoplankton have shapes characteristic of an inverted absorption spectrum with troughs corresponding to the absorption peaks (Fig. 6.6). The magnitude of backscattering varies widely among genera and between species dependent primarily on the cell size (particle size distribution) and cellular structure (real refractive index) of the cells (Svensen et al., 2007; Volten et al., 1998; Witkowski et al., 1998). Cells with considerable structure such as silica cell walls (e.g., diatom frustule), internal gas vacuoles (e.g., cyanobacteria), or scales (e.g., coccolithophores) have considerably higher scattering.

---

6. For treatment on atmospheric correction, see IOCCG Report No. 10.

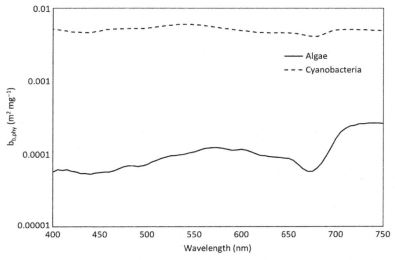

FIGURE 6.6   Chlorophyll-*a* specific backscattering spectra of a cyanobacteria and algal species modeled using a two-layered sphere. *From Matthews and Bernard (2013).*

In phytoplankton, the ratio of light scattered in the backward hemisphere to the total scattering coefficient ($\tilde{b}_b$) is typically less than 1% (e.g., Ahn et al., 1992; Whitmire et al., 2010; Zhou et al., 2012). Thus, phytoplankton generally scatter light poorly in the backward direction. As b_b is directly proportional to the observed reflectance ($R \propto b_b$), it is of primary interest.

According to Mie theory, larger cells exhibit enhanced forward scattering and consequently reduced $\tilde{b}_b$, but this is strongly variable as affected by the real refractive index and the cellular absorption (c_i) (Morel and Bricaud, 1986). This means that scattering varies widely independently of size. For instance, multiple studies present evidence indicating a weak positive relationship between $\tilde{b}_b$ and cell size (Vaillancourt et al., 2004; Whitmire et al., 2010; Zhou et al., 2012), seemingly contradicting the predictions of Mie theory. The enhanced backscattering ratios for larger cells are explained by increasing particulate organic carbon content and internal structures.

In particular, some authors argue for highest backscattering ratios from large dinoflagellates (e.g., Vaillancourt et al., 2004; Whitmire et al., 2010). However, this observation is not presently substantiated by empirical observations of high reflectance values from natural dinoflagellate blooms, and thus may result from bacteria or detached structures in cultures (see Whitmire et al., 2010). Low reflectance observed in dinoflagellate blooms may on the other hand result from their ability to regulate their position in the water column.

Although the backscattering properties of cyanobacteria are still poorly described, the occurrence of honeycomb-like arrangements of gas vacuoles towards the cell periphery is the cause of dramatically enhanced backscattering observed in cyanobacteria blooms (e.g., Dupouy et al., 2008; Matthews and Bernard, 2013). This scattering which results in high reflectance value can be an important factor for distinguishing cyanobacteria from algae, in addition to their unique pigmentation (e.g., Matthews et al., 2012). The other consequence of vacuoles is that cyanobacteria tend to float on or very near to the surface which has the further impact of increasing brightness. Thus, the vertical distribution of cyanobacteria blooms towards the surface in contrast to subsurface algal blooms is also a significant factor resulting in enhanced reflectance values (see Kutser et al., 2008, for an analysis; see also Chapter 8).

Backscattering normalized to chl-a ($b^*_{b,phy}$) ranges 4 orders of magnitude from 0.02 to 12×10^{-3} m^2 mg^{-1}, for poor and highly scattering species, respectively (e.g., Dupouy et al., 2008, Ahn et al., 1992). The largest values in the literature originate from vacuolate cyanobacteria (*Trichodesmium*), and algal species (e.g., dinoflagellates) containing unusual cellular structure (but see comment above). In general, small cells scatter more light than large cells per unit chl-a (see Vaillancourt et al., 2004). The intracellular chl-a concentration (c_i) has a strong influence on $b^*_{b,phy}$ as it normalizes the backscattering according to the number of cells.

There is a strong species-specific relationship between particulate backscattering and the concentration of chl-a (e.g., Dupouy et al., 2008, Fig. 6.2; Whitmire et al., 2010, Fig. 6.4; Antoine et al., 2011). This relationship is based on the particulate properties of the cells, rather than on any interaction with the pigment itself (absorbing components of the cell).[7] By way of reason, backscattering is caused by the cell wall and cellular inclusions which include the chloroplast and other cellular structures. Colonial arrangements or coenobium may also impart unique relationships between backscattering and chl-a. As the density of cells increases, the particulate backscattering from cellular material increases along with the concentration of chl-a.

The relationship between particulate backscattering and chl-a concentration has recently been experimentally examined in open ocean waters and may be described by a power law function (Huot et al., 2007; Antoine et al., 2011; Brewin et al., 2012). While it is admitted that $b_{b,p}$ is generally a relatively poor predictor of chl-a, the relationship is significant when considering fixed regimes and types. This is evidence that phytoplankton biomass plays a significant role in light backscattering in the oceans, despite the fact that they are presumed to scatter poorly. This holds even for very low biomass which chl-a concentrations not exceeding 0.1 mg m^{-3}. It has been deduced that in these conditions material (debris) that co-varies with phytoplankton might be responsible for this relationship, if not the phytoplankton itself

---

7. Although c_i can have a large impact on the intraspecies variability observed in $b^*_{b,phy}$.

(Huot et al., 2007). This relationship has yet to be fully investigated in inland waters.

The principle that chl-*a* is tightly linked to backscattering has been well demonstrated. In meso-eutrophic inland waters with chl-*a* concentrations regularly exceeding 10 mg m^{-3}, the contribution from phytoplankton backscattering to reflectance signals can be overwhelming. This is clearly observable from reflectance measurements made in eutrophic freshwaters that have distinct peaks near 560 and 700 nm from phytoplankton backscattering (see Fig. 6.7). Quibell (1992) noted, based on observations of increases in reflectance from higher concentrations of phytoplankton, that approaches based on algal backscattering should be preferred over those based on absorption effects. This is generally true for water with chl-*a* concentrations exceeding approx. 10 mg m^{-3}. A scattering-based approach to estimating chl-*a* utilizes the peak near 700 nm that becomes apparent at higher concentrations of algae and/or cyanobacteria.

Many similar studies in eutrophic inland waters demonstrate a significant relationship between the red-edge scattering signal near 700 nm and chl-*a* (e.g., Gitelson, 1992; Schalles et al., 1998; Rundquist et al., 1996; Gons, 1999). Chl-*a* is empirically correlated with both the height and position of this peak, which shifts towards increasing wavelengths as biomass increases. This peak results from increased particulate scattering offset by dramatically increasing water absorption at wavelengths larger than 700 nm, and the adjacent chl-*a* absorption band at 665 nm. Fluorescence may also contribute towards the tail of the peak in algae (680−690 nm) (Tao et al., 2013), but this is not usually observed in cyanobacteria, which rather has a distinct

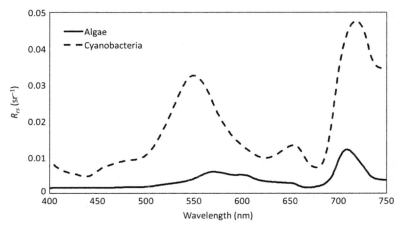

FIGURE 6.7 Reflectance spectra from natural populations of high biomass blooms (chl-*a* > 100 mg m^{-3}) of vacuolate cyanobacteria (*Microcystis Aeruginosa*) and algae (*Ceratium hirundinella*) demonstrating red-peak features and significantly enlarged reflectance.

trough in the fluorescence band around 680 nm, characteristic of reabsorption or state transition effects (Wynne et al., 2008; Matthews et al., 2012).

The scattering peak near 700 nm is preferred over that near 560 nm because of the influence of variable absorption from accessory pigments (e.g., carotenoids), and residual CDOM absorption at shorter wavelengths. In highly absorbing waters with very high phytoplankton biomass and/or CDOM there is typically very little light leaving the water in the blue. Thus red wavelength approaches provide more useable signal, despite the relatively high absorption by water, and the low depth of light penetration into the water (which at high biomass may only be a few centimeters). For this reason, the red-peak is most useful in well-mixed eutrophic water bodies, since it neglects the vertical distribution of phytoplankton. Red-wavelengths are also less prone to errors from atmospheric aerosol scattering than blue wavelengths, and techniques such as band ratios or peak height subtraction/derivatives have the advantage of effectively normalizing for atmospheric effects. There is evidence that even uncorrected top-of-atmosphere signals at 700 nm provide a good estimation of chl-a (e.g., Giardino et al., 2005; Matthews et al., 2012).

Similar to how the backscattering to chl-a relationship is species specific, the chl-a versus reflectance red peak height also displays specific relationships by genera (diatoms appear to be the most distinctive type) (Gitelson et al., 1999; Matthews and Bernard, 2013). The location of the red peak is also more consistent between genera than that of the peak near 560 nm, due to variable accessory pigmentation (Gitelson et al., 1999). Recent research has indicated that the chl-a specific absorption coefficient, and the fluorescence quantum yield significantly affects the shift in the peak position to longer wavelengths, with the result that various species possess unique relationships between peak position and height and chl-a (see Tao et al., 2013). The ability to model unique red-peak behavior resulting from species-specific backscattering and / or fluorescence effects presents an opportunity for distinguishing species based on peak height and position models, particularly in conjunction with accessory pigment related features (e.g., Tao et al., 2013, 2015).

Red-peak scattering-based algorithms may form simple reflectance ratios (e.g., 700/665 nm), more complex ratio forms (e.g., three band/four band variants), the normalized difference chlorophyll index (NDCI) (Mishra and Mishra, 2012), and also peak height models (e.g., the reflectance line height (Dierberg and Carriker, 1994); scattered line height (Schalles et al., 1998); the maximum chlorophyll index (MCI) (Gower et al., 2005); the maximum peak height (MPH) (Matthews et al., 2012); and the adaptive reflectance peak height algorithms (Ryan et al., 2014)). These empirical relationships can also be derived using radiative transfer models of reflectance utilizing the species-specific chl-a specific phytoplankton absorption and backscattering coefficients (e.g., Matthews and Bernard, 2013; Tao et al., 2013).

The simple two band model (using MERIS bands) is:

$$\text{Chl-}a \propto \frac{R_{rs}(708)}{R_{rs}(665)}$$

The three band model (using MERIS bands) is:

$$\text{Chl-}a \propto \left(R_{rs}(665)^{-1} - R_{rs}(708)^{-1}\right) \times R_{rs}(753)$$

The NDCI is written as follows (using MERIS bands):

$$\text{Chl-}a \propto \frac{R_{rs}(708) - R_{rs}(665)}{R_{rs}(708) + R_{rs}(665)}$$

Various relationships (e.g., linear, quadratic, power law, etc.) are used to correlate the concentration of chl-*a* with these variables.

The MPH algorithm searches for the position and height of the peak in the red between 680 and 750 nm, and is formulated for MERIS as follows (Matthews and Odermatt, 2015):

$$MPH = R_{max} - \left[R_0 + (R_1 - R_0) \times \frac{\lambda_{max} - \lambda_0}{\lambda_1 - \lambda_0}\right]$$

where

$$R_{max} = \max[R_{681}, R_{709}, R_{753}]$$

and where $0 = 665$ nm, $1 = 885$ nm, and R is ideally bottom-of-Rayleigh Reflectance.

The MPH principle can be applied to any sensor provided there are sufficient bands in the red. The MPH uses separate chl-*a* algorithms for cyanobacteria and algae dependent upon the identification of specific spectral features related to cyanobacteria.

It is also worth noting that the FLH algorithm has been used to indirectly measure the height of the 709 nm peak, or more directly the depth of the absorption trough at 681 nm (e.g., Palmer et al., 2014; Ampe et al., 2014). This results in a negative value which may be highly correlated with the concentration of chl-*a*. It is likely not appropriate to refer to FLH in these instances, since the lack of signal is not from fluorescence, but is related rather to scattering and absorption effects. This applies particularly to cyanobacteria-dominant waters which replace the fluorescence peak at 681 nm with a distinctive trough (Wynne et al., 2008; Matthews et al., 2012), where it has been referred to as the cyanobacteria index (CI) (Lunetta et al., 2014; Tomlinson et al., 2016).

In complex waters red-peak algorithm are negligibly affected by CDOM because of the spectral distance from blue absorption wavelengths. Variability in suspended sediment concentrations (deposited organic matter, clay and silt) is therefore likely to be the greatest source of error for these algorithms. Most sediment (perhaps with the exception of iron rich sediment) scatter light in a spectrally-invariant manner or with a gently sloping power law function dependent upon the particle size and refractive index (e.g., Stramski et al., 2007). Measured reflectance from sediment suspended in pure water do not exhibit a red peak at 700 nm the way phytoplankton do, due to the lack of a proximal absorption band at 665 nm (Quibell, 1991; Han et al., 1994; Hunter et al., 2008). The effect of sediment is primarily an increase in the magnitude of reflectance across the spectrum. Thus the red peak, which is very much determined by the chl-*a* absorption near 665 nm, is resilient (within limits) to variable suspended sediment concentrations (*ibid.*), and is therefore suitable for providing chl-*a* estimates in highly turbid coastal waters (see Kobayashi et al., 2011). Variations of the reflectance ratio algorithms have been developed to account for the effects of suspended sediment with some success, such as that of Gons, 1999, which is based on the reflectance approximation; that of Gitelson et al. (2008) which uses additional NIR band to account for scattering; and the four band model (Le et al., 2009).

A further case which presents a challenge to algorithms making use of red/NIR wavelengths is the case of optically shallow waters with sandy bottoms or submerged aquatic vegetation (SAV) (see Chapter 9 for more information on remote sensing of SAV). These may exhibit red peaks (e.g., Hill et al., 2014) and are an additional potential source of error when applying red-peak algorithms to satellite data in inland waters and shallow seas, particularly in areas adjacent to the shore. A method has been developed to flag out waters adjacent to the shore utilizing features in the red (see Matthews and Odermatt, 2015).

All things considered, the sources of error to red-peak derivations of chl-*a* can be summarized as being from variability in $b^*_{b,phy}$ and a^*_{phy}; to a lesser extent variability in the fluorescence quantum yield; scattering from suspended sediment; residual aerosol effects; and the presence of SAV and bottom effects. The likely sensitivity of red-peak/scattering approaches is limited to concentrations greater than approx. 10 mg m^{-3} at which biomass the red peak from scattering becomes apparent. It is important to note that most ocean color sensors lack a suitable narrow band near 710 nm for application of red-peak algorithms from satellite, making MERIS and OLCI particularly useful in eutrophic inland waters. In this respect future hyper/multispectral sensors (such as HICO) can fill an important gap for inland waters requiring higher spatial resolution. However, SNR remains problematic for many hyperspectral sensors (Kudela et al., 2015).

Scattering-based approaches using the red peak have the advantage of spectral proximity, and greater signal from typically higher biomass waters, for atmospheric correction. In this respect, red peak height approaches (MPH, MCI) provide the same advantages presented by FLH, and therefore can be more easily applied with partial atmospheric corrections, such as the bottom-of-Rayleigh reflectance (e.g., Matthews et al., 2012; Lunetta et al., 2014).

6.4 CONCLUSION

This chapter has provided an overview of the three pathways for estimating chl-*a* from remotely sensed measurements of the reflectance from natural waters. This should give the reader a better appreciation of the first principles involved in the estimation of chl-*a*, as well as an overview of the methods and applications of each approach. The three pathways discussed above represent alternative, complimentary, and sometimes divergent approaches to estimating chl-*a* from reflectance. Each pathway has its own set of advantages and weaknesses, limitations and constraints, and optimal conditions for use. Chapter 7 discussed fluorescence and its measurement and relationship to chl-*a* for readers requiring a more in-depth text.

Absorption approaches are optimal for oligotrophic waters with chl-*a* concentrations less than 5 mg m^{-3} due to the higher sensitivity than the other pathways at this biomass range. The available signal in the blue is reduced by higher concentrations of phytoplankton, and by CDOM and NAP absorption, and atmospheric interference remains a principal source of uncertainty. Fluorescence, whilst subject to wide variability due to physiology, and the influence of attenuation, provides a robust gross index for chl-*a* in the range 1 to ± 20 mg m^{-3}. A significant advantage is that FLH provides an effective atmospheric correction. The scattering approach is limited to concentrations above ± 10 mg m^{-3}, and is likely the most robust method for estimating chl-*a* concentrations above 20 mg m^{-3}, taking into account the higher relative signal from increasing biomass in the red (for additional sensitivity analysis on ranges of constituent concentrations and chl-*a* see Odermatt et al., 2012a, b). Peak-height calculations can also have the same advantages for atmospheric correction as the FLH. The development of hybrid approaches, such as combined absorption-scattering (Matsushita et al., 2015), and fluorescence-scattering approaches (Matthews et al., 2012) can optimize performance over wider biomass ranges. The optimal model for chl-*a* retrieval for inland and complex waters (yet to be developed) will probably combine all three approaches (absorption-florescence-scattering), with switching dependent on prior water type classification (e.g., Moore et al., 2014) or species-specific feature detection (e.g., Tao et al., 2015).

There are several areas for future research that will reduce uncertainty for chl-*a* estimates in complex waters. Better understanding of the fine resolution red features of backscattering from algae and cyanobacteria will enhance scattering algorithms (i.e., detailed measurements of the volume scattering function); further assessment of fluorescence variability, and attenuation effects will improve FLH corrections applied to turbid waters; better characterization of variables such as c_i, and further refinement of semi-analytical inversions will enhance absorption-based retrievals in complex waters.

It is clear that estimation of chl-a from space requires high thresholds for instrument sensitivity (SNR) (e.g., Hu et al., 2012; Moses et al., 2012) and precise placement of narrow spectral bands. The consequence is that only a few ocean color sensors are suitable for quantitative chl-a retrieval over its full scale of variability ($0.01-10\ 000$ mg m m^{-3}), which is likely to continue to be the case in future. From the perspective of inland waters, the implication is that, for water bodies less than 2 km^2, systematic quantitative chl-a estimation is likely to remain something for the future. However, current higher resolution sensors (e.g., Sentinel-2) can offer a partial solution in the interim. The outlook certainly indicates that in inland waters, provided that suitable instruments remain in space, chl-a estimates will continue to improve as these methods mature into generally accepted best-practice science, offering considerable advantages for use in a wide range of practical applications. These include services for monitoring threats from cyanobacteria blooms for recreational users and public authorities; systematic wide-scale monitoring of the impacts of eutrophication; and global assessments of lake bio-geochemistry and the long-term impacts of climate change on lake environments.

ACKNOWLEDGMENTS

The Water Research Commission project number K5/2458 is gratefully acknowledged for funding support.

REFERENCES

Abbott, M.R. & Letelier, R.M., 1999. Algorithm Theoretical Basis Document Chlorophyll Fluorescence (MODIS Product Number 20), NASA.

Ahn, Y.H., Bricaud, A., Morel, A., 1992. Light backscattering efficiency and related properties of some phytoplankters. Deep-Sea Res. 39, 1835–1855.

Ampe, E.M., et al., 2014. A wavelet approach for estimating chlorophyll-A from inland waters with reflectance spectroscopy. Geosci. Remote Sens. Lett., IEEE 11 (1), 89–93.

Antoine, D., 2010. Sentinel-3 optical products and algorithm definition. OLCI Level 2 Algorithm Theoretical Basis Document: Ocean Color Products in case 1 waters, Available at: https://sentinel.esa.int/documents/247904/349589/OLCI_L2_ATBD_Ocean_Colour_Products_Case-1_Waters.pdf.

Antoine, D., Siegel, D.A., Kostadinov, T., Maritorena, S., Nelson, N.B., Gentili, B., et al., 2011. Variability in optical particle backscattering in contrasting bio-optical oceanic regimes. Limnol. Oceanogr. 56 (3), 955–973.

Aurin, D.A., Dierssen, H.M., 2012. Advantages and limitations of ocean color remote sensing in CDOM-dominated, mineral-rich coastal and estuarine waters. Remote Sens. Environ. 125, 181–197.

Babin, M., et al., 2003. Variations in the light absorption coefficients of phytoplankton, nonalgal particles, and dissolved organic matter in coastal waters around Europe. J. Geophys. Res: Oceans 108 (C7), 4.1−4.20 (3211).

Bernard, S., Shillington, F.A., Probyn, T.A., 2007. The use of equivalent size distributions of natural phytoplankton assemblages for optical modeling. Opt. Express. 15 (5), 1995−2007.

Binding, C.E., et al., 2008. Spectral absorption properties of dissolved and particulate matter in Lake Erie. Remote Sens. Environ. 112 (4), 1702−1711.

Blondeau-Patissier, D., et al., 2014. A review of ocean color remote sensing methods and statistical techniques for the detection, mapping and analysis of phytoplankton blooms in coastal and open oceans. Prog. Oceanograph. 123, 123−144.

Bresciani, M., et al., 2016. Earth observation for monitoring and mapping of cyanobacteria blooms. Case studies on five Italian lakes. J. Limnol. 1565, 1−18.

Brewin, R.J.W., Dall'Olmo, G., Sathyendranath, S., Hardman-Mountford, N.J., 2012. Particle backscattering as a function of chlorophyll and phytoplankton size structure in the open-ocean. Opt. Express 20 (16), 17632.

Bricaud, A., Morel, A., 1986. Light attenuation and scattering by planktonic cells: a theoretical modeling. Appl. Opt. 25 (4), 571−580.

Bricaud, A., et al., 1995. Variability in the chlorophyll-specific absorption coefficients of natural phytoplankton: analysis and parameterization. J. Geophys. Res. 100 (C7), 13321−13332.

Bricaud, A., et al., 2004. Natural variability of phytoplanktonic absorption in oceanic waters: influence of the size structure of algal populations. J. Geophys. Res. 109 (C11010), 1−12.

Campbell, D., et al., 1998. Chlorophyll fluorescence analysis of cyanobacterial photosynthesis and acclimation. Microbiol. Mol. Biol. Rev. 62 (3), 667−683.

Carder, K.L., et al., 2004. Performance of the MODIS semi-analytical ocean color algorithm for chlorophyll-a. Adv. Space Res. 33 (7), 1152−1159.

Carder, K.L., Chen, F.R., Lee, Z.P., Hawes, S.K., Kamykowski, D., 1999. Semianalytic moderate-resolution imaging spectrometer algorithms for chlorophyll a and absorption with bio-optical domains based on nitrate-depletion temperatures. J. Geophys. Res. 104, 5403.

Codd, G.A., 2000. Cyanobacterial toxins, the perception of water quality, and the prioritisation of eutrophication control. Ecol. Eng. 16 (1), 51−60.

Defoin-Platel, M., Chami, M., 2007. How ambiguous is the inverse problem of ocean color in coastal waters? J. Geophys. Res. 112 (C03004), 1−16.

Desortová, B., 1981. Relationship between chlorophyll-a concentration and phytoplankton biomass in several reservoirs in Czechoslovakia. Int. Revue ges. Hydrobiol. 66 (2), 53−169.

Dierberg, F.E., Carriker, N.E., 1994. Field testing two instruments for remotely sensing water quality in the Tennessee Valley. Environ. Sci. Technol. 28 (1), 16−25.

Dierssen, H., et al., 2006. Red and black tides: quantitative analysis of water-leaving radiance and perceived color for phytoplankton, colored dissolved organic matter, and suspended sediments. Limnol. Oceanogr. 51 (6), 2646−2659.

Doerffer, R., Schiller, H., 2007. The MERIS Case 2 water algorithm. Int. J. Remote. Sens. 28 (3-4), 517−535.

Dupouy, C., et al., 2008. Bio-optical properties of the marine cyanobacteria *Trichodesmium* spp. J. Appl. Remote Sens. 2 (023503), 1−17.

Fell, F., et al., 2003. Retrieval of chlorophyll concentration from MERIS measurements in the spectral range of sun-induced chlorophyll fluorescence. Proc. SPIE 116.

Field, C.B., et al., 1998. Primary production of the biosphere: integrating terrestrial and oceanic components. Science 281 (5374), 237−240.

Giardino, C., Candiani, G., Zilioli, E., 2005. Detecting chlorophyll-a in Lake Garda using TOA MERIS radiances. Photogramm. Eng. Remote Sens. 71 (9), 1045−1051.

Gilerson, A., et al., 2007. Comparison of fluorescence retrieval algorithms using radiative transfer simulations and field measurements. IEEE International Conference on Geoscience and Remote Sensing, 2006. IEEE, pp. 1312−1315.

Gilerson, A., et al., 2008. Fluorescence component in the reflectance spectra from coastal waters. II. Performance of retrieval algorithms. Opt. Express. 16 (4), 2446−2460.

Gilerson, A., et al., 2010. Algorithms for remote estimation of chlorophyll-a in coastal and inland waters using red and near infrared bands. Opt. Express. 18 (23), 24109−24115.

Gitelson, A., 1992. The peak near 700 nm on radiance spectra of algae and water: relationships of its magnitude and position with chlorophyll concentration. Int. J. Remote. Sens. 13 (17), 3367−3373.

Gitelson, A.A., et al., 1999. Comparative reflectance properties of algal cultures with manipulated densities. J. Appl. Psychol. 11 (4), 345−354.

Gitelson, A.A., Dall'Olmo, G., Moses, W., Rundquist, D.C., Barrow, T., Fisher, T.R., et al., 2008. A simple semi-analytical model for remote estimation of chlorophyll-a in turbid waters: validation. Remote Sens. Environ. 112 (9), 3582−3593.

Gons, H.J., 1999. Optical teledetection of chlorophyll a in turbid inland waters. Environ. Sci. Technol. 33 (7), 1127−1132.

Gons, H.J., Auer, M.T., Effler, S.W., 2008. MERIS satellite chlorophyll mapping of oligotrophic and eutrophic waters in the Laurentian Great Lakes. Remote Sens. Environ. 112 (11), 4098−4106.

Gordon, H.R., 2002. Inverse methods in hydrologic optics. Oceanologia 44 (1), 9−58.

Gordon, H.R., Brown, O.B., Jacobs, M.M., 1975. Computed relationships between the inherent and apparent optical properties of a flat homogeneous ocean. Appl. Opt. 14 (2), 417−427.

Gower, J.F.R., Doerffer, R., Borstad, G.A., 1999. Interpretation of the 685 nm peak in water-leaving radiance spectra in terms of fluorescence, absorption and scattering, and its observation by MERIS. Int. J. Remote. Sens. 20 (9), 1771−1786.

Gower, J., King, S., Borstad, G., Brown, L., et al., 2005. Detection of intense plankton blooms using the 709 nm band of the MERIS imaging spectrometer. Int. J. Remote Sens 26 (9), 2005−2012.

Gregor, J., Marsalek, B., 2004. Freshwater phytoplankton quantification by chlorophyll a: a comparative study of in vitro, in vivo and in situ methods. Water. Res. 38 (3), 517−522.

Han, L., et al., 1994. The spectral responses of algal chlorophyll in water with varying levels of suspended sediment. Int. J. Remote. Sens. 15 (18), 3707−3718.

Hill, V.J., Zimmerman, R.C., Bissett, W.P., Dierssen, H., Kohler, D.D.R., 2014. Evaluating light availability, seagrass biomass, and productivity using hyperspectral airborne remote sensing in Saint Joseph's Bay, Florida. Estuaries Coasts, 1−23.

Hu, C., et al., 2012. Dynamic range and sensitivity requirements of satellite ocean color sensors: learning from the past. Appl. Opt. 51 (25), 6045−6062.

Hunter, P.D., et al., 2008. Spectral discrimination of phytoplankton colour groups: the effect of suspended particulate matter and sensor spectral resolution. Remote Sens. Environ. 112 (4), 1527−1544.

Huot, Y., Babin, M., 2010. In: Suggett, D.J., Prášil, O., Borowitzka, M.A. (Eds.), Chlorophyll a Fluorescence in Aquatic Sciences: Methods and Applications. Springer, Dordrecht, The Netherlands.

Huot, Y., et al., 2007. Does chlorophyll *a* provide the best index of phytoplankton biomass for primary productivity studies? Biogeosci. Discuss. 4 (2), 707−745.

IOCCG, 2012. Mission Requirements for Future Ocean-Colour Sensors. Report No. 13.

IOCCG, 1998. Minimum requirements for an operational ocean-colour sensor for the open ocean. Report No. 1.

Johnsen, G., et al., 1994. Chromoprotein- and pigment-dependent modeling of spectral light absorption in two dinoflagellates, Prorocentrum minimum and Heterocapsa pygmaea. Mar. Ecol. Prog. Ser. 114, 245−258.

Johnsen, G., Sakshaug, E., 2007. Biooptical characteristics of PSII and PSI in 33 species (13 pigment groups) of marine phytoplankton, and the relevance for pulse-amplitude-modulated and fast-repetition-rate fluorometry. J. Phycol. 43 (6), 1236−1251.

Kitchen, J.C., Zaneveld, J.R.V., 1992. A three-layered sphere model of the optical properties of phytoplankton. Limnol. Oceanogr. 37 (8), 1680−1690.

Kobayashi, H., Toratani, M., Matsumura, S., Siripong, A., Lirdwitayaprasit, T., Jintasaeranee, P., 2011. Optical properties of inorganic suspended solids and their influence on ocean colour remote sensing in highly turbid coastal waters. Int. J. Remote Sens. 32 (23), 8393−8420.

Kudela, R.M., et al., 2015. Application of hyperspectral remote sensing to cyanobacterial blooms in inland waters. Remote Sens. Environ. 167, 196−205.

Kutser, T., 2009. Passive optical remote sensing of cyanobacteria and other intense phytoplankton blooms in coastal and inland waters. Int. J. Remote. Sens. 30 (17), 4401−4425.

Kutser, T., Metsamaa, L., Dekker, A.G., 2008. Influence of the vertical distribution of cyanobacteria in the water column on the remote sensing signal. Estuar. Coast. Shelf Sci. 78, 649−654.

Le, C., Li, Y., Zha, Y., Sun, D., Huang, C., Lu, H., 2009. A four-band semi-analytical model for estimating chlorophyll a in highly turbid lakes: The case of Taihu Lake, China. Remote Sens. Environ. 113 (6), 1175−1182.

Lee, Z., et al., 2011. An inherent-optical-property-centred approach to correct the angular effects in water-leaving radiance. Appl. Opt. 50 (19), 3155−3167.

Letelier, M., Abbott, M.R., 1996. An analysis of chlorophyll fluorescence algorithms for the moderate resolution imaging spectrometer (MODIS). Remote Sens. Environ. 58 (2), 215−223.

Li, L., Li, L., Song, K., Li, Y., Tedesco, L.P., Shi, K., et al., 2013. An inversion model for deriving inherent optical properties of inland waters: Establishment, validation and application. Remote Sens. Environ. 135, 150−166.

Lu, Y., et al., 2016. Sunlight induced chlorophyll fluorescence in the near-infrared spectral region in natural waters: Interpretation of the narrow reflectance peak around 761 nm. J. Geophys. Res. Oceans 121 (7), 5017−5029.

Lunetta, R.S., et al., 2014. Evaluation of cyanobacteria cell count detection derived from MERIS imagery across the eastern USA. Remote Sens. Environ. 157, 24−34.

Lorenzen, C.J., 1966. A method for the continuous measurement of in vivo chlorophyll concentration. Deep Sea Res. Oceanogr. Abstr. 13 (2), 223−227.

Matsushita, B., et al., 2015. A hybrid algorithm for estimating the chlorophyll-a concentration across different trophic states in Asian inland waters. ISPRS J. Photogramm. Remote Sens. 102, 28−37.

Matthews, M.W., 2011. A current review of empirical procedures of remote sensing in inland and near-coastal transitional waters. Int. J. Remote. Sens. 32 (21), 6855−6899.

Matthews, M.W., 2014. Eutrophication and cyanobacterial blooms in South African inland waters: 10 years of MERIS observations. Remote Sens. Environ. 155, 161−177.

Matthews, M.W., Bernard, S., 2013. Characterizing the absorption properties for remote sensing of three small optically diverse South African reservoirs. Remote Sens. 5, 4370−4404.

Matthews, M.W., Odermatt, D., 2015. Improved algorithm for routine monitoring of cyanobacteria and eutrophication in inland and near-coastal waters. Remote Sens. Environ. 156, 374–382.

Matthews, M.W., Bernard, S., Robertson, L., 2012. An algorithm for detecting trophic status (chlorophyll-a), cyanobacterial-dominance, surface scums and floating vegetation in inland and coastal waters. Remote Sens. Environ. 124, 637–652.

Mckee, D., et al., 2007. Potential impacts of nonalgal materials on water-leaving Sun induced chlorophyll fluorescence signals in coastal waters. Appl. Opt. 46 (31), 7720–7729.

Meroni, M., et al., 2009. Remote sensing of solar-induced chlorophyll fluorescence: review of methods and applications. Remote Sens. Environ. 113 (10), 2037–2051.

Mishra, S., Mishra, D.R., 2012. Normalized Difference Chlorophyll Index: a novel model for remote estimation of chlorophyll-a concentration in turbid productive waters. Remote Sens. Environ. 117, 394–406.

Mishra, S., Mishra, D.R., Lee, Z., 2014. Bio-optical inversion in highly turbid and cyanobacteria dominated waters. IEEE Trans. Geosci. Remote Sens. 52 (1), 375–388.

Moore, T.S., et al., 2014. An optical water type framework for selecting and blending retrievals from bio-optical algorithms in lakes and coastal waters. Remote Sens. Environ. 143, 97–111.

Morel, A., Bricaud, A., 1986. Inherent optical properties of algal cells including picoplankton: theoretical and experimental results. Can. Bull. Fish. Aquat. Sci. 214, 521–559.

Morel, A., Prieur, L., 1977. Analysis of variations in ocean color. Limnol. Oceanogr. 22 (4), 709–722.

Moreno-Madriñán, M.J., Fischer, A.M., 2013. Performance of the MODIS FLH algorithm in estuarine waters: a multi-year (2003–2010) analysis from Tampa Bay, Florida (USA). Int. J. Remote Sens. 34 (19), 6467–6483.

Moses, W.J., et al., 2012. Impact of signal-to-noise ratio in a hyperspectral sensor on the accuracy of biophysical parameter estimation in case II waters. Opt. Express. 20 (4), 4309–4330.

Mouw, C.B., et al., 2015. Aquatic color radiometry remote sensing of coastal and inland waters: challenges and recommendations for future satellite missions. Remote Sens. Environ. 160, 15–30.

Neville, R., Gower, J.F.R., 1977. Passive remote sensing of phytoplankton via chlorophyll-α fluorescence. J. Geophys. Res. 82 (24), 3487–3493.

O'Reilly, J.E., et al., 1998. Ocean color chlorophyll algorithms for SeaWiFS. J. Geophys. Res. 103 (C11), 24937.

Odermatt, D., et al., 2012a. Review of constituent retrieval in optically deep and complex waters from satellite imagery. Remote Sens. Environ. 118, 116–126.

Odermatt, D., et al., 2012b. MERIS observations of phytoplankton blooms in a stratified eutrophic lake. Remote Sens. Environ. 126, 232–239.

Palmer, S., et al., 2013. Ultraviolet fluorescence LiDAR (UFL) as a measurement tool for water quality parameters in turbid lake conditions. Remote Sens. 5 (9), 4405–4422.

Palmer, S.C.J., et al., 2014. Validation of Envisat MERIS algorithms for chlorophyll retrieval in a large, turbid and optically-complex shallow lake. Remote Sens. Environ 157, 158–169.

Palmer, S.C.J., Kutser, T., Hunter, P.D., 2015. Remote sensing of inland waters: challenges, progress and future directions. Remote Sens. Environ. 157, 1–8.

Pan, H., Lyu, H., Wang, Y., Jin, Q., Wang, Q., Li, Y., et al., 2015. An improved approach to retrieve IOPs based on a quasi-analytical algorithm (QAA) for turbid eutrophic inland water. IEEE J. Sel. Top. Appl. Earth Obs. Remote Sens. 8 (11), 5177–5189.

Piskozub, J., McKee, D., 2011. Effective scattering phase functions for the multiple scattering regime. Opt. Express. 19 (5), 4786–4794.

Quibell, G., 1991. The effect of suspended sediment on reflectance from freshwater algae. Int. J. Remote. Sens. 12 (1), 177–182.

Quibell, G., 1992. Estimating chlorophyll concentrations using upwelling radiance from different fresh-water algal genera. Int. J. Remote. Sens. 13 (14), 2611–2621.

Quinby-Hunt, M.S., et al., 1989. Polarized-light scattering studies of marine chlorella. Limnol. Oceanogr. 34 (8), 1587–1600.

Quirantes, A., Bernard, S., 2006. Light-scattering methods for modeling algal particles as a collection of coated and/or nonspherical scatterers. J. Quant. Spectrosc. Radiat. Transfer 100 (1-3), 315–324.

Reynolds, C.S., 2006. The Ecology of Phytoplankton. Cambridge University Press, New York.

Riddick, C., Hunter, P.D., Tyler, A., Martinez-Vincente, V., Horvath, H., Kovacs, A., et al., 2015. Spatial variability of absorption coefficients over a biogeochemical gradiant in a large and optically complex shallow lake. J. Geophys. Res. Oceans JC011202, 1–16.

Ritchie, R.J., 2006. Consistent sets of spectrophotometric chlorophyll equations for acetone, methanol and ethanol solvents. Photosynth. Res. 89 (1), 27–41.

Robertson Lain, L., Bernard, S., Evers-King, H., 2014. Biophysical modelling of phytoplankton communities from first principles using two-layered spheres: equivalent Algal Populations (EAP) model. Opt. Express. 22 (14), 16745–16758.

Ruiz-Verdú, A., et al., 2008. An evaluation of algorithms for the remote sensing of cyanobacterial biomass. Remote Sens. Environ. 112 (11), 3996–4008.

Rundquist, D.C. et al., 1996. Remote measurement of algal chlorophyll in surface waters: the case for the first derivative of reflectance near 690 nm. Photogramm. Eng. Rem. S. 62(2), 195–200.

Ryan, J., et al., 2014. Application of the hyperspectral imager for the coastal ocean to phytoplankton ecology studies in Monterey Bay, CA, USA. Remote Sens. 6 (2), 1007–1025.

Salama, M.S., et al., 2009. Deriving inherent optical properties and associated inversion-uncertainties in the Dutch Lakes. Hydrol. Earth Syst. Sci. 13, 1113–1121.

Santini, F., et al., 2010. A two-step optimization procedure for assessing water constituent concentrations by hyperspectral remote sensing techniques: an application to the highly turbid Venice lagoon waters. Remote Sens. Environ. 114 (4), 887–898.

Sathyendranath, S., Platt, T., 1997. Analytic model of ocean color. Appl. Opt. 36 (12), 2620–2629.

Sayers, M.J., et al., 2015. A new method to generate a high-resolution global distribution map of lake chlorophyll. Int. J. Remote. Sens. 36 (7), 1942–1964.

Sayers, M., et al., 2016. Cyanobacteria blooms in three eutrophic basins of the Great Lakes: a comparative analysis using satellite remote sensing. Int. J. Remote. Sens. 37 (17), 4148–4171.

Schalles, J.F., et al., 1998. Estimation of chlorophyll a from time series measurements of high spectral resolution reflectance in an eutrophic lake. J. Phycol. 34 (2), 383–390.

Seppälä, J., et al., 2007. Ship-of-opportunity based phycocyanin fluorescence monitoring of the filamentous cyanobacteria bloom dynamics in the Baltic Sea. Estuar. Coast. Shelf Sci. 73 (3-4), 489–500.

Simis, S.G.H., et al., 2007. Influence of phytoplankton pigment composition on remote sensing of cyanobacterial biomass. Remote Sens. Environ. 106 (4), 414–427.

Simis, S.G.H., et al., 2012. Optimization of variable fluorescence measurements of phytoplankton communities with cyanobacteria. Photosynth. Res. 112 (1), 13–30.

Stramski, D., Bricaud, A., Morel, A., 2001. Modeling the inherent optical properties of the ocean based on the detailed composition of the planktonic community. Appl. Opt. 40 (18), 2929−2945.

Stramski, D., Babin, M., Wozniak, S.B., 2007. Variations in the optical properties of terrigenous mineral-rich particulate matter suspended in seawater. Limnol. Oceanogr. 52 (6), 2418−2433.

Svensen, Ø., Frette, Ø., Erga, S.R., 2007. Scattering properties of microalgae: the effect of cell size and cell wall. Appl. Opt. 46, 5762−5769.

Tao, B., et al., 2013. Influence of bio-optical parameter variability on the reflectance peak position in the red band of algal bloom waters. Ecol. Inform. 16 (April 2016), 17−24.

Tao, B., et al., 2015. A novel method for discriminating Prorocentrum donghaiense from diatom blooms in the East China Sea using MODIS measurements. Remote Sens. Environ. 158, 267−280.

Tomlinson, M.C., et al., 2016. Relating chlorophyll from cyanobacteria-dominated inland waters to a MERIS bloom index. Remote Sens. Lett. 7 (2), 141−149.

Vaillancourt, R.D., et al., 2004. Light backscattering properties of marine phytoplankton: relationships to cell size, chemical composition and taxonomy. J. Plankton Res. 26 (2), 191−212.

Verpoorter, C., et al., 2014. A global inventory of lakes based on high-resolution satellite imagery. Geophys. Res. Lett. 41 (18), 6396−6402.

Volten, A.H., et al., 1998. Laboratory measurements of angular distributions of light scattered by phytoplankton and silt. Limnol. Oceanogr. 43 (6), 1180−1197.

Wang, G., et al., 2016. Retrieving absorption coefficients of multiple phytoplankton pigments from hyperspectral remote sensing reflectance measured over cyanobacteria bloom waters. Limnol. Oceanogr. Methods 14 (7), 432−447.

Werdell, P.J., Bailey, S.W., 2005. An improved in-situ bio-optical data set for ocean color algorithm development and satellite data product validation. Remote Sens. Environ. 98 (1), 122−140.

Whitmire, A.L., et al., 2010. Spectral backscattering properties of marine phytoplankton cultures. Opt. Express. 18 (14), 1680−1690.

Witkowski, K., et al., 1998. A light-scattering matrix for unicellular marine phytoplankton. Limnol. Oceanographyol. 43 (5), 859−869.

Wynne, T.T., et al., 2008. Relating spectral shape to cyanobacterial blooms in the Laurentian Great Lakes. Int. J. Remote. Sens. 29 (12), 3665−3672.

Wynne, T., Stumpf, R., 2015. Spatial and temporal patterns in the seasonal distribution of toxic cyanobacteria in Western Lake Erie from 2002−2014. Toxins 7 (5), 1649−1663.

Yang, W., Matsushita, B., Chen, J., Yoshimura, K., Fukushima, T., 2013. Retrieval of inherent optical properties for turbid inland waters from remote-sensing reflectance. IEEE Trans. Geosci. Remote Sens. 51 (6), 3761−3773.

Zhang, Y., et al., 2012. Effect of phytoplankton community composition and cell size on absorption properties in eutrophic shallow lakes: field and experimental evidence. Opt. Express. 20 (11), 11882−11898.

Zhou, J., et al., 2008. Retrieving quantum yield of sun-induced chlorophyll fluorescence near surface from hyperspectral in-situ measurement in productive water. Opt. Express. 16 (22), 17468−17483.

Zhou, W., et al., 2012. Variations in the optical scattering properties of phytoplankton cultures. Opt. Express. 20 (10), 11189−11206.

Chapter 7

Bio-optical Modeling of Sun-Induced Chlorophyll-a Fluorescence

Alexander A. Gilerson[1] and Yannick Huot[2]
[1]City University of New York, New York, NY, United States, [2]Université de Sherbrooke, Québec, QC, Canada

7.1 INTRODUCTION, BASIC CONCEPTS, AND CURRENT KNOWLEDGE

Almost all surface waters harbor drifting photosynthesizing microbes, called phytoplankton. To carry out photosynthesis, phytoplankton contain several pigments. While the make-up of these pigments differs depending on the species and growth conditions, they all contain chlorophyll-a (chl-a). The chl-a molecule is strongly fluorescent with a peak *in vivo* centered near 685 nm. When phytoplankton photosynthetic pigments absorb light, a fraction (a few percent) of this absorbed light, referred to as the "quantum yield," is fluoresced. When the absorbed light is from the Sun, this fluorescence is termed Sun-induced chlorophyll-a fluorescence (SICF). While the quantum yield of fluorescence of chl-a in an organic solvent is constant at a given temperature, *in vivo* the effective quantum yield is variable and depends on the species present, the spectral distribution and amplitude of the excitation irradiance, and the physiological status of the cells.

SICF emission can be observed in the upwelling and downwelling light field as an additional contribution to sunlight. In the early 1970s, with measurements of reflectance spectra and the attenuation of irradiance in water becoming more available, it rapidly became clear that a significant and unusual process was occurring in the red part of the reflectance spectrum. Specifically, its discovery stems from observations in inland waters. Indeed, it is by multiplying measurements of reflectance by measurements of the

Bio-optical Modeling and Remote Sensing of Inland Waters.
DOI: http://dx.doi.org/10.1016/B978-0-12-804644-9.00007-0

diffuse attenuation coefficient taken in 1967 by Tyler and Smith (1970) in the San Vicente Reservoir (California, USA) that Howard Gordon (1974) identified the presence of an excess of light centered around 685 nm. In an otherwise very elegant contribution, the peak was misidentified as originating from anomalous dispersion due to the chlorophyll absorption band centered near 676 nm. In 1977, Morel and Prieur (1977) and Neville and Gower (1977) independently attributed the peak in reflectance spectra as originating from SICF. The latter even related the height of the fluorescence peak in reflectance measurements from overflights of Saanich Inlet (British Columbia, Canada) to the chl-a concentration ([chl-a], mg m^{-3}). Two years later, in a seminal paper Howard Gordon (Gordon, 1979), used the quasi-single scattering approximation to mathematically describe the enhancement of reflectance due to chlorophyll fluorescence and showed that (1) the quantum yield necessary to explain the observed enhancement of light near 685 nm was within the previously observed ranges, and (2) the derived spectra closely matched those of chlorophyll fluorescence, which strongly supported the findings reported two years earlier by Morel and Prieur (1977) and Neville and Gower (1977). Most of the following work (and the subject of this chapter) regarding SICF has focused on the upwelling radiance (or reflectance) at the water surface.

After the identification of the source of the red peak in reflectance spectra, two major research threads can be followed. On the one hand, the proportionality between the SICF emission and [chl-a] has led to several studies examining its potential as an estimator of phytoplankton [chl-a] or absorption. Many of these studies have taken place in coastal waters where the estimation of [chl-a] using ocean color algorithms is much more difficult due to the presence of uncorrelated (and often high) amounts of colored dissolved organic matter (CDOM) as well as detrital and mineral particles. Observations have shown that in these waters (many inland waters show similar optical properties), the contribution from fluorescence emission to the upwelling radiance can be used as an estimator of the phytoplankton concentration at least locally. In addition to the work of Neville and Gower (1977) mentioned above, Gower and Borstad (1990) have shown a good relationship between remotely measured [chl-a] and SICF in Barkley Sound (British Columbia, Canada), while Gower et al. (1999) (see also Sathyendranath et al., 2004) also presented several linear relationships between [chl-a] and SICF. In both cases, however, the relationships between [chl-a] and fluorescence emission showed different slopes depending on the location- and/or field-sampling campaign, which were interpreted as different efficiencies of the fluorescence emission. Beyond variability in the quantum yield, the influence of the optical properties of the water and, in particular, the attenuation coefficients can also be important. This has been noted by several authors and studied by Fischer and Kronfeld (1990), McKee et al. (2007) and Gilerson et al. (2007) who have discussed the potential impact of the presence of CDOM, and in some cases mineral particles, on the fluorescence emission. Largely due to different assumptions for the absorption and scattering coefficients, these authors, however, reached different conclusions concerning the impact it can have on the retrieval of [chl-*a*], with some suggesting that variations in the attenuation coefficient overwhelm the chlorophyll signal, while others observed limited impacts.

The other research thread has focused on the potential of the quantum yield of fluorescence to provide insights into the physiology of phytoplankton. This has been carried out mostly in open ocean waters where it is easier to estimate the amount of light absorbed by phytoplankton using ocean color algorithms. Indeed, in these waters, algorithms to obtain [chl-*a*] and other variables [e.g., the attenuation of scalar irradiance, $K(\lambda)$; the attenuation of upwelling irradiance, $K_{Lu}(685)$; downwelling irradiance just below the surface $E_o^-(\lambda)$] exist and are well validated. The main focus of using measurements of SICF in these waters has thus been the retrieval of the quantum yield of fluorescence as it contains physiological or taxonomic information about phytoplankton communities. In this case, the fluorescence quantum yield, ϕ_f, is obtained by constraining the optical effects in a model describing the emission (e.g., Babin et al., 1996; Behrenfeld et al., 2009; Huot et al., 2005). We will not examine this body of work in more detail but will mention that efforts have been made to link the variability observed in ϕ_f to primary productivity (Chamberlin et al., 1990; Chamberlin and Marra, 1992; Kiefer et al., 1989; an aspect that has been mostly left aside in recent years in view of the complex nature of the variability in ϕ_f), macro- and micronutrient limitation (Behrenfeld et al., 2009; Browning et al., 2014; Letelier et al., 1997; Schallenberg et al., 2008), and the photoacclimation or photoadaptation status (Morrison, 2003; Morrison and Goodwin, 2010; O'Malley et al., 2014). Similar sources of variability in ϕ_f likely occur in inland and coastal waters.

The results from many of the earlier studies described above have led to the inclusion of bands on satellite-based spectroradiometers, such as the MEdium Resolution Imaging Spectrometer (MERIS), the MODerate resolution Imaging Spectroradiometer (MODIS), the Geostationary Ocean Color Imager and very recently the Ocean and Land Colour Instrument to measure SICF from space. Current and past ocean color satellite sensors that allow the retrieval of SICF have a limited number of bands for this retrieval. While the positions of these bands differ slightly between sensors, two of these bands are located outside the fluorescence emission peak and one is located near the fluorescence emission peak (in general shifted toward shorter wavelengths to avoid an oxygen absorption band in the atmosphere). The radiance from the fluorescence emission ($L_f^-(\lambda_f)$, W m^{-2} μm^{-1} sr^{-1}) at the measurement wavelength, λ_f, is extracted by using the fluorescence line height (FLH) algorithm (Neville and Gower, 1977; Letelier and Abbott, 1996; Gower et al., 1999; see Fig. 7.11 described later) which estimates the excess radiance (or remote sensing reflectance) at the fluorescence peak band above a baseline connecting the radiance measured at the two bands outside the fluorescence band. This approach is appropriate in open ocean waters, where the FLH correlates well with the fluorescence magnitude (Huot et al., 2005). This retrieval is, however, much more difficult in waters with high [chl-*a*] or turbidity, as is often the case in inland and coastal waters, where the reabsorption by chl-*a* in the fluorescence band is significant and leads to a reduction in the fluorescence emission (and upwelling light from elastic scattering processes) observed at the surface. Together with the increasing water absorption in these bands, this leads to the appearance of a reflectance peak in the red and near infrared (NIR) spectral regions (Fischer and Kronfeld, 1990; Gower et al., 1999). These complications are further compounded in inland and coastal waters with the more difficult atmospheric correction as the assumption of negligible reflectance in the NIR, which is used in open ocean waters, fails (see IOCCG Report 10, 2010 and Chapter 3 of this book).

Therefore, the complex nature of inland waters, where the absorption and scattering by constituents in water are not well correlated with the amount of phytoplankton and they can be high and variable, implies that great care must be taken to account for their influence if inferences are to be made from the fluorescence emission. This chapter will examine a few key aspects of the use of fluorescence in inland and coastal waters.

7.2 MODELING OF REFLECTANCE SPECTRA WITH FLUORESCENCE

In this section, a simple approach for the forward modeling of reflectance spectra that includes fluorescence is presented. This approach is used to rapidly and efficiently create a large number of spectra—a synthetic dataset—based on inherent optical properties, or optical constituents' concentrations. In the following Sections (7.3 and 7.4), this dataset is used along with in situ and satellite data to highlight some properties of fluorescence emission in inland waters. This section starts by describing the reflectance components, which will be followed by detailing the parameterization of the inherent optical properties (IOPs).

7.2.1 Remote Sensing Reflectance

In this synthetic dataset, the remote sensing reflectance [$R_{rs}(\lambda)$, sr^{-1}] spectra were calculated as the sum of the elastic reflectance (R_{rs}^e, sr^{-1}) and the reflectance contribution from fluorescence (R_{rs}^f, sr^{-1}):

$$\begin{aligned} R_{rs}(\lambda) &= R_{rs}^e(\lambda) + R_{rs}^f(\lambda) \\ &= L_e(\lambda)/E_d(\lambda) + L_f(\lambda)/E_d(\lambda) \end{aligned} \quad (7.1)$$

The second row of Eq. (7.1) expresses the two remote sensing reflectances in terms of the elastically scattered $[L_e(\lambda)]$ and fluorescence $[L_f(\lambda)]$ radiance just above the surface and the downwelling planar irradiance just above the surface $[E_d(\lambda)]$.

7.2.2 Elastic Reflectance

The R_{rs}^e above water was calculated following Lee et al. (2002):

$$R_{rs}^e = 0.52 \frac{R_{rs}^-}{(1 - 1.7R_{rs}^-)}, \tag{7.2}$$

where R_{rs}^- is the remote sensing reflectance due to elastic scattering just below the surface, which is calculated as,

$$R_{rs}^- = g_1 u^2 + g_2 u, \tag{7.3}$$

$$u = b_b/(a + b_b), \tag{7.4}$$

where a (m^{-1}) and b_b (m^{-1}) are the total absorption and backscattering coefficients, respectively, and $g_1 = 0.125$ and $g_2 = 0.089$ are empirically derived parameters (Lee et al., 2009). This parameterization does not include inelastic scattering effects (i.e., Raman scattering) or the fluorescence contribution. Raman scattering is small in inland and coastal waters due to the generally strong absorption in the excitation bands and will be ignored here.

7.2.3 Fluorescence Reflectance

Since it is the focus of this chapter, the fluorescence model and the assumptions associated with it are described in detail.

The mathematical description of fluorescence upwelling radiance at a given depth is obtained by integrating with respect to depth the contribution from thin horizontal fluorescing layers below that depth. Its derivation in terms of incident irradiance and the inherent and apparent optical properties has been published by several authors (e.g., Gordon, 1979; Kattawar and Vastano, 1982; Maritorena et al., 2000) and it will not be derived here, but the resulting equation and the origin of the different variables is provided. At 685 nm, the upwelling radiance (W m^{-2} μm^{-1} sr^{-1}) just above the surface is given by:

$$L_f(685) = 0.54L_f^-(685) = 0.54\frac{1}{4\pi}\frac{\phi_f}{C_f}Q_a^*[\text{chl-}a]\int_{400}^{700}\frac{a_{phy}^*(\lambda)E_o^-(\lambda)}{K(\lambda)+K_{Lu}(685)}d\lambda. \tag{7.5}$$

The factor 0.54 accounts for the effect of reflectance (ρ) and refraction at the air-sea interface [i.e., $0.54=(1-\rho)/n_w^2$ where $\rho=0.025$ and $n_w=1.34$ is the index of refraction of water]; $L_f^-(685)$ is the upwelling radiance at 685 nm just below the surface. The factor $1/4\pi$ represents the fraction of the isotropic fluorescence emission in all directions in 1 steradian (in this case, for a radiance sensor pointing downward). Since ϕ_f (photons emitted x [photons absorbed]$^{-1}$) is defined as the emission over all wavebands (a roughly Gaussian peak) and the equation provides only the emission at 685 nm, ϕ_f is multiplied by the fraction of this emission within 1 nm at 685 nm to the total emission $(1/C_f)$. Since the fluorescence emitted within the cell can be reabsorbed by pigments (mostly chl-a) before leaving the cell (Babin et al. 1996; Collins et al. 1985), the fraction leaving the cell is given by the variable Q_a^*. At a given wavelength, λ (nm), just below the surface, the flux absorbed by phytoplankton is given by $a_{phy}(\lambda)E_o^-(\lambda)$, where $a_{phy}(\lambda)$ (m^{-1}) is the absorption coefficient of phytoplankton, and $E_o^-(\lambda)$ (μmol photon m^{-2} s^{-1} nm^{-1}) is the scalar irradiance (i.e., the excitation irradiance) just below the surface. The integration over the photosynthetically available radiation is necessary to obtain the absorption at all wavelengths. Eq. (7.5) conveniently requires only the subsurface irradiance as the resulting effect of attenuation after the depth integration is accounted for in the denominator. In Eq. (7.5), the absorption by phytoplankton has been further split into the chlorophyll-a specific absorption, $a_{phy}^*(\lambda)$ (m^2 mg^{-1}), and the [chl-a] according to $a_{phy}(\lambda) = [\text{chl-}a]a_{phy}^*(\lambda)$. The [chl-$a$] is taken outside the integral. Inside the integral, in the denominator, the attenuation of scalar irradiance, $K(\lambda)$ (m^{-1}), accounts for the attenuation of the excitation irradiance with depth, while the attenuation of upwelling irradiance, $K_{Lu}(685)$, accounts for the attenuation of upwelling fluorescence radiance (from the thin layer at depth z to the surface).

There are a few things that must be kept in mind when using Eq. (7.5). First, the quantum yield is defined as an effective quantum yield, that is, it is not the fraction of the absorbed photons by *photosynthetic* pigments that are reemitted (which is the quantum yield per se), but the fraction of the photons absorbed by *all* pigments within the cell that is reemitted. Thus, this is different from estimates that are derived by fluorescence lifetime measurements (e.g., Lin et al., 2016) or similar fluorescence-based measurements that observe only the result of the excitation of photosynthetic pigments. This approximation is, however, useful as it directly relates the emission to the absorption coefficient that is derived by ocean color algorithms. Second, in open ocean waters where this equation has been used the most, $K_{Lu}(685)$, is generally approximated as the total absorption coefficient (Kattawar and Vastano, 1982; Maritorena et al., 2000), which is a very good approximation due to the low backscattering to absorption ratio at 685 nm. In inland and coastal waters, when scattering (or backscattering) is important, this approximation can be less accurate; we found, however, that it applies well to all but the most turbid waters.

The fluorescence at 685 nm as a function of water parameters was modeled using Eq. (7.5). More accurate estimates of the fluorescence emission can be obtained by bio-optical modeling and radiative transfer (RT) programs like Hydrolight (Mobley, 1994, 2001). However, for the upwelling radiance at the surface observed at nadir and originating from waters with a uniform depth distribution of IOPs (including the quantum yield of fluorescence emission) RT programming can be effectively replaced by Eqs. (7.1) and (7.5). This approach is many orders faster and allows rapid analysis of the relationships between fluorescence emission and IOPs.

As just mentioned, the maximum of the peak fluorescence emission, $L_f(685)$ (W m^{-2} µm^{-1} sr^{-1}), was calculated using Eq. (7.5) by assuming a fluorescence quantum yield, ϕ_f, of 1%. The spectral shape of fluorescence was modeled as a Gaussian spectral profile centered at 685 nm, having a full width at half maximum (FWHM) of 25 nm (Gordon, 1979; Gower et al., 2004; Mobley, 1994)

$$L_f(\lambda) = L_f(685)\exp(-4\log(2)[\{\lambda - 685\}/25]). \qquad (7.6)$$

Finally, according to Eq. (7.1), $L_f(\lambda)$ was divided by $E_d(\lambda)$ to obtain R_{rs}^f. For all simulations, the $E_d(\lambda)$ (W m^{-2} nm^{-1}), just above the water surface was obtained from a RT simulation using the Hydrolight software (Mobley, 1994; Mobley and Sundman, 2001) with Gregg and Carder's (1990) solar irradiance model, assuming a cloud-free sky, a wind speed of 5 m/s, and a solar zenith angle, θ_s, of 30°. To obtain E_o^- (µmol photon m^{-2} s^{-1} nm^{-1}) in Eq. (7.5), $E_d(\lambda)$ was converted to quantum units and to scalar irradiance just below the surface.

7.2.4 Inherent Optical Properties and Attenuation Coefficients

The reflectance model that includes a fluorescence component as now been described. To use these models to obtain reflectance spectra, it is necessary to input inherent optical properties and describe how some of the attenuation coefficients are modeled. This is the focus of this section.

A four-component bio-optical model, similar to the one described in Chapter 2 of this book, was used to describe the contributions of the different water constituents to the IOPs (Babin et al., 2003; Bukata et al., 1995), which were then used as inputs to Eqs. (7.4)−(7.5) to generate reflectance spectra datasets. The constituents included were water, phytoplankton, CDOM, and nonalgal particulates (NAP, which include mineral and detrital material). The subscripts "w," "phy," "CDOM," and "NAP" are used, respectively, to identify these four components below. Similar assumptions to those used in Pozdnyakov et al. (2002), the IOCCG Report 5 (2006) and Gilerson et al. (2007) are followed with several differences. Datasets were generated for the spectral range of 400−800 nm with a 1-nm resolution. In this bio-optical model for the simulation of IOP spectra, [chl-*a*] is the only input; however, IOPs of the other constituents are allowed to vary from the mean statistical relationship through the application of random noise to their amplitude and, in some cases, their shape. In the main dataset, [chl-a] was in the range of $0.5-40$ mg m^{-3}. Results for a separate simulation where the optical variability in water was dominated by CDOM or NAP will also be presented.

7.2.4.1 Absorption Coefficient

The total spectral absorption coefficient, $a(\lambda)$, is modeled as

$$a(\lambda) = a_w(\lambda) + a_{phy}(\lambda) + a_{CDOM}(\lambda) + a_{NAP}(\lambda), \tag{7.7}$$

where the water absorption spectrum $a_w(\lambda)$ was obtained from Pope and Fry, 1997.

In coastal waters, $a_{phy}(443)$, $a_{CDOM}(443)$, and $a_{NAP}(443)$ are typically weakly correlated with each other (IOCCG, Report 5, 2006). There is not much information about such relationships in inland waters but, as a first approximation, relationships similar to coastal waters were assumed. Based on the data from the NASA bio-Optical Marine Algorithm Data (NOMAD) (Werdell and Bailey, 2005) and Chesapeake Bay field campaigns (Gilerson et al., 2015), we used the following relationships at 443 nm:

$$a_{phy}(443) = a_{phy}^*(443)[\text{chl-}a] = 0.042[\text{chl-}a]^{-0.2}[\text{chl-}a] = 0.042[\text{chl-}a]^{0.8}; \tag{7.8}$$

$$a_{CDOM}(443) = 0.7a_{phy}(443); \tag{7.9}$$

$$a_{NAP}(443) = 0.56a_{phy}(443). \tag{7.10}$$

A total of 500 [chl-a] values were modeled composed of two sets. The first one contained 300 values randomly distributed between 0.5 and 15.0 mg m^{-3}, while the second one comprised 200 values randomly distributed between 15.0 and 40 mg m^{-3}.

The spectral phytoplankton absorption coefficient is obtained by multiplying the [chl-a] by a chl-a specific absorption coefficient ($a_{phy}^*(\lambda)$, m^2 mg^{-1}),

$$a_{phy}(\lambda) = [\text{chl-}a]a_{phy}^*(\lambda). \tag{7.11}$$

The choice of $a_{phy}^*(\lambda)$ strongly influences the corresponding reflectance and the emission of fluorescence. We modeled the chl-a specific absorption coefficient according to Ciotti et al. (2002) as the weighted sum of the specific absorption coefficients of microplankton and picoplankton, with a weighting factor (referred to as a "size parameter" by Ciotti et al.), S_f (unitless),

$$a_{phy}^*(\lambda) = S_f \cdot a_{pico}^*(\lambda) + (1 - S_f) \cdot a_{micro}^*(\lambda). \tag{7.12}$$

S_f was determined from the combination of this equation with Eq. (7.8) for $a_{phy}^*(443)$.

Recent studies (Ficek et al., 2004; McKee et al., 2014) have shown that for most inland and coastal waters the chl-*a* specific absorption coefficient at 675 nm is higher than predicted by Ciotti et al. (2002). To account for this observation, $a_{phy}^*(675)$ ($\mathrm{m}^2\,\mathrm{mg}^{-1}$) was set to follow $a_{phy}^*(675) = 0.027[\text{chl-}a]^{-0.2}$ and spectra from 601 to 750 nm were modified to fit this relationship.

To simulate natural variability, $a_{phy}^*(443)$ and $a_{phy}^*(675)$ were multiplied by a random number drawn from a normal distribution $[N(\mu, \sigma^2)]$ with a mean $\mu=1$ and a variance $\sigma^2=0.0625$: $X_1 \sim N(1, 0.0625)$. In a similar manner, $a_{CDOM}(443)$ and $a_{NAP}(443)$ in Eqs. (7.9) and (7.10) were multiplied by $X_2 \sim N(1, 0.1225)$. The ranges of variability here and below were based primarily on the published values from IOCCG Report 5 (2006), NOMAD and the authors' data (Gilerson et al., 2015).

The spectral absorption coefficients of both CDOM and NAP were modeled as having an exponentially decreasing shape with wavelength and referenced to 443 nm (Bukata et al., 1995; Stramski et al., 2001):

$$a_{CDOM}(\lambda) = a_{CDOM}(443)e^{-S_{CDOM}(\lambda - 443)}, \tag{7.13}$$

$$a_{NAP}(\lambda) = a_{NAP}(443)e^{-S_{NAP}(\lambda - 443)}. \tag{7.14}$$

S_{CDOM} was modeled as a normal distribution $0.017N(1, 0.01^2)$ and S_{NAP} as $0.010N(1, 0.01^2)$. Eq. (7.10) was also used to determine the concentration of NAP, [NAP] ($\mathrm{g\,m^{-3}}$):

$$[NAP] = a_{NAP}(443)/a_{NAP}^*(443), \tag{7.15}$$

where $a_{NAP}^*(443)$ ($\mathrm{m}^2\,\mathrm{g}^{-1}$) is the mass-specific absorption coefficient of NAP at 443 nm, which was simulated as a uniformly distributed random number $0.02 \le a_{NAP}^*(443) \le 0.05$ ($\mathrm{m}^2\,\mathrm{g}^{-1}$). The [NAP] was in the range of $0-30\,\mathrm{g\,m^{-3}}$ with most values below $15\,\mathrm{g\,m^{-3}}$. Examples of the resulting absorption coefficients at 443 nm as a function of [chl-*a*] are shown in Fig. 7.1.

7.2.4.2   Scattering and Backscattering Coefficients

The total scattering coefficient ($b(\lambda)$, $\mathrm{m^{-1}}$) was simulated as a sum of three components (applicable to lakes; e.g., Peng et al., 2009):

$$b(\lambda) = b_w(\lambda) + b_{phy}(\lambda) + b_{NAP}(\lambda). \tag{7.16}$$

Scattering by NAP was modeled using a power law function (Stramski et al., 2001; Twardowski et al., 2001) as follows:

$$b_{NAP}(\lambda) = b_{NAP}(550)\left(\frac{550}{\lambda}\right)^{\gamma_2}, \tag{7.17}$$

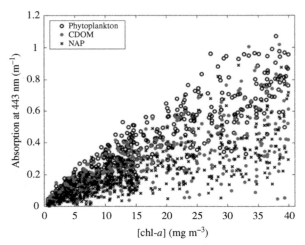

FIGURE 7.1   Example of the distributions of the absorption coefficients of CDOM, NAP, and phytoplankton used to obtain the total absorption spectra for simulations where CDOM and NAP covary with the phytoplankton absorption coefficient.

$$b_{NAP}(550) = b^*_{NAP}(550)[NAP], \tag{7.18}$$

where $b^*_{NAP}(550) = 0.75N(1, 0.01)$ (m^2 g^{-1}) is the mass-specific scattering of nonalgal particles at 550 nm, and $\gamma_2 = 0.8N(1, 0.0049)$.

The scattering by phytoplankton was calculated as the difference between their attenuation and absorption coefficients (Voss, 1992; Roesler and Boss, 2003):

$$b_{phy}(\lambda) = c_{phy}(\lambda) - a_{phy}(\lambda). \tag{7.19}$$

The attenuation coefficient itself was modeled as a power law function (Voss, 1992),

$$c_{phy}(\lambda) = c_{phy}(550)\left(\frac{550}{\lambda}\right)^{\gamma_1}, \tag{7.20}$$

where $c_{phy}(550) = 0.3[chl\text{-}a]^{0.57}$ and $\gamma_1 = 0.8$.

In the simulations, the backscattering coefficient [$b_b(\lambda)$, m^{-1}] was modeled as the sum of the contributing components,

$$b_b(\lambda) = b_{bw}(\lambda) + \tilde{b}_{b\ phy}b_{phy}(\lambda) + \tilde{b}_{b\ NAP}b_{NAP}(\lambda), \tag{7.21}$$

where $b_{bw}(\lambda)$ is obtained according to Morel (1974) and $\tilde{b}_{b\ phy}$ and $\tilde{b}_{b\ NAP}$ are backscattering ratios for phytoplankton and non-algal particles assumed to be independent of wavelength (Twardowski et al., 2001; Sydor and Arnone, 1997). Typical values were used as $\tilde{b}_{b\ phy}(\lambda) = 0.006$ and $\tilde{b}_{b\ NAP}(\lambda) = 0.0183$ (Petzold, 1972) (assuming a strong mineral component in NAP scattering).

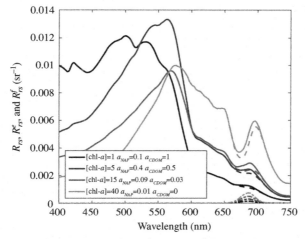

FIGURE 7.2   Examples of remote sensing reflectance spectra with [$R_{rs}(\lambda)$, continuous lines] and without [$R_{rs}^e(\lambda)$, dashed lines] fluorescence for various [chl-*a*]; $R_{rs}^f(\lambda)$ is also shown in dot-dashed lines. [chl-*a*] is in units of mg m^{-3}; a_{NAP} and a_{CDOM} are in units of m^{-1} at 443 nm.

7.2.4.3   Diffuse and Radiance Attenuation Coefficients

The attenuation coefficient of scalar irradiance, $K(\lambda)$, was assumed equal to that of the planar irradiance and modeled as $K(\lambda) = 1.0547[a(\lambda) + b_b(\lambda)]/\cos\theta_s'$, (Albert and Mobley, 2003; Gordon, 1989) where θ_s' is the solar zenith angle in water. The attenuation coefficient of upwelling irradiance, $K_{Lu}(685)$, was considered equal to the total absorption coefficient at 685 nm.

Some examples of spectra resulting from the use of this model are presented in Fig. 7.2.

7.3   RELATIONSHIPS BETWEEN THE FLUORESCENCE MAGNITUDE AND THE CONCENTRATIONS OF CHLOROPHYLL AND OTHER WATER CONSTITUENTS

7.3.1   Simplified Fluorescence Model—Theoretical Considerations

Gower et al. (2004) proposed a simplification of Eq. (7.5) for open ocean waters by assuming that the optical properties of water were solely driven by the absorption of water and phytoplankton and using an average emission "efficiency,"

$$L_f^G(685) = 0.15[\text{chl-}a]\{a_w^{ex+em}/(a_w^{ex+em} + a_{phy}^{*ex+em}[\text{chl-}a])\}. \qquad (7.22)$$

Here, $a_w^{ex+em} = a_w^{ex} + a_w^{em}$ and $a_{phy}^{*ex+em} = a_{phy}^{*ex} + a_{phy}^{*em}$; a_w^{ex} is the average absorption of water in the excitation wavelength region 400–700 nm; a_w^{em} is the average absorption of water in the emission wavelength region around 685 nm; and a_{phy}^{*ex} and a_{phy}^{*em} are the chl-a specific absorption coefficients of algae in the excitation and emission bands. Summing the absorptions in the excitation and emission bands is required since the downwelling light is attenuated by absorption in the excitation band, while the upwelling fluorescence is similarly attenuated in the emission band.

All other parameters in Eq. (7.5) are included in the proportionality coefficient (0.15 W m^{-2} sr^{-1} μm^{-1} [mg m^{-3}]$^{-1}$) in Eq. (7.22), which implicitly includes the chl-a specific absorption coefficient in the excitation band, a_{phy}^{*ex}, as well as the quantum yield. This coefficient was chosen (Gower et al., 2004) to be consistent with the relationship $L_f^G(685) = 0.15$[chl-a], which is obtained from field measurements for clear waters with low [chl-a] where the attenuation is dominated by water absorption (i.e., the factor in the curly brackets in Eq. (7.22) is approximately equal to 1). In waters with higher [chl-a], the attenuation due to phytoplankton absorption must be accounted for by multiplying 0.15[chl-a] with $a_w^{ex+em}/(a_w^{ex+em} + a_{phy}^{*ex+em}$[chl-a]). Further estimation of a_w^{ex+em} and a_{phy}^{*ex+em} (Gower et al., 2004) in the excitation and emission bands results in the following parameterization for the fluorescence magnitude suitable for open ocean waters

$$L_f^G(685) = 0.15[\text{chl-}a]/(1 + 0.2[\text{chl-}a]). \tag{7.23}$$

Here, $L_f^G(685)$ has radiance units (W m^{-2} μm^{-1} sr^{-1}), the coefficient 0.2 in the denominator has units of m^3 mg^{-1} and the coefficient 0.15 has units of radiance per [chl-a] (W m^{-2} sr^{-1} μm^{-1} [mg m^{-3}]$^{-1}$). Note that the latter was later scaled by 0.9 (Gower et al., 2004, Fig. 7.7) to fit MODIS data; the 0.15 value will be used herein.

This relationship will be used throughout as it provides a simple model, grounded in remotely sensed data, which allows comparisons among the different results presented. With increasing [chl-a] the $L_f^G(685)$ per [chl-a] decreases because of stronger light absorption in both the excitation and fluorescence emission bands. However, it should be kept in mind that, in addition to spectral effects, many aspects are neglected in this parameterization such as covariation of a_{phy}^{*ex+em} (Bricaud et al., 1995) and $a_{CDOM}(\lambda)$ with [chl-a] in open ocean waters, and the diffuse attenuation of light. In addition, the coefficients in Eq. (7.23) are based on simplified case 1 water assumptions, in particular high a_{phy}^{*ex+em}. Therefore, Eq. (7.23) will be progressively less accurate for productive waters.

Eq. (7.23) is based on experimental data; by comparing the fluorescence calculated by Hydrolight using standard case 1 water assumptions the best match with Eq. (7.23) was obtained when Hydrolight simulations were carried out with a fluorescence quantum yield of $\phi_f = 1\%$ for

$1 \geq [\text{chl-}a] \leq 15 \text{ mg m}^{-3}$ (Gilerson et al., 2007). This is consistent with observations of the quantum yield in the Baltic Sea (Ostrowska et al. 1997) and the coastal waters of the United Kingdom (Morrison, 2003), and slightly higher than those found in the equatorial Pacific and west coast of South America (Maritorena et al., 2000), where surface values were closer to 0.5%.

The simple model [Eqs. (7.22) and (7.23)] described above is not appropriate for inland and coastal waters where, in the presence of CDOM and mineral particles, attenuation is not generally dominated by phytoplankton unlike the open ocean. In this case, the attenuation of upwelling radiance and downwelling irradiance is caused by absorption of CDOM and NAP as well as by backscattering by NAP in addition to absorption by phytoplankton. Thus, with similar assumptions to those above, but including the influence of CDOM and NAP in the development of Eq. (7.22) as additional attenuation mechanisms, one can find,

$$L_f^{C2}(685) \propto [\text{chl-}a]\{a_w^{ex+em}/(a_w^{ex+em} + a_{CDOM}^{ex+em} + a_{NAP}^{ex+em} + b_{bNAP}^{ex+em} + a_{phy}^{*ex+em}[\text{chl-}a])\},$$
(7.24)

or

$$L_f^{C2}(685) \propto [\text{chl-}a]/\{1 + (a_{CDOM}^{ex+em} + a_{NAP}^{ex+em} + b_{bNAP}^{ex+em} + a_{phy}^{*ex+em}[\text{chl-}a])/a_w^{ex+em}\}.$$
(7.25)

Here the superscript C2 in $L_f^{C2}(685)$ refers to case 2 waters, $a_w^{ex+em} \approx 0.5 \text{ m}^{-1}$ [Gower et al., 2004] and a_{CDOM}^{ex+em}, a_{NAP}^{ex+em}, and b_{bNAP}^{ex+em} are respectively the sums of CDOM absorption, NAP absorption and NAP backscattering in the excitation and emission spectral regions. Since the CDOM and NAP absorption and NAP backscattering decrease rapidly with the wavelength [Eqs. (7.13), (7.14) and (7.17)], they can generally be neglected in the emission band.

Using the simulated reflectance spectra, taking into account CDOM and NAP absorption and assuming that NAP absorption and backscattering are proportional to the concentration of NAP and are thus correlated with each other, we used a least-squares fit to estimate the coefficients $X=[x_1, x_2, x_3]$ in

$$L_f^{C2}(685) = x_1[\text{chl-}a]/(1 + x_2 a_{CDOM+NAP}(443) + x_3[\text{chl-}a]) \quad (7.26)$$

where $a_{CDOM+NAP}(443)$ is the sum of CDOM and NAP absorption at 443 nm, and the term $x_2 a_{CDOM+NAP}(443)$ approximately accounts for the effect of CDOM absorption, NAP absorption, and backscattering.

We found that for the main dataset, with $0.5 < [\text{chl-}a] < 40 \text{ mg m}^{-3}$ and $\phi_f = 1\%$, the best fit was achieved with the following coefficients:

$$L_f^{C2}(685) = 0.092[\text{chl-}a]/(1 + 0.40 a_{CDOM+NAP}(443) + 0.078[\text{chl-}a]). \quad (7.27)$$

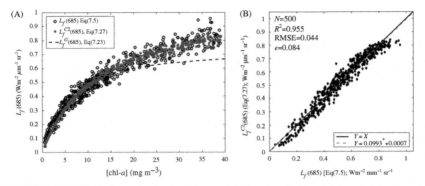

FIGURE 7.3 $L_f(685)$ determined using the empirical approach (Eq. 7.27) and full spectral model (Eq. 7.5), $\phi_f = 1\%$ and $0.5 < [\text{chl-}a] < 40$ mg m^{-3} (A) [chl-a] − $L_f(685)$ relationship, (B) regression between $L_f^{C2}(685)$ and $L_f(685)$. The coefficient of determination, R^2, the root mean square error (RMSE), and the coefficient of variation e are also provided in panel B.

In this expression, the coefficient in the numerator is reduced from 0.15 in Eq. (7.23) to 0.092, and the coefficient 0.2 in front of [chl-a] in the denominator is reduced to 0.078. This is due, in part, to the lower a_{phy}^{*ex+em} in inland and coastal waters for high [chl-a] compared with a_{phy}^{*ex+em} for open ocean waters (Bricaud et al., 1995) and the presence of CDOM and NAP. This relationship is illustrated in Fig. 7.3A for $\phi_f = 1\%$, which shows that Eq. (7.27) can be successfully used to quantify the fluorescence magnitude with a coefficient of determination of $R^2 = 0.944$ (Fig. 7.3B), where the fluorescence radiance is functionally related to the optical effects of CDOM, NAP and chl-a.

As a first approximation, the relationship (Eq. 7.27) can be used to evaluate the impact of CDOM and NAP on the fluorescence component as shown in Fig. 7.4. If $a_{CDOM+NAP}(443)$ and [chl-a] are not known in advance, the model can also be used in a statistical way to define a confidence region in the $L_f^{C2}(685)$ versus [chl-a] plane, where we can expect the bulk of the measurements to occur and where the boundaries of these regions are formed by taking appropriate limits. Depending on the attenuation by CDOM and detrital material in the water, the $L_f(685)$ value can be up to 40% lower for the same [chl-a] compared to clear waters (Fig. 7.4). At the same time, for a typical $a_{CDOM+NAP}(443) < 1$ m^{-1}, variability in $L_f^{C2}(685)$ for the fixed quantum yield is small. As described below, this can make Eq. (7.27) an important additional constraint in retrieval algorithms. It should be noted that, in addition to the absolute reduction in $L_f(685)$ due to the absorption of downwelling irradiance and upwelling radiance described in Eq. (7.27), the relative contribution of fluorescence to the reflectance further decreases with increasing [NAP] due to the increase of elastic scattering reflectance (Gilerson et al., 2007). For [NAP] > 10 g/m^3, the contribution of fluorescence to the total reflectance peak drops below 5−10%, which makes it very difficult to retrieve.

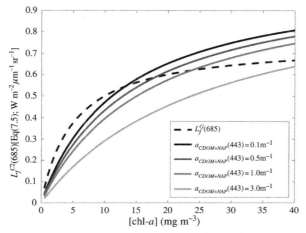

FIGURE 7.4 Impact of the combined absorption of CDOM and NAP on the fluorescence emission, according to the empirical model in Eq. (7.27) [$L_f^{C2}(685)$]. See the legend for the modeled values of $a_{CDOM+NAP}$.

7.3.2 Simplified Fluorescence Model—Comparison with Field Measurements

Results of simulations are compared with field data (Table 7.1). This data was collected in the Chesapeake Bay in 2013 by researchers from the City College of New York with collaborators (Gilerson et al., 2015), referred to as CB-CCNY, researchers from the University of South Florida (PI—C. Hu) in 2011 (Le et al., 2013), referred to as CB-USF, and from several lakes in Nebraska by researchers from the Center for Advanced Land Management Information Technologies (CALMIT) at the University of Nebraska-Lincoln in 2008 and 2009 (Gurlin et al., 2011), referred to as NE08 and NE09, respectively. The [chl-*a*] and hyperspectral $R_{rs}(\lambda)$ were measured in all field campaigns. Some absorption and, more often, scattering measurements were not fully available in the CB-USF and Nebraska datasets. For those cases, the modeled values were used as described in Section 7.2.4.

To obtain the contribution from fluorescence, the elastic fraction of the remote sensing reflectance spectrum was subtracted from the total $R_{rs}(\lambda)$. To do this, $R_{rs}^e(\lambda)$ was modeled [$R_{rs\,mod}^e(\lambda)$, sr^{-1}] for each station based on the attenuation and absorption coefficient spectra measured with a WET Labs AC-S instrument. The following relationship was used to model $u_{mod}(\lambda)$ as,

$$u_{mod}(\lambda) = \frac{\tilde{b}_{b\,mod}b_m(\lambda) + 0.5b_w(\lambda)}{a_m(\lambda) + \tilde{b}_{b\,mod}b_m(\lambda) + 0.5b_w(\lambda) + a_w(\lambda)}, \qquad (7.28)$$

TABLE 7.1 Ranges of Main Water Quality Parameters for Datasets from the Chesapeake Bay and Nebraska Lakes

	Number of Stations	[chl-a] (mg m^{-3})	[NAP] (g m^{-3})	a_{CDOM}(443) (m^{-1})	$a_{CDOM+NAP}$ (443) (m^{-1})
Chesapeake Bay, CB-CCNY, 2013	43	0.74–95.3	0.76–19.2	0.125–0.90	0.16–1.86
Chesapeake Bay, CB-USF, 2011	46	9.4–67.5	3.9–21.6	0.236–1.3	0.38–2.07
Nebraska lakes, CALMIT, 2008	89	2.26–200	0.15–5.8	0.46–1.46	0.47–1.75
Nebraska lakes, CALMIT, 2009	63	3.97–196.4	0.1–6.2	0.35–1.35	0.35–1.66

where $b_m(\lambda)$ and $a_m(\lambda)$ are the measured scattering and absorption spectra without water. As a first approximation, it was assumed that the backscattering ratio $\tilde{b}_{b\,\mathrm{mod}}$ was spectrally flat and equal to 0.01. Eqs. (7.2)–(7.4) were then applied to obtain the corresponding modeled remote sensing reflectance (usually lower than the measured spectra due to the $\tilde{b}_{b\,\mathrm{mod}} = 0.01$ assumption). They were multiplied by a constant to match the measured $R_{rs}(665)$ at 665 nm and arrive at the modeled elastic reflectance $R^e_{rs\,\mathrm{mod}}(\lambda)$. This approach works well for turbid waters where particulate backscattering is significantly greater than the backscattering of water. The fluorescence reflectance, $R^f_{rs}(\lambda)$, was determined as the difference between the measured and modeled reflectance spectra. For the field experiment data, the fluorescence radiance $L^{\mathrm{exp}}_f(685)$ was calculated by multiplying $R^f_{rs}(\lambda)$ by the measured downwelling irradiance (when available) and by $E_d(685) = 1100 \ \mathrm{W \ m}^{-2} \ \mu m^{-1}$ from the Hydrolight model spectrum when E_d measurements were not available. Despite variations in data availability, a very good fit was achieved for most of the stations especially in the green-red parts of the spectrum (Fig. 7.5). This modeling approach was used to provide the best possible spectral model for the elastic part of the reflectance $R^e_{rs\,\mathrm{mod}}(\lambda)$ in the fluorescence band, instead of using a linear baseline that may not represent the appropriate fluorescence shape, especially in high-chlorophyll waters where phytoplankton absorption causes a significant curvature.

The relationships between $L^{\mathrm{exp}}_f(685)$ and [chl-*a*] from the field measurements are compared with $L^G_f(685)$ from Eq. 7.22 and $L^{C2}_f(685)$ from Eq. 7.27 with the lowest and highest $a_{CDOM+NAP}(443)$ (Fig. 7.6) measured from the

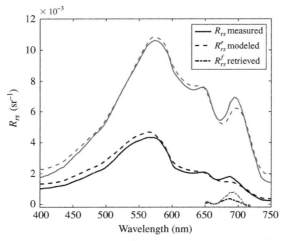

FIGURE 7.5 Examples of the retrieval of $R^f_{rs}(\lambda)$ from the CB-CCNY dataset: black lines are for [chl-*a*] = 5.54 mg m^{-3}, [NAP] = 3.2 g m^{-3} and $a_{CDOM}(443) = 0.386$ m^{-1}; gray lines are for [chl-*a*] = 22.8 mg m^{-3}, [NAP] = 9.3 g m^{-3}, and $a_{CDOM}(443) = 0.58$ m^{-1}.

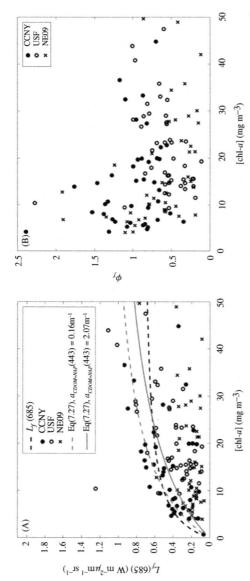

FIGURE 7.6 Fluorescence radiance and quantum yield estimated from the Chesapeake Bay and Nebraska lakes datasets compared with modeled functions: (A) $L_f^{exp}(685)$ and (B) fluorescence quantum yield.

field data (see Table 7.1). A few outliers with negative $L_f^{exp}(685)$ and quantum yields as well as high $\phi_f > 2.5$ were excluded. It can be seen that the modeled impact of CDOM and NAP absorption can only explain a small fraction of the variability in the measured $L_f^{exp}(685)$. Other sources of variability, which can include variability in the chl-*a* specific absorption coefficient or variations in the fluorescence quantum yield (estimates are shown in Fig. 7.6B), likely contributed to the observed variability. Quantum yields for all four datasets are mostly in the range of $\phi_f = 0.2-1.5\%$ with higher values found mostly in the CB-CCNY dataset and the lower values mostly in the NE08 and NE09 datasets. In the NE08 dataset, the fluorescence radiance was very small and almost irretrievable, while in the NE09 dataset the fluorescence component was noticeably lower than that in the Chesapeake Bay datasets, most likely due to a lower quantum yield. There is a generally decreasing trend in ϕ_f as [chl-*a*] increases. This trend could originate from the well-described shift toward cyanobacteria-dominated communities at higher chlorophyll concentrations in lakes (Downing et al., 2001), though data on phytoplankton community structure were not available to verify this hypothesis. This is something that should be tested with future datasets as it could provide a diagnostic tool for the presence of cyanobacteria. These results are consistent with the decrease in quantum yield with trophic status as observed in the Baltic Sea (Ostrowska, 2011). Thus, for inland and coastal waters, the high variability observed in the fluorescence magnitude for a given [chl-*a*] value appears to originate from variability in the quantum yield with a smaller impact of absorption by phytoplankton, CDOM and NAP.

7.4 RETRIEVAL OF THE FLUORESCENCE COMPONENT FROM REFLECTANCE SPECTRA

The synthetic datasets developed herein can be used to test various algorithms for the retrieval of IOPs, water parameters and the fluorescence component from reflectance spectra and examine how the number of available spectral bands affects these retrievals. In this section, some examples of retrievals are presented for hyperspectral data as well as multispectral data typical of MERIS and MODIS satellite sensors. The performance of simple algorithms like the FLH is also discussed.

7.4.1 Combined Retrieval of the Fluorescence and Water Constituents

An inversion algorithm (inspired by the work of Roesler and Perry, 1995; Culver and Perry, 1997; Roesler and Boss, 2003) was applied to the synthetic dataset for the simultaneous retrieval of the following seven parameters: (i) [chl-*a*]; (ii) the sum of CDOM and NAP absorption $a_{CDOM+NAP}(443)$; (iii) the spectral slope of $a_{CDOM+NAP}$, $S_{CDOM+NAP}$; (iv) the backscattering coefficient of particulates at 550 nm $b_b(550)$; (v) the spectral slope t of $b_b(550)$; (vi) the S_f factor which describes the chl-a specific absorption; and (vii) the quantum yield, ϕ_f. The chl-a specific absorption coefficients were assumed to be the sum of micro- and picoplankton specific absorption spectra (Ciotti et al., 2002) with S_f as a weighting factor as in Eq. (7.12). To take into account higher values of the chl-a specific absorption in the red part of the spectrum in inland and coastal waters in comparison with open ocean waters (Ficek et al., 2004; McKee et al., 2014), the picoplankton specific absorption spectrum was "stretched" from 601 to 750 nm by multiplying the original a^*_{pico} by $\alpha_1 = [1 + 0.008(\lambda - 651)]$ for $601 \leq \lambda \leq 675$ and by $\alpha_2 = [1 + 0.008(750 - \lambda)]$ for $676 \leq \lambda \leq 750$. The calculated $a^*_{pico}(675)$ of $0.036 \text{ m}^2/\text{mg}$ is 1.6 times higher than that in the original model (Ciotti et al., 2002). The fluorescence maximum is modeled using Eq. (7.27) multiplied by the quantum yield, which served as an additional constraint, assuming a Gaussian fluorescence shape as in Eq. (7.6). The parameter $u(\lambda)$, which was defined in Eq. (7.4), and its components were thus described as

$$u(\lambda) = b_b(\lambda)/[a(\lambda) + b_b(\lambda)], \tag{7.29}$$

$$a_{phy}(\lambda) = [S_f \cdot a^*_{pico}(\lambda) + (1 - S_f) \cdot a^*_{micro}(\lambda)][\text{chl-}a], \tag{7.30}$$

$$b_b = b_{bw} + b_b(550) \cdot (550/\lambda)^t, \tag{7.31}$$

$$a(\lambda) = a_w(\lambda) + a_{phy}(\lambda) + a_{CDOM+NAP}(443)\exp[-S_{CDOM+NAP}(\lambda - 443)]. \tag{7.32}$$

Other equations, which are based on Eqs. (7.2), (7.3), (7.27), (7.6), and (7.1) and which are included in the model, are repeated to help the reader:

$$R^-_{rs}(\lambda) = g_1 u(\lambda)^2 + g_2 u(\lambda), \tag{7.33}$$

$$R^e_{rs} = 0.52 \frac{R^-_{rs}}{(1 - 1.7R^-_{rs})}, \tag{7.34}$$

$$L_f(685) = \phi_f^* 0.092\,[\text{chl-}a]/(1 + 0.40 a_{CDOM+NAP}(443) + 0.078\,[\text{chl-}a]), \tag{7.35}$$

$$L_f(\lambda) = L_f(685)\exp\left(-4\ln(2)[\{\lambda - 685\}/25]^2\right), \tag{7.36}$$

$$R_{rs}(\lambda) = R^e_{rs}(\lambda) + L_f(\lambda)/E_d(\lambda). \tag{7.37}$$

It should be noted that ϕ_f from Eq. (7.35), which was retrieved during the inversion, is a multiplication factor to the functional form established by Eq. (7.27). Conveniently, because Eq. (7.27) was built for $\phi_f = 1\%$, the multiplication factor becomes equivalent to the quantum yield, expressed in percent. Eqs. (7.29)−(7.37) were applied for the retrieval of the fluorescence component and water parameters for each given $R_{rs}(\lambda)$ spectrum simulated by Eqs. (7.1)−(7.21).

Parameter estimation was carried out in MATLAB using a non-linear least-squares data fitting procedure (Levenberg, 1944; Marquardt, 1963). Results for the retrieval of [chl-*a*], $a_{CDOM+NAP}(443)$, $L_f(685)$ [obtained by multiplying $R_{rs}^f(\lambda)$ with $E_d(685)$] and ϕ_f are shown in Fig. 7.7 for $R_{rs}(\lambda)$ in the spectral range of 400−800 nm and with a 5-nm spectral resolution (as is planned for the sensor in NASA's Plankton, Aerosols, Clouds, ocean Ecosystem mission (PACE, 2012). These results together with retrieval results for MERIS (412, 442, 490, 510, 560, 620, 665, 681, 709, and 754 nm) and MODIS (412, 442, 466, 488, 530, 547, 554, 645, 666, 678, and 747 nm) band sets are given in Table 7.2.

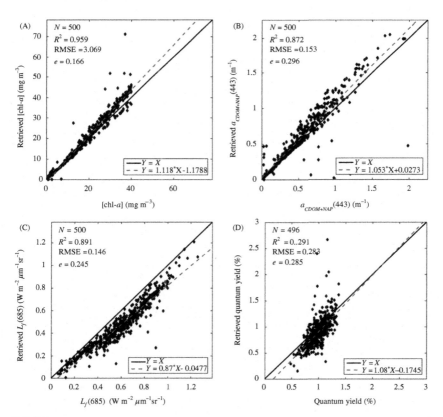

FIGURE 7.7 Comparison between inputs and retrieved values for the synthetic reflectance spectra in the 400−800 nm range with 5-nm steps and inversion for (A) [chl-*a*], (B) $a_{CDOM+NAP}(443)$, (C) $L_f(685)$, and (D) ϕ_f. Points with quantum yields above 3% ($N = 2$) were removed from the statistics.

TABLE 7.2 Retrieval Statistics for [chl-*a*], $a_{CDOM+NAP}$(443),L_f(685), and ϕ_f from the Simulated Reflectance Spectra in the 400–800 nm Range at 5-nm Spectral Resolution and at the Band Locations of MERIS and MODIS

	400–800 nm by 5 nm R^2 (e)	MERIS Bands R^2 (e)	MODIS Bands R^2 (e)
[chl-*a*]	0.959 (0.166)	0.887 (0.527)	0.894 (0.406)
L_f(685)	0.915 (0.23)	0.884 (0.291)	0.683 (0.358)
$a_{CDOM+NAP}$(443)	0.872 (0.296)	0.885 (0.509)	0.155 (1.03)
ϕ_f	0.231 (0.338)	0.169 (0.387)	0.117 (0.473)

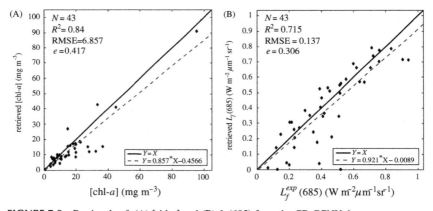

FIGURE 7.8 Retrievals of: (A) [chl-*a*] and (B) L_f(685) from the CB-CCNY dataset.

The quality of the retrieval degrades with the number of bands used. L_f(685) was also always underestimated by ~25% in these retrievals. This is most likely due to the simple backscattering model used in Eq. (7.31) that does not resolve spectral features of backscattering, especially in the red-NIR region, which are more properly accounted for in the more detailed model for the backscattering coefficient in Eqs. (7.16)–(7.21). Retrievals of the fluorescence quantum yield ϕ_f had some outliers with very high retrieved ϕ_f values, but, generally, they were in the correct range and consistent with inputs. It should be noted that similar retrieval procedures for all band sets, which were performed without the constraint of Eq. (7.27), resulted in a much lower quality of retrieval for all parameters.

Similar retrieval tests were conducted on the CB-CCNY dataset (Fig. 7.8) for the MERIS spectral bands, which produced results similar to the ones from the synthetic dataset also for MERIS bands as shown in Table 7.2.

TABLE 7.3 Retrieval Statistics for [chl-*a*], $a_{CDOM+NAP}(443)$, and $L_f(685)$ from the CB-CCNY Dataset for Reflectance in the 400−700 nm Range for a 5-nm Spectral Resolution and at the Band Locations of MERIS and MODIS

	400−700 nm by 5 nm R^2 (e)	MERIS Bands R^2 (e)	MODIS Bands R^2 (e)
[chl-*a*]	0.83 (0.44)	0.84 (0.42)	0.49 (0.96)
$L_f(685)$	0.91 (0.30)	0.72 (0.31)	0.67 (0.35)
$a_{CDOM+NAP}(443)$	0.51 (0.24)	0.55 (0.36)	0.70 (0.40)

More detailed statistics (Table 7.3) for the [chl-*a*], $L_f(685)$ and $a_{CDOM+NAP}(443)$ retrievals for the three spectral configurations (PACE, MERIS and MODIS) show the moderate quality of the retrievals. Here, $L_f^{exp}(685)$ is the fluorescence magnitude estimated from the field data as described in Section 7.3.2. These results also highlight that the results of the inversion procedure are highly dependent on the choice of wavebands used during the inversion, often in ways that are not expected. For example, the 5-nm resolution set produced better retrievals for $L_f(685)$ and [chl-*a*] but worse retrievals for $a_{CDOM+NAP}(443)$ than the MODIS bands. More advanced approaches can be considered which can include, for example, a separate retrieval of [chl-*a*] using red-NIR bands (Gitelson, 1992; Gilerson et al., 2010; Gurlin et al., 2011).

7.4.2 Fluorescence Line Height Algorithms and Their Limitations

The operational FLH algorithms (Letelier and Abbott, 1996; Gower et al., 1999) for MODIS and MERIS satellite sensors are based on measurements of reflectance in three bands. Two "outer" bands are meant to estimate the baseline or elastic reflectance (mostly outside the fluorescence spectrum), while the middle band determines the height. The baseline approach is graphically shown in Fig. 7.9A. FLH is defined (Letelier and Abbott, 1996; Gower, et al., 2004) as

$$FLH = L_2 - [L_3 + (L_1 - L_3)(\lambda_3 - \lambda_2)/(\lambda_3 - \lambda_1)], \qquad (7.38)$$

$$FLH = L_2 - kL_1 - (1 - k)L_3, \qquad (7.39)$$

where L_1, L_2, and L_3 are radiances at appropriate spectral bands centered at wavelengths λ_1, λ_2 and λ_3 and $k = (\lambda_3 - \lambda_2)/(\lambda_3 - \lambda_1)$. On current satellite sensors, these bands have bandwidths (FWHM) between 7 and 10 nm. The central band is chosen at a location shorter than 685 nm to avoid oxygen absorption at 687 nm (Letelier and Abbott, 1996).

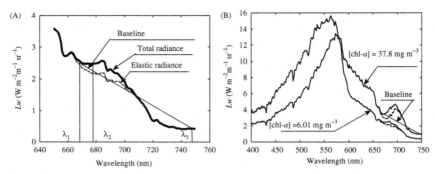

FIGURE 7.9 FLH approach to separate the fluorescence contribution from the upwelling radiance. (A) Fluorescence height over baseline and (B) overlapping of fluorescence and elastic radiance peaks in red-NIR for [chl-a] = 6.01 mg m^{-3}, where elastic radiance is close to the baseline and [chl-a] = 37.8 mg m^{-3} with a significant difference between elastic radiance and the baseline.

The FLH approach is appropriate for fluorescence retrieval in the open ocean (e.g., Huot et al., 2005), where atmospheric correction algorithms work well and the elastic reflectance in the fluorescence band is well approximated by the baseline curve due to the relatively flat and low elastic scattering signal. However, while recent advances in atmospheric corrections suitable for coastal waters (IOCCG, Report 10, 2010, Chapter 3) allow better separation of the water-leaving radiance in the red-NIR spectral region, application of the FLH algorithms in inland and coastal waters with higher [chl-a] is significantly hindered by a peak in the underlying elastic reflectance, which spectrally overlaps and contaminates any fluorescence retrieval (see, for example, Fig. 7.9B).

This reflectance peak occurs at a wavelength beyond the chl-a absorption feature at 675 nm and is caused by the combined effect of decreasing absorption by chl-a, increasing absorption by water, and scattering by other constituents in water (Gitelson, 1992; Schalles, 2006; and references therein). With increasing [chl-a], the peak in total $R_{rs}(\lambda)$ or radiance shifts from the fluorescence maximum at 685 nm to longer wavelengths and can reach 705 nm at [chl-a] of about 100 mg m^{-3} (Schalles, et al., 1998; Zimba and Gitelson, 2006). The impact on the fluorescence retrieval using FLH can be easily seen by imagining a baseline that connects R_{rs} or radiance in two bands outside the fluorescence bands (e.g., 665 and 750 nm) as shown in Fig. 7.5, see also Fig. 7.9B for the radiance. In Fig. 7.5 (black lines, lower [chl-a]), this baseline is quite close to the elastic reflectance (or radiance); however, it is very different from the elastic reflectance for high [chl-a] (gray lines), see also Fig. 7.9B for the radiance, which can result in significant biases in the retrieval of L_f values (at 678 nm on MODIS and 681 nm on MERIS) using the FLH algorithms. Indeed, in some cases, the depression due to chl-a absorption overwhelms the fluorescence emission in satellite data as observed by several authors (Gons et al., 2008; Binding et al., 2010; Palmer et al., 2015) and negative FLH can be obtained.

To improve the extraction of the fluorescence component of reflectance for inland and coastal waters, it is clear that suitable models that take into account the spectral variation of the underlying elastic reflectance peak must be developed. These can then be coupled with studies on the new optimized positioning of spectral channels to improve retrievals and reduce errors that cannot be totally compensated for by modeling alone.

Synthetic data, field data and satellite imagery were used to analyze the performance of FLH algorithms based on MODIS and MERIS bands. Three bands on each of these sensors are used to determine the FLH. Since fluorescence is not equal to zero at λ_1, only part of the fluorescence can be captured by the FLH approach, even in an ideal case and related measurements should be scaled appropriately (see Abbott and Letelier, 1999). The bands used are centered at 666, 678, and 747 nm for MODIS and 665, 681, and 709 nm for MERIS (Gower, et al., 2004). An additional MERIS band exists at 754 nm and the combination of bands centered at 665, 681, and 754 nm is also considered here and referred to as the MERIS2 configuration.

In case 1 waters, the FLH algorithms work reasonably well because the elastic radiance for low [chl-*a*] < 10 mg/m³ is a close approximation to the baseline between λ_1 and λ_3 (Gower et al., 2004; Huot et al., 2005). For [chl-*a*] = 6.01 mg m^{-3} (Fig. 7.9A and B), the baseline almost matches the elastic radiance. However, for [chl-*a*] = 37.8 mg/m³, the peak in the elastic radiance described earlier is also observed within the same wavelength range as the fluorescence emission and, therefore, overlaps with the fluorescence spectrum (Gilerson et al., 2008; Gower et al., 1999). The elastic spectrum clearly deviates more from a simple baseline, which affects the FLH retrieval accuracy.

Using theoretical band profiles, Gower et al. (2004) estimated that the relative signals due to fluorescence are 0.20 for the 666 nm band and 0.74 for the central band (676.8 nm was used as the central band location for these calculations). That is, the 676.8 nm band detects only 74% of the peak radiance of the fluorescence signal. The reference band at 666 nm, which would ideally be outside the fluorescence range, in fact detects 20% of the peak signal. The position of this band was chosen with the goal of avoiding the influence of absorption bands and minimizing the spectral interval between the fluorescence peak and the reference band location. This also minimizes the impact of variations between the bands due to atmospheric effects on the radiance. MODIS has a FLH reduction factor 0.57, which means that it responds only to 57% of the fluorescence signal. The FLH reduction factor for MERIS is 0.78 (Gower et al., 2004). The spectral locations of the red and NIR bands of MODIS and MERIS together with the fluorescence and atmospheric absorption spectra are shown in Fig. 7.10.

The combination of such a reduction of the fluorescence signal and the deviation of the elastic radiance from the straight line creates significant

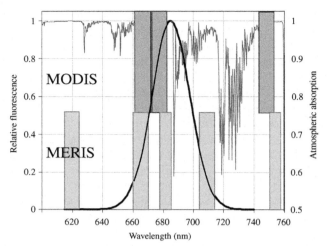

FIGURE 7.10 Theoretical phytoplankton fluorescence, showing positions and widths of MODIS and MERIS bands and atmospheric absorption features (from Gower et al., 2004). *Used with permission from the Canadian Aeronautics and Space Institute.*

difficulties in the interpretation of FLH signals at high [chl-*a*] values. FLH-based retrievals of fluorescence from the synthetic data are shown in Fig. 7.11A and B for the MODIS bands, Fig. 7.11C and D for the MERIS bands centered at 665, 681 and 709 nm, and Fig. 7.11E and F for the MERIS bands centered at 665, 681, and 754 nm.

The FLH is correlated with [chl-*a*] up to $5-7$ mg m^{-3}. At higher [chl-*a*], correlations are very low and values are much more dispersed than for the modeled $L_f(685)$. In Fig. 7.11C, for the MERIS band configuration, FLH values quickly become negative. Attempts have been made to use this relationship where FLH decreases with [chl-*a*] (Palmer et al., 2015); however, such algorithms would be applicable to very special regional conditions. The best performance is achieved for the 665, 681, and 754 nm band combination (MERIS2), where the band centered at 681 nm is positioned closer to the fluorescence maximum at 685 nm than in the MODIS algorithm. The significantly lower performance of the standard MERIS FLH algorithm in comparison with MERIS2 is probably due to the proximity of the 709 nm band to the 665 nm band which makes the baseline a less accurate replacement of the elastic radiance in comparison with other band combinations. This is clearly an additional challenge associated with high [chl-*a*] in inland or coastal waters that is rarely, if ever, encountered in offshore waters. For waters where optical variability is dominated by CDOM or NAP, there is a significant change in the patterns of the relationships between FLH and [chl-*a*] as shown in Fig. 7.12 for MODIS and MERIS algorithms (7.12 A, B for CDOM and 7.12 C, D for NAP).

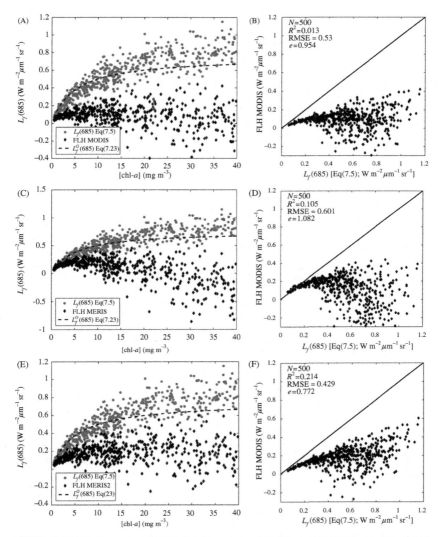

FIGURE 7.11 Performance of FLH algorithms using MODIS and MERIS bands: (A and B) MODIS; (C and D) MERIS using 665, 681, and 709 nm; and (E and F) MERIS using 665, 681, and 754 nm.

Retrievals of FLH from the CB-CCNY dataset are shown in Fig. 7.13A−C for the MODIS band combination and the two MERIS band combinations and from the NE09 dataset in Fig. 7.15D for the MODIS band combination.

The trends are similar to those observed for the synthetic dataset: a change of slope around $L_f(685)=0.35−0.4$ W m^{-2} μm^{-1} sr^{-1}; better retrievals from

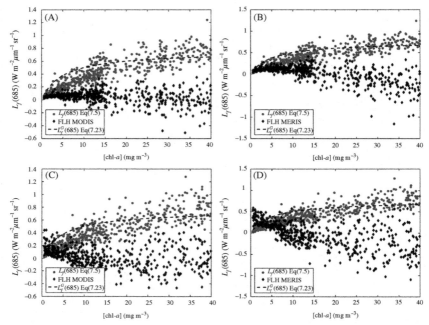

FIGURE 7.12 Performance of FLH algorithms based on MODIS and MERIS bands for waters with high optical variability arising from NAP and CDOM absorption: (A) $0 \leq a_{CDOM}(443) \leq 2.5$ m^{-1}, MODIS; (B) $0 \leq a_{CDOM}(443) \leq 2.5$ m^{-1} MERIS; (C) $0 < [NAP] < 40$ g m^{-3}, MODIS; and (D) $0 < [NAP] < 40$ g m^{-3}, MERIS.

the MERIS2 band combination; and a very poor retrieval for the main MERIS band combination. Nevertheless, all three FLH algorithms performed better on the field data than on the synthetic data, capturing the initial portion of the fluorescence magnitude without the need for any corrections. This can be partially due to the more complex absorption and reflectance spectra in the field data than in the model caused by the presence of additional pigments that were not considered in the model or because the quantum yield was higher than in the model and therefore there was less spectral curvature in the baseline for the same $L_f(685)$. Notably, the Chesapeake Bay reflectance spectra shown in Fig. 7.5 have pronounced troughs around 620 nm, which indicate the presence of phycocyanin typical of cyanobacteria (Chapter 8) with its fluorescence features possibly affecting radiance near 665 nm.

7.4.3 Performance of Fluorescence Algorithms with Satellite Data

The standard MODIS FLH product in the NASA Ocean Color Level 2 processing is the normalized fluorescence line height, *nflh*, which includes a

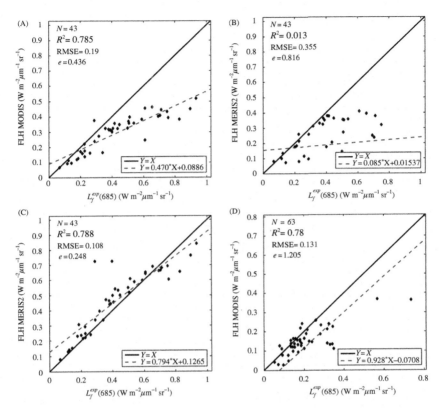

FIGURE 7.13 Performance of FLH algorithms for the CB-CCNY dataset using: (A) MODIS, (B) MERIS, and (C) MERIS2 band combinations, and for NE09 dataset using (D) MODIS bands.

correction for bidirectional effects (Behrenfeld et al., 2009). Fig. 7.14A shows a plot of *nflh* versus [chl-*a*] retrieved through the standard processing for a MODIS image acquired over Chesapeake Bay on September 15, 2015. All pixels with $R_{rs}(412) < 0$ were excluded which reduced the number of points, especially in the Upper and Middle Bay, but also eliminated retrievals where the atmospheric correction certainly failed. Comparison of *nflh* shown in Fig. 7.14A with the FLH retrievals shown in Fig. 7.11A indicates higher variability in *nflh* than observed in the synthetic data. However, the main trends are not dissimilar, with a very pronounced change in direction near [chl-*a*]=5 mg m^{-3}. It should be noted that while some patches of high [chl-*a*] associated with algal blooms exist in the bay in the summer and fall, considering the 1-km spatial resolution of MODIS, it is unrealistic to expect many pixels with [chl-*a*] > 15 mg m^{-3} (Magnuson et al., 2004; Werdell et al., 2009). Therefore, [chl-*a*] values higher than 15 mg m^{-3} in Fig. 7.14A are most likely the result of overestimation of [chl-*a*] by the standard MODIS

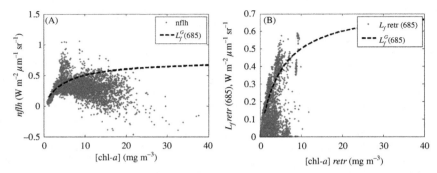

FIGURE 7.14 Retrieval of the fluorescence component from the MODIS image of Chesapeake Bay acquired on September 15, 2015: (A) [chl-*a*] and *nflh* satellite products and (B) retrieval using Section 7.4.1 approach.

algorithm (which could, in part, be due to residual effects of the atmospheric correction and high CDOM concentration). Following the reasonable retrievals of both [chl-*a*] and $L_f(685)$ for MODIS bands in Section 7.4.1, we applied the same algorithm to the retrieval of these parameters from MODIS reflectance spectra for each pixel. Results of these retrievals, denoted as [chl-*a*] *retr* and $L_f(685)$ *retr*, are shown in Fig. 7.14B demonstrating a significant change in [chl-*a*] and $L_f(685)$ in comparison with Fig. 7.14A with both parameters staying within the expected ranges: [chl-*a*] *retr* < 15 mg m^{-3} and $0 < L_f(685)$ *retr* < 0.6 Wm^{-2} μm^{-1} sr^{-1}. The [chl-*a*] *retr* vs. $L_f(685)retr$ relationship is more akin to what was observed with field data as shown in Fig. 7.6A. Maps of MODIS [chl-*a*] from the standard OC3M algorithm, [chl-*a*] *retr*, MODIS *nflh*, $L_f(685)retr$, and the quantum yield ϕ_f *retr* (all parameters with "*retr*" were retrieved with Section 7.4.1 methodology) are shown in Fig. 7.15, illustrating very noticeable differences between the approaches. The retrieved quantum yield ϕ_f *retr* was found to be within the expected range of 0%−1.5%.

A MERIS image from September 15, 2011 was used to test the performance of the fluorescence algorithms for Chesapeake Bay. FLH MERIS (665, 681, and 709 nm bands) and FLH MERIS2 (665, 681, and 754 nm bands) together with the OC4E [chl-*a*] algorithm produced the patterns shown in Fig. 7.16A and B, which are different from those observed with the synthetic data as discussed in the previous Section 7.4.2. It must be noted that the retrieved [chl-*a*] could be biased in these coastal waters. Application of the approach described in Section 7.4.1 to MERIS data produced the [chl-*a*] *retr*-$L_f(685)$ *retr* pattern, shown in Fig. 7.16C, which is similar to the one for MODIS data (Fig. 7.14A) for the same time of the year, with the relationship primarily following the $L_f^G(685)$ curve. Clearly, the results from the approach described in Section 7.4.1 are different from the standard OC and FLH retrievals.

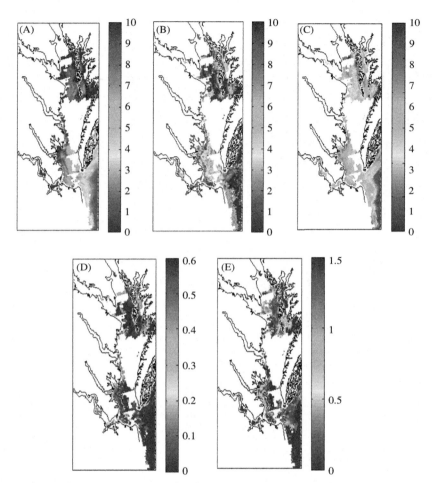

FIGURE 7.15 Maps of parameters retrieved from the satellite image of Chesapeake Bay on September 15, 2015: (A) MODIS [chl-*a*] (mg m^{-3}) from the OC3 algorithm; (B) [chl-*a*] *retr* (mg m^{-3}) using the approach described in Section 7.4.1; (C) MODIS *nflh* (W m^{-2} μm^{-1} sr^{-1}); (D) $L_f(685)$ *retr*, (W m^{-2} μm^{-1} sr^{-1}) using Section 7.4.1 approach; and (E) ϕ_f *retr* (%) using Section 7.4.1 approach.

Applying the same algorithms to MODIS imagery of three of the Great Lakes (Lake Superior, Lake Michigan and Lake Huron) resulted in a complex [chl-*a*]-*nflh* pattern (Fig. 7.17A, while the relationship between [chl-*a*] *retr* and $L_f(685)$ *retr* followed the $L_f^G(685)$ curve with some deviations (Fig. 7.17B). This confirmed, once more, the potential of the approach described in Section 7.4.1 for the retrieval of water quality parameters and the fluorescence magnitude, $L_f(685)$. The current section is not meant to suggest that the approach developed in Section 7.4.1 is universally

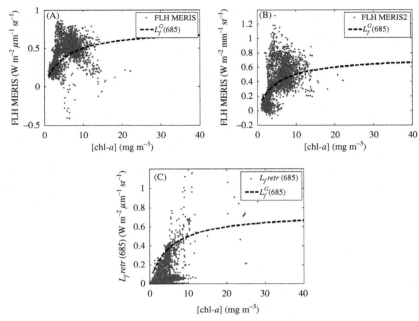

FIGURE 7.16 Retrieval of the fluorescence component from the MERIS image of Chesapeake Bay acquired on September 15, 2011: (A) [chl-a] and FLH MERIS, (B) [chl-a] and FLH MERIS2, and (C) retrieval of [chl-a] and $L_f(685)$ *retr* using the approach described in Section 7.4.1.

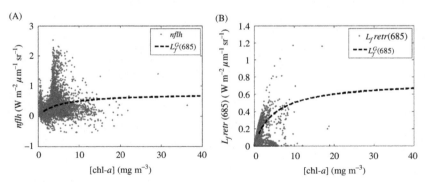

FIGURE 7.17 Retrieval of the fluorescence component from the MODIS image of the Great Lakes acquired on September 30, 2015: (A) [chl-a] and *nflh* satellite products and (B) retrieval using Section 7.4.1 approach.

applicable, as it is certainly not validated yet. However, the results shown here, together with recent advances in the atmospheric correction for inland and coastal waters (see Chapter 3), suggest that more advanced approaches to separate fluorescence from the elastic scattering contribution have the potential to make fluorescence a useful product even in waters with high phytoplankton absorption. Realizing the full potential of such approaches may, however, require satellites capable of collecting radiance at a larger number of bands, which will help to better constrain the spectral inversion for fluorescence retrieval.

The *nflh* algorithm, based on the water-leaving radiance just above the sea-surface measured at three bands, is influenced by the quality of the atmospheric correction (Chapter 3). Alternatively, a similar algorithm can be applied to radiance at the top of the atmosphere (TOA, level 1 data) (Gower et al., 2004). In this case, an offset can be expected between these two approaches, depending on the atmospheric correction quality, aerosol loading and impact of the linear baseline (instead of elastic radiance on the underlying spectra at the surface level and at TOA). In addition, absorption by the atmosphere is about 10% at 685 nm (Gower et al., 2004). The *flh* calculated in accordance with Eq. (7.38) at the surface level and at TOA are compared in Fig. 7.18 showing an offset of about $0.2 \, \mathrm{Wm}^{-2} \, \mu\mathrm{m}^{-1} \, \mathrm{sr}^{-1}$. While it is unlikely that either of them is an appropriate measure of fluorescence emission, due to the baseline issue highlighted above, they provide very similar results and are affected in similar ways by phytoplankton absorption in water. The TOA *flh* was indeed successfully used with some corrections for the analysis of chlorophyll fluorescence, algal blooms and matchups with in situ [chl-*a*] (Blondeau-Patissier et al., 2014). There were no noticeable differences between the *nflh* product and *flh* calculated using Eq. (7.38) for this image.

The last sections have examined the retrieval of the fluorescence signal from the magnitude of the emission. Another approach that uses the polarimetric nature of the light field to discriminate the fluorescence signal will now be described.

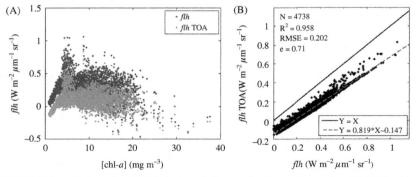

FIGURE 7.18 *flh* from the MODIS image of Chesapeake Bay on September 15, 2015 calculated from near surface and TOA radiances: (A) as a function of [chl-*a*] and (B) correlation between two *flh*.

7.4.4 Retrieval of the Fluorescence Component from Polarimetric Hyperspectral Observations

When entering the Earth's atmosphere, solar radiation is initially unpolarized. Sunlight is then scattered by aerosols and atmospheric molecules, refracted and reflected at the atmosphere−ocean interface, and further scattered by hydrosols and water molecules. As a result of these interactions, the solar radiation becomes partially polarized and the resulting polarization field depends on the optical properties of the atmosphere-ocean system (AOS). Thus, the polarization state of light carries information about the AOS that can be utilized in remote sensing applications (Hansen and Travis, 1974; Chami et al., 2001).

Hyperspectral polarimetric measurements of reflectance spectra below and above the water surface are becoming increasingly available (Tonizzo et al., 2009; Gilerson et al., 2013; Foster et al., 2015) and can provide important information about macro- and microphysical properties of particulates in water. In addition, these spectra can be used for retrieving the fluorescence component, especially in inland and coastal waters with [chl-a] $>5-7$ mg m^{-3} where, as shown earlier, FLH algorithms are inaccurate. Such retrievals are based on the isotropic property of chlorophyll fluorescence (Ahmed et al., 2004; Gilerson et al., 2006), which means that the contribution of the fluorescence component to the polarized reflectances is the same for measurements taken with different orientations of the polarizer. This contribution is equal to $0.5 L_f(\lambda) / E_d$.

Thus, if a polarizer is placed in front of a probe used to measure the radiance of scattered light (or remote sensing reflectance, if normalized by the downwelling irradiance) then rotating the polarizer will cause the reflectance to vary from a minimum, $R_{rs\min}(\lambda)$, when the radiance polarization is most perpendicular to the axis of the polarizer with maximum transmission, to a maximum, $R_{rs\max}(\lambda)$, when the radiance polarization is most parallel to this axis of the polarizer. These reflectances can be decomposed into signal components, $R_{rs\,\parallel}^{e}(\lambda)$ and $R_{rs\,\perp}^{e}(\lambda)$, that are polarized parallel and perpendicular to the scattering plane, respectively, plus half of the $L_f(\lambda) / E_d$ caused by the chlorophyll fluorescence feature centered at 685 nm with a FWHM of 25 nm. That is,

$$R_{rs\max}(\lambda) = R_{rs\perp}^{e}(\lambda) + 0.5 L_f(\lambda)/E_d, \tag{7.40}$$

$$R_{rs\min}(\lambda) = R_{rs\,\parallel}^{e}(\lambda) + 0.5 L_f(\lambda)/E_d. \tag{7.41}$$

The total signal $R_{rs}(\lambda)$ is given by

$$R_{rs}(\lambda) = R_{rs\,\max}(\lambda) + R_{rs\,\min}(\lambda) = R_{rs\,\perp}^{e}(\lambda) + R_{rs\,\parallel}^{e}(\lambda) + L_f(\lambda)/E_d, \tag{7.42}$$

while the elastically scattered polarized component is expressed as

$$R_{rs}^{e}(\lambda) = R_{rs\,\perp}^{e}(\lambda) + R_{rs\,\parallel}^{e}(\lambda). \tag{7.43}$$

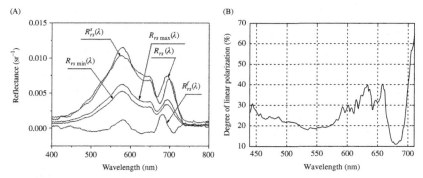

FIGURE 7.19 Retrieval of the fluorescence component using polarized reflectance: (A) measured and processed spectra for field measurements collected from the Chesapeake Bay in July 2005 ([chl-*a*]=41 mg m^{-3}). (B) Spectrum of *DoP* processed from above-water measurements taken on July 31, 2014, near the Chesapeake Bay COVE AERONET station (Foster et al., 2015).

It was shown (Ahmed et al., 2004; Gilerson et al., 2006) that this elastic spectrum is well correlated with the difference signal, $R_{rs}^e{}_d(\lambda)$, given by

$$R_{rs}^e{}_d(\lambda) = R_{rs\,max}(\lambda) - R_{rs\,min}(\lambda) = R_{rs\,\perp}^e(\lambda) - R_{rs\,\|}^e(\lambda). \qquad (7.44)$$

Using this assumption of a correlation, a linear regression can be used to fit $R_{rs}^e{}_d(\lambda)$ into $R_{rs}(\lambda)$ in the 450−660 nm and 710−750 nm ranges (i.e., outside the fluorescence spectral region). The regression parameters are then used to extract $R_{rs}^e(\lambda)$ from $R_{rs}^e{}_d(\lambda)$ in the wavelength range of 660−710 nm (fluorescence spectral region), after which $R_{rs}^f(\lambda)$ can be retrieved using $R_{rs}^f(\lambda) = [R_{rs}(\lambda) - R_{rs}^e(\lambda)]$. An example of such a retrieval for spectra measured in the Chesapeake Bay in July 2005 is shown in Fig. 7.19A. The spectra were collected by the GER spectroradiometer (SpectraVista) with the fiber probe and the polarizer in front of the probe. The solar zenith angle was 20° and the probe was positioned off-nadir at 45° just below the water surface. Radiances were measured just below the water surface and propagated to just above the surface to obtain remote sensing reflectance using a factor of 0.54.

The spectral degree of polarization *DoP*(λ) (assuming negligible circular polarization) can be calculated as

$$DoP(\lambda) = R_{rs}^e{}_d(\lambda)/R_{rs}(\lambda). \qquad (7.45)$$

DoP(λ) depends strongly on the constituents in water and spectrally follows the trend of the absorption-to-attenuation ratio (a/c) (Timofeeva, 1970; Ibrahim et al., 2012) because higher absorption is associated with a smaller number of scattering events and correspondingly higher polarization of light, and vice versa. This results in a significant increase in *DoP*(λ) for wavelengths higher than 600 nm due to the increase in pure water absorption. Since the fluoresced light is unpolarized, it has a strong impact on *DoP*(λ), causing a drop in $DoP(\lambda)$ in the fluorescence spectral region as can be seen in Fig. 7.19B. It can be shown (Tonizzo et al., 2010) that this effect can be related to the fluorescence component in the reflectance spectrum:

$$L_f(\lambda)/E_d(\lambda) = [DoP_{elastic}(\lambda) - DoP(\lambda)]\frac{R_{rs}(\lambda)}{DoP_{elastic}(\lambda)}. \tag{7.46}$$

7.4.5 Application of SICF to the Detection of Algal Blooms

Algal blooms or harmful algal blooms (HABs) are observed more and more frequently in inland and coastal waters because of the increased anthropogenic effects and enhanced monitoring capabilities in situ and from space. These blooms usually manifest higher [chl-a] in comparison with typical water conditions in the area, and the fluorescence signal can be used as an additional proxy for this [chl-a] in productive waters when standard [chl-a] retrievals are unreliable. For some types of algae, the application of the fluorescence emission is less straightforward and is discussed in this section. The examples are related to HABs of *Karenia brevis*, which are frequent along the Western Florida Shelf (WFS). Blooms of these algae become toxic at [chl-a] even below 1 mg m^{-3}, which complicates the detection. Together with several other approaches, two fluorescence techniques proved to be successful (Hu et al., 2011). These techniques exploit the fact that these algae have particle sizes of $20-40 \text{ }\mu\text{m}$ (Cannizzaro et al., 2009), which are larger than most typical phytoplankton particles. This results in lower backscattering properties for the particulates, which correspond to lower R_{rs}^e and thus a more pronounced fluorescence signal for the same [chl-a] than for other chlorophyllous particles in water. The MODIS FLH algorithm (Hu et al., 2005) was efficiently used in a very challenging nearshore WFS environment to detect *K. brevis* blooms. Another technique, the so-called Red Band Difference, RBD (Amin et al., 2009), which is also often used in *K. brevis* detection, is based on an unusual combination of only two MODIS bands in the fluorescence spectral region:

$$RBD = nLw(678) - nLw(667). \tag{7.47}$$

Examples of two blooms monitored using this technique are shown in Fig. 7.20.

The shift of the NIR peak in R_{rs} (combined effect of the elastic reflectance and fluorescence emission) to longer wavelengths with increasing [chl-a], which was discussed in Section 7.4.2, makes the 709 nm band quite sensitive to [chl-a]. This effect was used in the development of the Maximum Chlorophyll Index product for the MERIS sensor (Blondeau-Patissier et al., 2014) primarily for the detection of blooms with high [chl-a] $> 30 \text{ mg m}^{-3}$ and was successfully used for the global monitoring of phytoplankton blooms in the world's oceans (Gower and King, 2011).

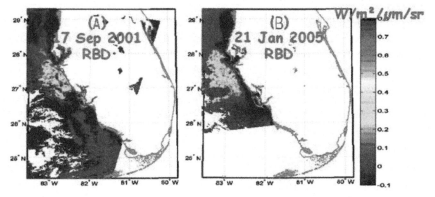

FIGURE 7.20   *K. brevis* blooms detected using the RBD technique on the West Florida Shelf on (A) September 17, 2001 and (B) Jan 21, 2005 (Amin et al., 2009).

7.5 SUMMARY

Bio-optical models for the main water optical properties—absorption, scattering, and backscattering coefficients—used to simulate remote sensing reflectance spectra combined with Eq. (7.5) can be successfully applied to estimate the magnitude of SICF in inland and coastal waters. In typical sunny conditions, the effective fluorescence quantum yield is close to 1%, with most values in the range of 0%−1.5%. Based on the analysis of the dependence of the fluorescence magnitude on the optical properties of water using synthetic datasets, an empirical relationship Eq. (7.27) has been developed, which can be used to estimate fluorescence $L_f^{C2}(685)$ as a function of [chl-*a*] and $a_{CDOM+NAP}(443)$ for various water conditions. This relationship is close to $L_f^{G}(685)$ (Gower et al., 2004) determined for relatively clear waters, with usually smaller values for $L_f^{C2}(685)$ below [chl-*a*]=15 mg m^{-3} and higher values for [chl-*a*] > 15 mg m^{-3}. For most conditions in inland and coastal waters with $a_{CDOM+NAP}(443) < 1$ m^{-1}, this relationship varies by no more than 20% for a given [chl-*a*] value and thus can be used as an additional constraint in retrieval models. In the case of NAP-dominated waters, the relative fluorescence contribution becomes smaller as $R_{rs}(\lambda)$ becomes higher, and this can lead to difficulty in detecting this contribution. In inland or coastal waters with high [chl-*a*], the performance of MODIS and MERIS FLH algorithms is diminished by the overlap with the chl-*a* fluorescence with an elastic reflectance peak in the same spectral region, which is due to the combined absorption features of phytoplankton and water. This overlap results in the deviation of the FLH baseline from the elastic reflectance and leads to inaccurate retrievals. This also makes it difficult to relate $L_f(685)$ determined from the FLH algorithms to [chl-*a*] in inland and coastal waters, taking into account significant uncertainties in [chl-*a*] retrievals using standard ocean color algorithms. One of the possible approaches for the retrieval of the fluorescence component together with other water parameters is based on models similar to that developed by Roesler and Perry (1995), together with Eq. (7.27) as an additional relationship for the fluorescence component and its connection to other water parameters. Applying such an approach to synthetic, field and satellite data from different inland and coastal water bodies, covering a wide range of bio-optical properties and a number of geographic locations, has shown interesting potential for retrieving SICF from remotely sensed data.

ACKNOWLEDGMENTS

We would like to thank Anatoly Gitelson and CALMIT (University of Nebraska-Lincoln) for the NE08 and NE09 datasets, the group led by Chuanmin Hu at the Optical Oceanography Laboratory (University of South Florida) for the CB-USF dataset and NASA OBPG for the satellite imagery, the anonymous reviewers for their insightful remarks and suggestions and Catherine A. Brown from the University of Sherbrook, Canada for editorial efforts A. Gilerson is grateful for the valuable collaboration with Sam Ahmed, Fred Moshary, and Barry Gross, at the Optical Remote Sensing Laboratory, Department of Electrical Engineering, CCNY; for the help from the researchers and students in the laboratory who participated in the field measurements and data processing (Ioannis Ioannou, Robert Foster, Carlos Carrizo, Ahmed El-Habashi, Anna McGilloway); and for the assistance from Maria Tzortziou, CCNY, and Michael Ondrusek, NOAA, NESDIS, who co-organized the Chesapeake Bay campaign in 2013. A. Gilerson's contribution was partially supported by NOAA CREST. Y. Huot's contribution was funded by the Canada Research Chair and NSERC Discovery Grants programs.

REFERENCES

Abbott, M.R., & Letelier, R.M., 1999. Algorithm theoretical basis document chlorophyll fluorescence (MODIS product number 20).

Albert, A., Mobley, C.D., 2003. An analytical model for subsurface irradiance and remote sensing reflectance in deep and shallow case-2 waters. Opt. Expr. 11 (22), 2873–2890.

Ahmed, S., Gilerson, A., Gill, A., Gross, B.M., Moshary, F., Zhou, J., 2004. Separation of fluorescence and elastic scattering from algae in seawater using polarization discrimination. Opt. Commun. 235, 23–30.

Amin, R., Zhou, J., Gilerson, A., Gross, B., Moshary, F., Ahmed, S., 2009. Novel optical techniques for detecting and classifying toxic dinoflagellate *Karenia brevis* blooms using satellite imagery. Optics Express 17, 9126–9144.

Babin, M., Morel, A., Gentili, B., 1996. Remote sensing of sea surface sun-induced chlorophyll fluorescence: consequences of natural variations in the optical characteristics of phytoplankton and the quantum yield of chlorophyll a fluorescence. Int. J. Remote. Sens. 17, 2417–2448.

Babin, M., Stramski, D., Ferrari, G.M., Claustre, H., Bricaud, A., Obolensky, G., et al., 2003. Variations in the light absorption coefficients of phytoplankton, non-algal particles, and dissolved organic matter in coastal waters around. Eur. J. Geophys. Res. 108 (C7)), 3211. Available from: http://dx.doi.org/10.1029/2001JC000882.

Behrenfeld, M.J., Westberry, T.K., Boss, E.S., O'Malley, R.T., Siegel, D.A., Wiggert, J.D., et al., 2009. Satellite-detected fluorescence reveals global physiology of ocean phytoplankton. Biogeosciences 6, 779–794.

Binding, C.E., Greenberg, T.A., Jerome, J.H., Bukata, R.P., Letourneau, G., 2010. An assessment of MERIS algal products during an intense bloom in lake of the woods. J. Plankton Res 33 (5), 793–806. Available from: http://dx.doi.org/10.1093/plankt/fbq13.

Blondeau-Patissier, D., Gower, J.F.R., Dekker, A.G., Phinn, S.R., Brando, V.E., 2014. A review of ocean color remote sensing methods and statistical techniques for the detection, mapping and analysis of phytoplankton blooms in coastal and open oceans. Prog. Oceanogr. 123, 123–144.

Bricaud, A., Babin, M., Morel, A., Claustre, H., 1995. Variability in the chlorophyll-specific absorption coefficients of natural phytoplankton: analysis and parameterization. J. Geophys. Res. 100, 13321–13332.

Browning, T.J., Bouman, H.A., Moore, C.M., 2014. Satellite-detected fluorescence: decoupling nonphotochemical quenching from iron stress signals in the south Atlantic and Southern ocean. Global Biogeochem. Cy 28 (5), 510–524. Available from: http://dx.doi.org/10.1002/2013gb004773.

Bukata, R.P., Jerome, J.H., Kondratyev, K.Y., Pozdnyakov, D.V., 1995. Optical Properties and Remote Sensing of Inland and Coastal Waters. CRC Press, Boca Raton, FL.

Cannizzaro, J.P., Hu, C., English, D.C., Carder, K.L., Heil, C.A., Muller-Karger, F.E., 2009. Detection of Karenia brevis blooms on the west Floridashelf using in situ backscattering and fluorescence data. Harmful. Algae 8, 898–909.

Chamberlin, S., Marra, J., 1992. Estimation of photosynthetic rate from measurements of natural fluorescence: analysis of the effects of light and temperature. Deep-Sea Res. 39 (10), 1695–1706.

Chamberlin, W.S., Booth, C.R., Kiefer, D.A., Morrow, J.H., Murphy, R.C., 1990. Evidence for a simple relationship between natural fluorescence, photosynthesis and chlorophyll in the sea. Deep-Sea Res. 37 (6), 951–973.

Chami, M., Santer, R., Dilligeard, E., 2001. Radiative transfer model for the computation of radiance and polarization in an ocean-atmosphere system: polarization properties of suspended matter for remote sensing. Appl. Opt. 40, 2398–2416.

Ciotti, A.,M., Lewis, M.R., Cullen, J.J., 2002. Assessment of the relationships between dominant cell size in natural phytoplankton communities and the spectral shape of the absorption coefficient. Limnol. Oceanogr. 47 (2), 404–417.

Collins, D.J., Kiefer, D.A., SooHoo, J.B., McDermid, I.S., 1985. The role of reabsorption in the spectral distribution of phytoplankton fluorescence emission. Deep-Sea Res. 32 (8), 983–1003.

Culver, M.E., Perry, M.J., 1997. Calculation of solar-induced fluorescence in surface and subsurface waters. J. Geophys. Res. 102 (C5), 10563–10572.

Downing, J.A., Watson, S.B., McCauley, E., 2001. Predicting cyanobacteria dominance in lakes. Can. J. Fish. Aquat. Sci. 58 (10), 1905–1908.

Ficek, D., Kaczmarek, S., Stoń-Egiert, J., Woźniak, B., Majchrowski, R., Dera, J., 2004. Spectra of light absorption by phytoplankton pigments in the Baltic; conclusions to be drawn from a Gaussian analysis of empirical data. Oceanologia 46 (4), 533–555.

Fischer, J., Kronfeld, U., 1990. Sun-stimulated chlorophyll fluorescence. 1. Influence of oceanic properties. Int. J. Rem. Sens. 11, 2125–2147.

Foster, R., Ibrahim, A., Gilerson, A., El-Habashi, A., Carrizo, C., Ahmed, S., 2015. Characterization of sun and sky glint from wind ruffled sea surfaces for improved estimation of polarized remote sensing reflectance. Proc. SPIE 9613.

Gilerson, A., Zhou, J., Oo, M., Chowdhary, J., Gross, B., Moshary, F., et al., 2006. Retrieval of fluorescence from reflectance spectra of algae in sea water through polarization discrimination: modeling and experiments. Appl. Opt. 45, 5568–5581.

Gilerson, A., Zhou, J., Hlaing, S., Ioannou, I., Schalles, J., Gross, B., et al., 2007. Fluorescence component in the reflectance spectra from coastal waters. Dependence on water composition. Opt. Express. 15, 15702–15721.

Gilerson, A., Zhou, J., Hlaing, S., Ioannou, I., Gross, B., Moshary, F., et al., 2008. Fluorescence component in the reflectance spectra from coastal waters. Performance of retrieval algorithms. Opt. Express. 16, 2446–2460.

Gilerson, A.A., Gitelson, A.A., Zhou, J., Gurlin, D., Moses, W.J., Ioannou, I., et al., 2010. Algorithms for remote estimation of chlorophyll-a in inland and coastal waters using red and near infrared bands. Opt. Express. 18, 24109–24125.

Gilerson, A.A., Stepinski, J., Ibrahim, A., You, Y., Sullivan, J., Twardowski, M., et al., 2013. Benthic effects on the polarization of light in shallow waters. Appl. Opt. 52 (36), 8685–8705.

Gilerson, A., Ondrusek, M., Tzortziou, M., Foster, R., El-Habashi, A., Tiwari, S.P., et al., 2015. Multi-band algorithms for the estimation of chlorophyll concentration in the Chesapeake Bay. Proc. SPIE 9638. Available from: http://dx.doi.org/10.1117/12.2195725.

Gitelson, A., 1992. The peak near 700 nm on reflectance spectra of algae and water: relationships of its magnitude and position with chlorophyll concentration. Int. J. Remote Sens. 13, 3367–3373.

Gons, H.J., Auer, M.T., Effler, S.W., 2008. MERIS satellite chlorophyll mapping of oligotrophic and eutrophic waters in the laurentian great lakes. Rem. Sens. of Environ. 112 (11), 4098–4106. Available from: http://dx.doi.org/10.1016/j.rse.2007.06.02.

Gordon, H.R., 1974. Spectral variations in the volume scattering function at large angles in natural waters. JOSA 64 (6), 773–775.

Gordon, H.R., 1979. Diffuse reflectance of the ocean: the theory of its augmentation by chlorophyll a fluorescence at 685 nm. Appl. Opt. 18 (8), 1161–1166.

Gordon, H.R., 1989. Can the Lambert-Beer law be applied to the diffuse attenuation coefficient of ocean water?. Limnol. Oceanogr. 34 (8), 1389–1409.

Gower, J., King, S., 2011. A global survey of intense surface plankton blooms and floating vegetation using MERIS MCI. In: Tang, D. (Ed.), Remote Sensing of the Changing Oceans. Springer, London, New York, p. 396.

Gower, J.F.R., Borstad, G.A., 1990. Mapping of phytoplankton by solar-stimulated fluorescence using an imaging spectrometer. Int. J. Rem. Sens. 11 (2), 313–320. Available from: http://dx.doi.org/10.1080/01431169008955022.

Gower, J.F.R., Doerffer, R., Borstad, G.A., 1999. Interpretation of the 685 nm peak in water-leaving radiance spectra in terms of fluorescence, absorption and scattering, and its observation by MERIS. Int. J. Rem. Sens. 20 (9), 1771–1786.

Gower, J.F.R., Brown, L., Borstad, G.A., 2004. Observation of chlorophyll fluorescence in west coast waters of Canada using the MODIS satellite sensor. Can. J. Rem. Sens. 30, 17–25.

Gregg, W.W., Carder, K.L., 1990. A simple spectral solar irradiance model for cloudless maritime atmospheres. Limnol. Oceanogr. 35 (8), 1657–1675.

Gurlin, D., Gitelson, A.A., Moses, W.J., 2011. Remote estimation of chl-a concentration in turbid productive waters—return to a simple two-band NIR-red model?. Rem. Sens. Environ. 115, 3479–3490.

Hansen, J.E., Travis, L.D., 1974. Light-scattering in planetary atmospheres. Space. Sci. Rev. 16, 527–610.

Hu, C., Muller-Karger, F.E., Taylor, C., Carder, K.L., Kelble, C., Johns, E., et al., 2005. Red tide detection and tracing using MODIS fluorescence data: a regional example in SW Florida coastal waters. Remote Sens. Environ. 97, 311–321.

Hu, C., Cannizzaro, J., Carder, K.L., Lee, Z., Muller-Karger, F.E., Soto, I., 2011. Red tide detection in the eastern Gulf of Mexico using MODIS imagery. In: Morales, J., Stuart, V., Platt, T., Sathyendranath, S. (Eds.), Handbook of Satellite Remote Sensing Image Interpretation: Applications for Marine Living Resources Conservation and Management, EU PRESPO and IOCCG. Dartmouth, Canada, pp. 95–110.

Huot, Y., Brown, C.A., Cullen, J.J., 2005. New algorithms for MODIS sun-induced chlorophyll fluorescence and a comparison with present data products. Limnol. Oceanogr. 3, 108–130.

Ibrahim, A., Gilerson, A., Harmel, T., Tonizzo, A., Chowdhary, J., Ahmed, S., 2012. The relationship between upwelling underwater polarization and attenuation/absorption ratio. Opt. Express. 20, 25662–25680.

IOCCG Report 5, 2006: Remote Sensing of Inherent Optical Properties: Fundamentals, Tests of Algorithms, and Applications. Edited by Z. P. Lee, pp. 126.

IOCCG Report 10, 2010. Atmospheric Correction for Remotely-Sensed Ocean-Colour Products. Edited by: M. Wang, pp. 78.

Kattawar, G.W., Vastano, J.C., 1982. Exact 1-D solution to the problem of chlorophyll fluorescence from the ocean. Appl. Opt. 21, 2489–2492.

Kiefer, D.A., Chamberlin, W.S., Booth, C.R., 1989. Natural fluorescence of chlorophyll a: Relationship to photosynthesis and chlorophyll concentration in the western south pacific gyre. Limnol. Oceanogr. 34, 868–881.

Lee, Z., Carder, K.L., Arnone, R.A., 2002. Deriving inherent optical properties from water color: a multiband quasi-analytical algorithm for optically deep water. Appl. Opt. 41, 5755–5772.

Lee, Z.P., Lubac, B., Werdell, J., Arnone, R., 2009. An Update of the Quasi-Analytical Algorithm (QAA_v5). www.ioccg.org/groups/Software_OCA/QAA_v5.pdf.

Letelier, R.M., Abbott, M.R., 1996. An analysis of chlorophyll fluorescence algorithms for the moderate resolution imaging spectrometer (MODIS). Rem. Sens. Environ., 58, 215–223.

Letelier, R.M., Abbott, M.R., Karl, D.M., 1997. Chlorophyll fluorescence response to upwelling events in the southern ocean. J. Geophys. Res. 24 (4), 409–412.

Levenberg, K., 1944. A method for the solution of certain non-linear problems in least squares. Q. Appl. Math. 2, 164–168.

Lin, H., Kuzminov, F.I., Park, J., Lee, S., Falkowski, P.G., Gorbunov, M.Y., 2016. The fate of photons absorbed by phytoplankton in the global ocean. Science 351, 264–267. Available from: http://dx.doi.org/10.1126/science.aab2213.

Magnuson, A., Harding Jr., L.W., Mallonee, M.E., Adolf, J.E., 2004. Bio-optical model for Chesapeake Bay and the middle Atlantic bight. Estuar. Coast. Shelf Sci., 61, 403–424.

Marquardt, D., 1963. An algorithm for least-squares estimation of nonlinear parameters. SIAM J. Appl. Math. 11 (2), 431–441.

Maritorena, S., Morel, A., Gentili, B., 2000. Determination of the fluorescence quantum yield by oceanic phytoplankton in their natural habitat. Appl. Opt. 39 (36), 6725–6737.

McKee, D., Cunningham, A., Wright, D., Hay, L., 2007. Potential impacts of nonalgal materials on water-leaving sun induced chlorophyll fluorescence signals in coastal waters. Appl. Opt. 46, 7720–7729.

McKee, D., Rottgers, R., Neukermans, G., Calzado, V.S., Trees, C., Ampolo-Rella, M., et al., 2014. Impact of measurement uncertainties on determination of chlorophyll-specific absorption coefficient for marine phytoplankton. J. Geophys. Res. Oceans 119, 9013–9025. Available from: http://dx.doi.org/10.1002/2014JC009909.

Mobley, C.D., 1994. Light and Water. Radiative Transfer in Natural Waters. Academic Press, New York.

Mobley, C.D., Sundman, L.K., 2001. Hydrolight 4.2. Sequoia Scientific, Inc.

Morel, A., 1974. Optical properties of pure water and pure sea water. In: Jerlov, N.G., Nielsen, E.S. (Eds.), Optical aspects of oceanography. Academic Press, New York, pp. 1–24.

Morel, A., Prieur, L., 1977. Analysis of variations in ocean color. Limnol. Oceanogr. 22, 709–722.

Morrison, J.R., 2003. In situ determination of the quantum yield of phytoplankton chlorophyll a fluorescence: a simple algorithm, observations, and a model. Limnol. Oceanogr. 48 (2), 618–631.

Morrison, J.R., Goodwin, D.S., 2010. Phytoplankton photocompensation from space-based fluorescence measurements. Geoph. Res. Let 37 (6), L06603. Available from: http://dx.doi.org/10.1029/2009GL04179.

Neville, R.A., Gower, J.F.R., 1977. Passive remote sensing of phytoplankton via chlorophyll fluorescence. J. Geophys. Res. 82, 3487–3493.

O'Malley, R.T., Behrenfeld, M.J., Westberry, T.K., Milligan, A.J., Shang, S., Yan, J., 2014. Geostationary satellite observations of dynamic phytoplankton photophysiology. Geoph. Res. Let. 41 (14), 5052–5059. Available from: http://dx.doi.org/10.1002/2014gl06024.

Ostrowska, M., 2011. Dependence between the quantum yield of chlorophyll a fluorescence in marine phytoplankton and trophicity in low irradiance level. Optica Applicata 41, 567–577.

Ostrowska, M., Darecki, M., Wozniak, B., 1997. An attempt to use measurements of sun-induced chlorophyll fluorescence to estimate chlorophyll a concentration in the baltic sea. Proc. SPIE 3222, 528–537.

PACE, 2012. Pre-Aerosol, Clouds, and ocean Ecosystem (PACE) Mission Science Definition Team Report.

Palmer, S.C.J., Odermatt, D., Hunter, P.D., Brockmann, C., Presing, M., Balzter, H., et al., 2015. Satellite remote sensing of phytoplankton phenology in lake Balaton using 10 years of MERIS observations. Rem. Sens. Environ. 158, 441–452.

Peng, F., Effler, S.W., O'Donnell, D., Weidemann, A.D., Auer, M.T., 2009. Characterizations of minerogenic particles in support of modeling light scattering in lake superior through a two-component approach. Limnol. Oceanogr. 54 (4), 1369–1381.

Petzold, T.J., 1972. Volume Scattering Functions for Selected Ocean Waters. Scripps Institution of Oceanography, San Diego, CA.

Pope, R., Fry, E., 1997. Absorption spectrum (380–700 nm) of pure waters: II. Integrating cavity measurements. Appl. Opt. 36, 8710–8723.

Pozdnyakov, D., Lyaskovsky, A., Grassl, H., Pettersson, L., 2002. Numerical modelling of trans-spectral processes in natural waters: implications for remote sensing. Int. J. Rem. Sens. 23, 1581–1607.

Roesler, C.S., Boss, E., 2003. Spectral beam attenuation coefficient retrieved from ocean color inversion. Geophys. Res. Lett. 30, 1468. Available from: http://dx.doi.org/10.1029/2002GL016185.

Roesler, C.S., Perry, M.J., 1995. In situ phytoplankton absorption, fluorescence emission, and particulate backscattering spectra determined from reflectance. J. Geophys. Res. 100, 13279–13294. Available from: http://dx.doi.org/10.1029/95JC00455.

Sathyendranath, S., Platt, T., Irwin, B., Horne, E., Borstad, G.A., Stuart, V., et al., 2004. A multispectral remote sensing study of coastal waters off Vancouver Island. Int. J. Rem. Sens. 25, 893–919.

Schalles, J.F., 2006. Optical remote sensing techniques to estimate phytoplankton chlorophyll-a concentrations in coastal waters with varying suspended matter and CDOM concentrations.

In: Richardson, L.L., LeDrew, E.F. (Eds.), Remote Sensing of Aquatic Coastal Ecosystem Processes: Science and Management Applications. Springer, The Netherlands, pp. 27–79.

Schalles, J.F., Gitelson, A.A., Yacobi, Y.Z., Kroenke, A.E., 1998. Estimation of chlorophyll-a from time series measurements of high spectral resolution reflectance in an eutrophic lake. J. Phycol., 34, 383–390.

Schallenberg, C., Lewis, M.R., Kelley, D.E., Cullen, J.J., 2008. The inferred influence of nutrient availability on the relationship between sun-induced fluorescence and incident irradiance in the Bering sea. J. Geophys. Res., 113, C07046. Available from: http://dx.doi.org/10.1029/2007JC00435.

Stramski, D., Bricaud, A., Morel, A., 2001. Modeling the inherent optical properties of the ocean based on the detailed composition of the planktonic community. Appl. Opt. 40, 2929–2945.

Sydor, M., Arnone, R.A., 1997. Effect of suspended particulate and dissolved organic matter on remote sensing of coastal and riverine waters. Appl. Opt. 36, 6905–6912.

Timofeeva, V.A., 1970. Degree of light polarization in turbid media. Izvestiya Akademii Nauk Sssr Fizika Atmosfery I Okeana 6, 513.

Tonizzo, A., Zhou, J., Gilerson, A., Twardowski, M.S., Gray, D.J., Arnone, R.A., et al., 2009. Polarized light in coastal waters: hyperspectral and multiangular analysis. Opt. Express. 17 (7), 5666–5682.

Tonizzo, A., Ibrahim, A., Zhou, A., Gilerson, A., Gross, B., Moshary, F., et al., 2010. The impact of algal fluorescence on the underwater polarized light field. Proc. SPIE 7678.

Twardowski, M.S., Boss, E., Macdonald, J.B., Pegau, W.S., Barnard, A.H., Zaneveld, J.V., 2001. A model for estimating bulk refractive index from the optical backscattering ratio and the implications for understanding particle composition in case I and case II waters. J. Geoph. Res., 106, 14129–14142.

Tyler, J.E., Smith, R.C., 1970. Ocean Sciences: Measurements of Spectral Irradiance under Water. Gordon and Breach Science Publishers, New-York, p. 103.

Voss, K.J., 1992. A spectral model of the beam attenuation coefficient in the ocean and coastal areas. Limnol. Oceanogr. 37, 501–509.

Werdell, P.J., Bailey, S.W., 2005. An improved in-situ bio-optical data set for ocean color algorithm development and satellite data product validation. Rem. Sens. Environm. 98 (1), 122–140.

Werdell, P.J., Bailey, S.W., Franz, B.A., Harding Jr., L.W., Feldman, G.C., McClain, C.R., 2009. Regional and seasonal variability of chlorophyll-a in Chesapeake Bay as observed by SeaWiFS and MODIS-Aqua. Rem. Sens. Environm. 113, 1319–1330.

Zimba, P.V., Gitelson, A.A., 2006. Remote estimation of chlorophyll concentration in hypereutrophic aquatic systems: model tuning and accuracy optimization. Aquaculture. 256(1-4), 272–286.

Chapter 8

Bio-optical Modeling of Phycocyanin

Linhai Li[1] and Kaishan Song[2]
[1]*University of California San Diego, La Jolla, CA, United States,*
[2]*Chinese Academy of Sciences, Changchun, China*

8.1 INTRODUCTION

Cyanobacteria (blue-green algae) are the oldest oxygenic photoautotrophs on Earth and have exerted major impacts on shaping the biosphere, which had led to the evolution of the higher terrestrial plant and animal lives (Paerl and Otten, 2013). Through the long evolutionary history (around 3.5 billion years), cyanobacteria have acquired the ability to adapt to geochemical and climatic changes (Paerl and Paul, 2012) and diverse aquatic environments with different nutrient levels, water diversion, and salinization (Reynolds, 1987; Paerl and Otten, 2013). Global warming and eutrophication in water bodies have increased cyanobacterial bloom potentials due to cyanobacterial preferential growth conditions of higher temperature and nutrient levels (Smith, 1986; Paerl and Huisman, 2008). Such trend is unfavorable because the blooms produce surface scums and earthy smell in water bodies for drinking water supplies (Codd et al., 1999) and it was reported that cyanobacteria produce a variety of toxic compounds, for example, microcystins and saxitoxins (Baker et al., 2002; Chorus and Bartram, 1999; Sivonen, 1996). These cyanobacteria-generated toxins may cause severe health issues; for example, it was suggested that microcystins caused acute liver injury or even death (Falconer et al., 1983; Jochimsen et al., 1998; Carmichael et al. 2001; Azevedo et al. 2002) and chronic exposure to low-level microcystins increased the tumor risk (Falconer and Humpage, 1996; Humpage et al., 2000). Thus, the adverse cyanobacterial blooms result in serious issues in lakes and reservoirs for drinking water supplies and/or recreational activities. It is critical to monitor cyanobacterial blooms and determine the toxicity in drinking water sources from early stage (Baker et al., 2002; Codd et al., 2005; Song et al., 2014).

Bio-optical Modeling and Remote Sensing of Inland Waters.
DOI: http://dx.doi.org/10.1016/B978-0-12-804644-9.00008-2
© 2017 Elsevier Inc. All rights reserved.

Therefore, determining cyanobacterial abundance helps water resource managers to take measures to decrease the health accidents caused by cyanobacteria. The first approach is to count cyanobacterial cells using microscopy, which also provides the identification of cyanobacterial genus. The polymerase chain reaction was also adapted to directly identify cyanobacteria from other organisms in water samples by amplifying the phycocyanin intergenic spacer region, which as well provides direct implication of toxigenicity (Baker et al., 2002). Application of high-pressure liquid chromatography (HPLC) is another standard protocol for identification and quantification of cyanobacteria (Lawton et al., 1994). Although HPLC does not have the capability to identify cyanobacterial species, it measures various pigments and provides a robust measure of cyanobacterial abundance. On the other hand, *in vivo* fluorometric approaches usually provide indirect measure of cyanobacterial abundance by determining concentration of one or two pigments such as chlorophyll-*a* (chl-*a*) and phycocyanin (PC). For instance, various companies (e.g., Hydrolab and YSI) manufacture submersible multiparameter sondes equipped with fluorescence sensors for quantifying chl-*a*, PC, and phycoerythrin within a water column. Despite that laboratory and field approaches described above provide useful data to infer cyanobacterial growth, these traditional approaches are often limited due to the extensive efforts of water sampling (Hunter et al., 2009) or small spatial scale of coverage (Guanter et al., 2010).

As a result, the traditional approaches are not suitable to monitor cyanobacterial blooms which are ephemeral in time and widely dispersed in space (Hunter et al., 2010; Huang et al., 2015). The limitations of traditional approaches are overcome by remote sensing techniques, which provide a synoptic view at a short temporal scale. For example, remote sensing was used to monitor the spatiotemporal dynamics of chl-*a* concentration ([chl-*a*]) in inland waters (e.g., Gons, 1999; Gons et al., 2008; Li et al., 2013 and references therein; also see Chapters 6 and 7 in this book for bio-optical modeling and fluorescence of chl-*a* in inland waters, respectively), which was subsequently used to infer cyanobacterial biomass (Kutser, 2004). Nevertheless, [chl-*a*] only implies the total algal biomass but not specific biomass of cyanobacteria. Quantification of chl-*a* is not a robust means to accurately estimate cyanobacterial abundance because all phytoplankton contains chl-*a* (Randolph et al., 2008; Song et al., 2014). A quantification of a pigment contained by freshwater cyanobacteria is needed in order to more accurately determine the cyanobacterial abundance in the water bodies. It was suggested that PC is a unique pigment of freshwater cyanobacteria and depicts a distinctive optical feature around 620−630 nm (see Section 8.2), which makes the remote detection (i.e., bio-optical modeling) of cyanobacteria possible (e.g., Dekker, 1993; Simis et al., 2005; Mishra et al., 2013). Therefore, remote sensing of freshwater cyanobacteria has been largely focused on developing deferent bio-optical algorithms to estimate PC within last more than two decades (e.g., Dekker, 1993; Schalles and Yacobi, 2000; Simis et al., 2005, 2007; Li et al., 2015 and references therein) although a few works on remote quantification of cyanobacterial cells were also found (e.g., Lunetta et al., 2015).

As discussed above, traditional approaches for monitoring cyanobacteria that rely on field sampling are time-consuming and labor intensive; thus, this chapter is focused on the emerging remote sensing techniques that provide a fast and efficient method to provide us both intensity and spatial distribution of cyanobacterial blooms. In Section 8.2, the absorption feature of PC and its corresponding response on remote sensing reflectance will be discussed, as which is the fundamental basis for bio-optical modeling of phyocyanin from remote sensing measurements. We then in Section 8.3 comprehensively reviewed existing remote sensing algorithms of [PC] in literature. A few representative algorithms are subsequently evaluated and compared using the same large field dataset in Section 8.4. In Section 8.5, maps of [PC] were generated from airborne images to demonstrate the capability of remote sensing to present a synoptic view of [PC] in a water body. At last, a summary of this chapter and future research direction of bio-optical modeling of [PC] are noted in Section 8.6.

8.2   THEORETICAL BASIS FOR REMOTE SENSING OF PHYCOCYANIN

The inherent optical properties (IOPs), for example, absorption and scattering, control how light propagates within and leaves a water column (Mobley, 1994). Radiative transfer theory (see Chapter 2) describes the complete relationship between IOPs and radiance within a water column, based on which Gordon et al. (1988) derived an alternative relationship between remote sensing reflectance and IOPs using Monte Carlo simulations:

$$R_x(\lambda) = D(\lambda)\frac{b_b(\lambda)}{a(\lambda) + b_b(\lambda)}, \tag{8.1}$$

where $R_x(\lambda)$ represents remote sensing reflectance either above $[R_{rs}(\lambda)]$ or below $[R_{rs}^-(\lambda)]$ water subsurface at light wavelength λ (in vacuo), $D(\lambda)$ is a spectral constant but different for $R_{rs}(\lambda)$ and $R_{rs}^-(\lambda)$, total absorption coefficient $a(\lambda)$ is the sum of phytoplankton $[a_{phy}(\lambda)]$, colored detritus matter $[a_{CDM}(\lambda)]$, and water $[a_w(\lambda)]$ absorption, and total backscattering coefficient $b_b(\lambda)$ is the sum of backscattering by particulate matter $[b_{bp}(\lambda)]$ and water $[b_{bw}(\lambda)]$.

As $R_x(\lambda)$, an apparent optical property (AOP), does not straightforwardly reflects the abundance of optically active water constituents, absorption coefficients otherwise maintain direct relationship with the concentrations of individual constituent in the water column. For example, PC absorption

$[a_{pc}(\lambda)]$, a portion of $a_{phy}(\lambda)$, is generally suggested to be linearly correlating with [PC]:

$$a_{pc}(\lambda) = [PC]a_{pc}^*(\lambda), \tag{8.2}$$

where $a_{pc}^*(\lambda)$ is phycocyanin-specific absorption in unit of m^2 (mg PC)$^{-1}$. Therefore, it is expected that $a_{pc}(\lambda)$ increases with [PC], which in turn causes an observed response on $R_x(\lambda)$ according to Eq. 8.1.

PC has an absorption peak around 620−630 nm and its absorption decreases fast from the spectral peak position (Dekker, 1993; Jupp et al., 1994; Simis and Kauko, 2012; Mishra et al., 2013). Because $R_x(\lambda)$ inversely relates to absorption coefficient (see Eq. 8.1), the absorption feature of PC and chl-a results in a trough around 620−630 nm and a shoulder at $\sim$650 nm on $R_x(\lambda)$ (Metsamaa et al., 2006). Mishra et al. (2009) further found that the ratio of [PC] to [chl-a] controls the appearance of the shoulder at $\sim$650 nm and the concentration of chlorophyll-b affects the exact position of this spectral shoulder. Fig. 8.1 shows an example spectrum of the measured $R_{rs}(\lambda)$ and $a_{phy}(\lambda)$, in which the inverse relationship between $R_{rs}(\lambda)$ and $a_{phy}(\lambda)$ is well illustrated. The features caused by PC and chl-a absorption on $R_{rs}(\lambda)$ are easily identified around 620−630 and 670−680 nm, respectively, where their spectral absorption peaks are located. These physical relationships between $R_{rs}(\lambda)$ and PC (and chl-a) are the basis for majority of remote sensing algorithms of [PC] reviewed in Section 8.3.

Except for empirical algorithms, the physical basis of all other algorithms in Section 8.3 can be understood through IOPs by replacing $R_x(\lambda)$ in the algorithms with Eq. 8.1. In general, the key to developing a PC algorithm is

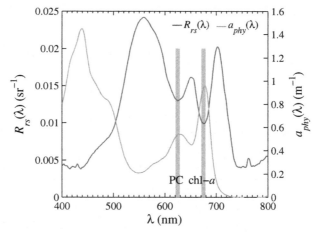

FIGURE 8.1 An example spectra of remote sensing reflectance $R_{rs}(\lambda)$ and phytoplankton absorption $a_{phy}(\lambda)$ to demonstrate how $a_{phy}(\lambda)$ causes distinctive optical features on $R_{rs}(\lambda)$.

to either determine the radiometric index that is sensitive to PC absorption (e.g., semi-empirical algorithms) or directly derive $a_{pc}(\lambda)$ from radiometric measurements (e.g., semi-analytical algorithms). Because $R_x(\lambda)$ is the bulk signal that has contributions from optical processes such as $a(\lambda)$, $b_b(\lambda)$, and even fluorescence of all constituents and pigments including PC, the performance of a PC algorithm highly depends on how the interference from other optical processes (e.g., scattering and fluorescence) and from other constituents is addressed. For example, the influence of $b_b(\lambda)$ can be reduced by taking ratio of $R_x(\lambda)$ at different spectral bands that are not spectrally far away (e.g., band ratio and three-band algorithms); the effect of PC fluorescence around 645 nm (Pizarro et al., 2001; Singh et al., 2015) can be avoided by not using the spectral bands within this spectral range. Furthermore, Mishra et al. (2009) proved that absorption of chl-*a* and chlorophyll-*b* affected the remote estimation of cyanobacteria, and Simis et al. (2007) suggested that the accuracy of remotely estimated [PC] was also affected by accessory pigments such as chlorophylls *b* and *c*, and pheophytin. In fact, a major interference on the performance of PC algorithms comes from other pigments and constituent such as colored detritus material (CDM) (Li et al., 2015), which is discussed in detail in Sections 8.3.3 and 8.4.6 when describing semi-analytical algorithms.

8.3 LITERATURE REVIEW OF REMOTE SENSING ALGORITHMS OF PHYCOCYANIN

According to the recommendations of Ogashawara (2015), remote sensing algorithms for applications in aquatic ecosystems can be classified into different types such as empirical, semi-empirical, and semi-analytical algorithms. The remote sensing algorithms for [PC] reviewed below are classified mainly by following Ogashawara's recommendations. Here it is restated that the algorithms are empirical or semi-empirical if statistical relationships are required directly between radiometric data (e.g., remote sensing reflectance and water-leaving radiance) and [PC], and the algorithms are semi-analytical if there is no direct statistical relationship between radiometric data and [PC], while limited empirical assumptions are only made during derivation of IOPs such as $a_{pc}(\lambda)$. It is possible that an algorithm can be classified into a different category when using different criteria, and the approach in this chapter is not necessary the only way to classify the reviewed PC algorithms.

Ruiz-Verdu et al. (2008) compared performance of a empirical, semi-empirical, and semi-analytical algorithm in Spanish and Netherlands lakes. A relatively complete list of remote sensing algorithms for [PC] until 2014 is found in Table 1 of Qi et al. (2014). More details about those algorithms and newly developed ones are reviewed in Sections 8.3.1–8.3.3.

8.3.1 Empirical Algorithms

Empirical algorithms for estimating [PC] usually focus on establishing statistical regressions between [PC] measured *in situ* and radiometric data determined by various optical remote sensing platforms. This type of algorithms generally has no or very limited physical basis. They typically use most or all spectral bands of radiometric measurements to determine the statistical relationship between the radiometric data of the selected spectral bands and the *in situ* measured [PC]. Empirical algorithms are easy to implement, but these algorithms are often site-specific and require recalibration for applications in different sites or datasets.

For example, much efforts were made to map [PC] using Landsat TM and ETM+ satellite images. Vincent et al. (2004) firstly reported that [PC] could be estimated statistically using spectral bands 1, 3, 5, and 7 and their band ratios from Landsat 7 ETM+ in Lake Erie. Similar approach was recently applied by Ogashawara et al. (2014) to Guarapiranga Reservoir in Brazil using Landsat 5 TM and 7 ETM+ images as well as by Sun et al. (2015) to Lake Dianchi in China using Landsat 7 ETM+ images. This group of algorithms using Landsat satellite images is purely statistical, while the semi-empirical algorithms reviewed below have more underlying physical basis.

8.3.2 Semi-empirical Algorithms

Semi-empirical algorithms utilize more physical assumptions (e.g., absorption features of PC) to determine the spectral bands to construct the band combinations of radiometric data. Thus, semi-empirical algorithms theoretically have better site-transferability than empirical algorithms. However, semi-empirical algorithms often involve the procedure to select the best spectral bands of radiometric data result in the highest correlation with [PC]. The selection of best spectral bands is accomplished by methods of iterative searching, stepwise regression analysis, and other iteration approaches. It also requires an additional step to determine the statistical relationship between the radiometric data of the selected spectral bands and the *in situ* measured [PC]. Therefore, statistical relationships between band combinations of radiometric data and [PC] are still needed to be determined through regression analysis. Semi-empirical algorithms often lead to satisfactory performance for the dataset where they are developed, but a recalibration is also required when applying semi-empirical algorithms to different sites and datasets.

According to such definition, the first group of semi-empirical remote sensing algorithms for [PC] is band ratio algorithms, which has the form of

$$[PC] \propto R_x(\lambda_1)/R_x(\lambda_2), \tag{8.3}$$

where $R_x(\lambda)$ can be $R_{rs}(\lambda)$, $R_{rs}^-(\lambda)$, or irradiance reflectance $R(\lambda)$, and λ_1 and λ_2 are different light wavelengths (i.e., spectral bands). The underlying physical basis of band ratio algorithms is typically that λ_2 should be located around PC absorption peak and λ_1 is away from PC absorption bands, so that the band ratio (Eq. 8.3) is overall related to $a_{pc}(\lambda)$, when replacing $R_x(\lambda)$ with Eq. 8.1, and thus correlated with [PC], although the selected bands by iterative processes sometimes do not follow such principle. The first band ratio algorithm for [PC] that can be found in literature was developed by Schalles and Yacobi (2000), which correlated [PC] to $R_{rs}(650)/R_{rs}(625)$ for lakes in Nebraska, USA. Later Hunter et al. (2008b, 2009) proposed new algorithm using $L_w(710)/L_w(620)$ for Norfolk Broads in England. Other different band ratios used to estimate [PC] include $R_{rs}(700)/R_{rs}(600)$ (Mishra et al., 2009), $R_{rs}(709)/R_{rs}(600)$ (Mishra, 2012), and $R_{rs}(595)/R_{rs}(660)$ (Woźniak et al., 2016). Sun et al. (2012) and Woźniak et al. (2016) also suggested that a better estimation accuracy was achieved when combining several band ratios into a single statistical relationship with [PC]. The comparison of performance of different band ratio algorithms was found in Mishra et al. (2009), Li et al. (2010), Sun et al. (2012), Ogashawara et al. (2013), Woźniak et al. (2016) for laboratory experiments or different study sites with various limnological conditions.

A large group of semi-empirical algorithms is called three-band algorithms which were inspired by the pioneering work of Gitelson et al. (2003) for remote estimation of [chl-a] of terrestrial plants. The three-band algorithm has the following form:

$$[PC] \propto \left[R_x(\lambda_1)^{-1} - R_x(\lambda_2)^{-1}\right] \times R_x(\lambda_3), \tag{8.4}$$

where λ_1 and λ_2 are typically within visible spectral range and λ_3 is within the near-infrared. λ_1 is the spectral band most sensitive to PC absorption, λ_2 is the spectral band not sensitive to PC absorption, and the spectral band λ_3 has negligible influence from PC. Compared to band ratio algorithms, three-band algorithms further reduce the influence of $a_{CDM}(\lambda)$ and $b_b(\lambda)$ by introducing a third spectral band, which is important for turbid waters, assuming that $a_{CDM}(\lambda)$ does not vary much within spectral range between λ_1 and λ_2 and $b_b(\lambda)$ is wavelength-independent within the spectral range between λ_1 and λ_3. Hunter et al. (2008a) firstly found that [PC] correlated well with $[R_{rs}(630)^{-1} - R_{rs}(660)^{-1}] \times R_{rs}(750)$ for laboratory experiments using cultured cyanobacteria and later they (Hunter et al., 2010) modified the algorithm to $[R_{rs}(615)^{-1} - R_{rs}(600)^{-1}] \times R_{rs}(725)$ for two lakes in United Kingdom. Upon the success of three-band algorithm for [PC], subsequent efforts were made by different groups to modify the algorithm for improved performance. For example, Le et al. (2011) introduced a fourth spectral band into the algorithm to eliminate the backscattering term in the denominator, which exists in the standard three-band algorithm when substituting Eq. 8.1

into Eq. 8.4 and having the assumption of spectrally flat $b_b(\lambda)$, and thus further reduce the influence from the scattering of other water constituents in highly turbid Lake Taihu:

$$[PC] \propto [R_{rs}(625)^{-1} - R_{rs}(650)^{-1}] \times [R_{rs}(730)^{-1} - R_{rs}(695)^{-1}]^{-1}; \qquad (8.5)$$

Mishra and Mishra (2014) introduced a correction factor Ψ, an proxy for the ratio of chl-a absorption coefficient between 665 and 620 nm, to deal with influence of accessory pigments (particularly chl-a) suggested by Simis et al. (2007):

$$[PC] \propto [R_{rs}(620)^{-1} - (\psi R_{rs}(665))^{-1}] \times R_{rs}(778), \qquad (8.6a)$$

with

$$\psi = 0.709*[R_{rs}(560)/R_{rs}(665)]^{0.874}, \qquad (8.6b)$$

based on the assumption that $R_{rs}(560)/R_{rs}$ (665) is highly correlated with [chl-a].

Another well-known semi-empirical algorithm was the baseline algorithm developed by Dekker (1993). The physical basis of the baseline algorithm is that the PC absorption at 624 nm, $a_{pc}(624)$, can be determined from phytoplankton absorption, $a_{phy}(\lambda)$, using

$$a_{pc}(624) = a_{phy}(624) - 0.5 \times [a_{phy}(648) + a_{phy}(600)]. \qquad (8.7)$$

As shown in Fig. 8.1, remote sensing reflectance is nearly reciprocal to a_{phy} around the spectral absorption peaks of pigments such as around 620–630 nm by PC absorption. Therefore, the baseline algorithm based on radiometric measurements was expressed as

$$[PC] \propto 0.5 \times [R_x(648) + R_x(600)] - R_x(624). \qquad (8.8)$$

This algorithm was originally developed for various lakes, rivers, and canal in Netherlands. Dekker (1993) found that the statistical relationship between [PC] and $0.5 \times [R(648) + R(600)] - R(624)$ [here $R(\lambda)$ is irradiance reflectance] were not stable due to the change of the absorption or scattering properties of the study sites. It indicates that the optical processes within the inland water bodies are complex and optical substances other than PC mask the optical response of the PC on radiometric signal of the water column.

Li et al. (2012) reported that the original baseline algorithm (Eq. 8.8) could not achieve a satisfactory performance for three central Indiana reservoirs which may have very different optical characteristics from those lakes, rivers, and canal in Netherlands. They, thus, proposed a new baseline algorithm using two three-band indices:

$$R31 = [R_x(624)^{-1} - R_x(600)^{-1}] \times R_x(725) \qquad (8.9a)$$

and

$$R32 = [R_x(648)^{-1} - R_x(600)^{-1}] \times R_x(725). \qquad (8.9b)$$

The two three-band indexes were then inversed to absorption $a(\lambda)$ and backscattering $b_b(\lambda)$ using Eq. 8.1. Combining Eqs. 8.1, 8.7, and 8.9, the final form of the algorithm in Li et al. (2012) is

$$[PC] = \frac{1}{2} \frac{[a_w(725) + b_b(725)](R31 + R32) - 2a_w(624) + a_w(600) + a_w(648)}{a^*_{pc}(624)},$$

(8.10)

where

$$b_b(725) = \frac{a_w(725)R^-_{rs}(725)}{0.082 - R^-_{rs}(725)},$$

(8.11)

with remote sensing reflectance below water surface $R^-_{rs}(\lambda)$ converted from remote sensing reflectance above water surface $R_{rs}(\lambda)$ using $R^-_{rs}(\lambda) = R_{rs}(\lambda)/0.54$ (Mobley, 1999) and $a_w(\lambda)$ from Buiteveld et al. (1994). The band 725 nm was used to correct the effect of $b_b(\lambda)$ at shorter wavelengths and can be theoretically replaced with any spectral bands in the near-infrared spectral region where $a(\lambda) \approx a_w(\lambda)$. Compared to Dekker (1993), the improvement by Li et al. (2012) result from the explicit attempts to remove the interference from $a_{CDM}(\lambda)$, $b_b(\lambda)$, and $a_w(\lambda)$. In principle, Eq. 8.10 is a semi-analytical algorithm because no separate statistical regression is needed to derive [PC]. However, it is still recommended to have a statistical regression analysis to determine $a^*_{pc}(624)$ due to two reasons: (1) the assumption of flat spectral $a_{CDM}(\lambda)$ and $b_b(\lambda)$ and 2) $a^*_{pc}(624)$ was reported to vary under different nutrient and light conditions (Tandeau de Marsac, 1977).

Two more semi-empirical algorithms for [PC] published recently are Dash et al. (2011) and Qi et al. (2014). Dash et al. (2011) found that [PC] could be derived from the remote sensing reflectance slope between 556.4 and 510.6 nm:

$$[PC] \propto \frac{R_{rs}(556.4) - R_{rs}(510.6)}{556.4 - 510.6}.$$

(8.12)

This algorithm was used to map [PC] in Lac des Allemands, Louisiana, USA using Oceansat-1 satellite Ocean Color Monitor data. Qi et al. (2014) suggested that [PC] related to the spectral band difference around 620 nm:

$$[PC] \propto R_{rs}(560) + \frac{620 - 560}{665 - 560}[R_{rs}(665) - R_{rs}(560)] - R_{rs}(620).$$

(8.13)

They assumed that the band difference at 620 nm was mainly caused by PC absorption, and the algorithm was used to map [PC] in Lake Taihu and Dianchi in China using MEdium Resolution Imaging Spectrometer (MERIS) images.

8.3.3 Semi-analytical Algorithms

While empirical and semi-empirical algorithms estimate [PC] directly from radiometric measurements, semi-analytical algorithms otherwise firstly derive

IOPs [e.g., $a_{pc}(\lambda)$] and then determine [PC] from $a_{pc}(\lambda)$. Therefore, semi-analytical algorithms are also called indirect or two-step algorithms. The first step of semi-analytical algorithms of [PC] generally depends on the radiative transfer theory (Eq. 8.1, Gordon et al. 1988) and other assumptions to explicitly separate $a_{pc}(\lambda)$ from absorption of other constituents and pigments, and the second step is to compute [PC] by simply assuming that [PC] physically (either linearly or nonlinearly) correlates with $a_{pc}(\lambda)$ (e.g., Eq. 8.2). Based on such definition, quasi-analytical algorithms are also reviewed in this section due to the insignificant difference between quasi-analytical and semi-analytical algorithms. Because semi-analytical algorithms use minimal number of empirical assumptions and maximal physical basis during their derivation, this type of algorithms often has best spatiotemporal transferability although the usage of empirical assumption(s) may degrade such transferability.

The first and widely used semi-analytical algorithm for remote estimation of [PC] was initially derived by Simis et al. (2005) for Lake Loosdrecht and Lake IJsselmeer in Netherlands and later applied to both Spanish and Netherlands lakes (Simis et al., 2007). The Simis' algorithm was formulated as the following:

$$[PC] = a_{pc}(620)/a_{pc}^*(620), \tag{8.14a}$$

with $a_{pc}^*(620) = 0.007 \text{ m}^2 \text{ (mg PC)}^{-1}$ and

$$a_{pc}(620) = a_{phy}(620) - \varepsilon \times a_{phy}(665), \tag{8.14b}$$

where ε is a constant 0.24 to calculate the contribution of chlorophylls absorption to $a_{phy}(620)$,

$$a_{phy}(620) = \left(\frac{R_x(709)}{R_x(620)} \times (0.727 + b_b) - b_b - 0.281\right) \times \delta^{-1}, \tag{8.14c}$$

and

$$a_{phy}(665) = \left(\frac{R_x(709)}{R_x(665)} \times (0.727 + b_b) - b_b - 0.401\right) \times \gamma^{-1}, \tag{8.14d}$$

in which b_b is the backscattering coefficients at 778 nm which is determined using Eq. 8.11 by changing light wavelength from 725 to 778 nm, and δ and γ are two constants, 0.84 and 0.68, respectively. The Simis' algorithm so far has been validated or modified in central Indiana reservoirs in the United States (Randolph et al., 2008), Spanish, and Netherlands lakes (Ruiz-Verdu et al., 2008), European Lakes (Guanter et al., 2010), Lake Taihu, and Dianchi in China (Lyu et al., 2013).

It is worth noting that Simis et al. (2005, 2007) assumed flat spectral $b_b(\lambda)$ and ignored $a_{CDM}(\lambda)$, which was suggested to be potential issues for application of this algorithm to water bodies with spectrally-variant $b_b(\lambda)$ and high $a_{CDM}(\lambda)$ (Mishra et al., 2013; Li et al., 2015). Therefore, Lyu et al.

(2013) attempted to accommodate such influence by empirically adjusting ε, δ, and γ using locally measured dataset in highly turbid water bodies. Another potential factor affecting the performance of the Simis' algorithm is the absorption of other accessory pigments such as chlorophyll-b and chlorophyll-c (Simis et al., 2007). In order to address all these potential issues, two semi-analytical (or quasi-analytical) algorithms were recently developed by Mishra et al. (2013) and Li et al. (2015).

Mishra et al. (2013) firstly derived $a_{phy}(\lambda)$ from $R_{rs}(\lambda)$ using a reparameterized quasi-analytical algorithm (QAA) for turbid waters (Mishra et al., 2014). Upon the derivation of $a_{phy}(\lambda)$ at 665 and 620 nm, the PC absorption at 620 nm was determined using

$$a_{pc}(620) = \frac{\psi_1 a_{phy}(620) - a_{phy}(665)}{\psi_1 - \psi_2}, \qquad (8.15a)$$

where

$$\psi_1 = 2.867 \ln \frac{R_x(560)}{R_x(665)} + 2.214, \qquad (8.15b)$$

and

$$\psi_2 = 0.254 \left[\frac{R_x(620)}{R_x(665)} \right]^{2.219}. \qquad (8.15c)$$

[PC] was subsequently derived from $a_{pc}(620)$ using Eq. 8.14a by assuming $a_{pc}^*(620) = 0.0048 \text{ m}^2 (\text{mg PC})^{-1}$. In general, Mishra et al. (2013) used the same principle as in Simis et al. (2005, 2007) to retrieve $a_{pc}(620)$ from $a_{phy}(665)$ and $a_{phy}(620)$ except that the contribution of $a_{phy}(665)$ to $a_{phy}(620)$ was not assumed to be a constant but varied with remote sensing reflectance through ψ_1 and ψ_2. It could explain why better estimation accuracy was achieved than the Simis' algorithm. However, the limitation of semi-analytical algorithm by Mishra et al. (2013) may lie on the component QAA, which was initially developed for oceanic waters (Lee et al., 2002), if the model is applied to non-case 1 waters. Many studies (e.g., Le et al., 2009; Aurin and Dierssen, 2012; Li et al., 2013) found that QAA does not work for water bodies that are turbid or rich in colored dissolved organic matter, and reparameterization of QAA was required in order to achieve satisfactory performance to retrieve absorption coefficients in inland waters. Mishra et al. (2013, 2014) shared the same dataset for calibrating QAA and deriving Eq. 8.15. It is likely that the QAA component in Mishra et al. (2013) needs be recalibrated again when applying to different water bodies.

Li et al. (2015), on the other hand, used different approach to derive $a_{phy}(\lambda)$ and $a_{pc}(\lambda)$. The complete steps to derive [chl-a] and [PC] by the Li's IIMIW model are shown in Table 8.1. The Li's IIMIW model firstly derives $b_b(778)$ from $R_x(778)$ using Eq. 8.1 with assumption of $a(778) \approx a_w(778)$. The $b_b(778)$ was then extrapolated to shorter wavelengths by a power

TABLE 8.1 Steps to Estimate IOPs, [chl-a], and [PC] for Turbid and Productive Waters Using IIMIW Model Developed by Li et al. (2013, 2015)

Step	Variable	Formula
1	$b_b(778)$	$b_b(778) = \dfrac{R^-_{rs}(778)a_w(778)}{0.082 - R^-_{rs}(778)}$ If necessary, $R^-_{rs}(\lambda) = R_{rs}(\lambda)/0.54$
2	Y	$Y = 2.0 \times [1 - 1.2\exp(-0.9R^-_{rs}(443)/R^-_{rs}(560))]$
3	$b_{bp}(560)$	$b_{bp}(560) = [b_b(778) - b_{bw}(778)]/0.7198^Y$
4	$b_b(\lambda)$	$b_b(\lambda) = b_{bp}(560)(560/\lambda)^Y + b_{bw}(\lambda)$
5	$a_{nw}(\lambda)$	$a_{t-w}(\lambda) = \dfrac{R_x(709)b_b(\lambda)[a_w(709) + b_b(709)]}{R_x(\lambda)b_b(709)} - b_b(\lambda) - a_w(\lambda),$ where $R_x(\lambda)$ represents either $R_{rs}(\lambda)$ or $R^-_{rs}(\lambda)$
6	[chl-a] (mg m^{-3})	$[chl - a] = [-0.3319a_{sol}(630) - 1.7485a_{sol}(647)$ $\qquad\qquad + 11.9442a_{sol}(665) - 1.4306a_{sol}(691)] \times 4.34$ where $a_{sol}(665) = a_{nw}(665)/0.68$ and $a_{sol}(\lambda_i) = a_{t-w}(\lambda_i)$ ($\lambda_i = 630, 647$ and 691, respectively)
	[chl-a] (mg m^{-3})	$[chl-a] = a_{t-w}(665)/a^*_{ph}(665),$ $a^*_{ph}(665) = 0.016$ m^2 (mg chl-a)$^{-1}$
7[a]	$a_{phy\text{-}pc}(\lambda)$	$a_{ph-pc}(\lambda) = 1.1872C1(\lambda)a_{nw}(665) + C2(\lambda)$
8	$a_{CDM+pc}(\lambda)$	$a_{CDM+pc}(\lambda) = a_{nw}(\lambda) - a_{phy-pc}(\lambda)$
9	S_{CDM}	$a_{CDM}(412) = a_{CDM+pc}(412), a_{CDM}(510) = a_{CDM+pc}(510),$ $S_{CDM} = \dfrac{1}{98}\ln\dfrac{a_{CDM}(412)}{a_{CDM}(510)}$
10	$a_{CDM}(\lambda)$	$a_{CDM}(\lambda) = a_{CDM}(412)\exp[-S_{CDM} \times (\lambda - 412)]$ and $a_{CDM}(709)$ is forced to 0.
11	$a_{phy}(\lambda)$	$a_{phy}(\lambda) = a_{nw}(\lambda) - a_{CDM}(\lambda)$
12	[PC] (mg m^{-3})	$[PC] = a_{pc}(620)/a^*_{pc}(620)$, where $a^*_{pc}(620) = 0.006$ m^2 (mg PC)$^{-1}$ and $a_{pc}(620) = a_{phy}(620) - a_{phy-pc}(620)$
13[b]	[PC] (mg m^{-3})	$[PC] = a_{pc}(620)/0.0043$, where $a_{pc}(620) = a_x(620) - a_x(600) - \frac{20}{48}[a_x(648) - a_x(600)]$

Refer to Li et al. (2015) for Description of Symbols and Acronyms Not Included in This Book.
[a]*C1 and C2 can be found in Li et al. (2015).*
[b]*IOP-baseline algorithm, a_x can be either a_{phy} or a_{nw}.*

function. Total absorption coefficients $a(\lambda)$ was subsequently determined from band ratio $R_x(709)/R_x(\lambda)$ according to Eq. 8.1. The band ratio was utilized to minimize the influence of variation in the spectral constant $D(\lambda)$ which has only weak spectral dependence but varies much for different limnological conditions. The fact that only chl-*a* and accessory pigments such as chlorophyll-*b*, chlorophyll-*c*, and carotenoids are soluble in acetone but not PC was used to determine phytoplankton absorption without PC contribution, $a_{phy-pc}(\lambda)$, from non-water absorption coefficient $a_{nw}(\lambda)$ [Note: $a_{nw}(\lambda) = a(\lambda) - a_w(\lambda)$], at 665 nm:

$$a_{phy-pc}(\lambda) = 1.1872 C1(\lambda) a_{nw}(665) + C2(\lambda) \qquad (8.16)$$

where $C1(\lambda)$ and $C2(\lambda)$ are spectral constants derived from $a_{phy-pc}(\lambda)$ measured in pigment extraction using acetone in laboratory and shown in the Appendix B of Li et al. (2015). The next step was to derive $a_{CDM}(\lambda)$ from $a_{nw}(\lambda) - a_{phy-pc}(\lambda)$ at 412 and 510 nm based on the fact that PC has very weak absorption at wavelengths shorter than 550 nm. Thereafter, $a_{pc}(620)$ was calculated by

$$a_{pc}(620) = a_{nw}(620) - a_{CDM}(620) - a_{phy-pc}(620) \qquad (8.17)$$

Finally, [PC] was determined using Eq. 8.14a by assuming $a_{pc}^*(620) = 0.006 \text{ m}^2 (\text{mg PC})^{-1}$ (see discussion in Section 8.4.6). The advantages of the Li's IIMIW model come from: (1) it was specifically derived for turbid and productive inland waters and not restrained by models derived for oceanic waters such as QAA; (2) it explicitly address the potential influence of spectrally variant $a_{CDM}(\lambda)$ and $b_b(\lambda)$; and (3) it was also attempted to address effects by chl-*a* and accessory pigments. As a result, better performance was achieved by Li's IIMIW, especially for low [PC], compared to Simis' algorithm (Li et al., 2015). However, limitation of Li's IIMIW model is that $C1(\lambda)$ and $C2(\lambda)$ using in the model were determined *in vitro* and only for three central Indiana reservoirs. Further efforts may be needed to derive *in vivo* $C1(\lambda)$ and $C2(\lambda)$ for different water bodies using the approach developed by Simis and Kauko (2012).

8.4 EVALUATION OF REPRESENTATIVE ALGORITHMS USING A LARGE FIELD DATASET

A few algorithms discussed above are selected and evaluated using a large field dataset for the purpose of further illustrating their advantages and limitations for remote sensing of [PC].

8.4.1 Description of the Field Dataset

The field dataset was collected in three central Indiana reservoirs, Eagle Creek Reservoir (ECR, 39°51′ N, 86°18.3′ W), Geist Reservoir (GR, 39°55′ N, 85°56.7′ W), and Morse Reservoir (Mr, 40°6.4′ N, 86°2.3′ W). These reservoirs have similar depth (3.2–4.7 m), surface area (5–7.5 km²), volume (21–28 million m³), and residence time (55–70 days). They were selected for the study because they are important sources for supplying drinking water to residents surrounding the Indianapolis metropolitan area and all exhibit toxic cyanobacterial blooms as a result of severe eutrophication. Field campaigns were conducted on these three reservoirs between 2005 and 2010 to determine remote sensing reflectance as well as [PC], [chl-*a*], and total suspended matter concentrations ([TSM]). The detailed description of how these parameters were measured can be found in Li et al. (2015) while it is only briefly described below.

Remote sensing reflectance: In 2005 and 2006, an ASD FieldSpec ultraviolet/visible and near-infrared (UV/VNIR) spectroradiometer (Analytical Spectral Devices, Inc., Boulder, CO, USA) was used to measure the remote sensing reflectance above the water surface by following the NASA's protocol (Mueller and Fargion, 2002) except no sky radiance was determined. In order to remove the effects of reflected sunglint and sky radiance by the water surface, the method by Kutser et al. (2013) was applied to derive the final above water-surface remote sensing reflectance, $R_{rs}(\lambda)$. In 2007, 2008, and 2010, an Ocean Optics USB4000 unit (Ocean Optics, Inc., Dunedin, FL, USA) with dual radiometers was used to measure remote sensing reflectance below water surface $R_{rs}^-(\lambda)$ by following the procedure described in Gitelson et al. (2007). The measured raw data were converted to $R_{rs}^-(\lambda)$ using the manufacturer software CALMIT Data Acquisition Program (CDAP), during which process the immersion factor for upwelling radiance was applied. The total 649 $R_{rs}(\lambda)$ and $R_{rs}^-(\lambda)$ spectra are shown in Fig. 8.2. We note R_{rs} and R_{rs}^- can be converted to each other by $R_{rs}(\lambda)=0.54 \, R_{rs}^-(\lambda)$ (Mobley, 1999).

PC, chl-a and total suspended matter concentrations: Water samples were collected at multiple depths from surface to the euphotic depth at all sampling locations where *in situ* reflectance was measured. Water at different depths was mixed in a big bucket, filled into 1-liter amber high-density polyethylene bottle, and then temporarily stored on ice in coolers. Once shipped back to laboratory, samples were filtered and frozen immediately to prevent pigments denaturalization. For the determination of [PC] of the samples collected in all years, a homogenization method with a tissue grinder was modified to improve the accuracy of measurements (Sarada et al., 1999), which is described in Randolph et al. (2008). Samples were concentrated onto 0.7 μm pore size glass fiber filter (Millipore APFF). The filter was then grinded in 50 mmol L^{-1} sodium phosphate buffer (pH 7.0 ± 0.2) using a stainless steel spatula and centrifuged under temperature of 5°C and speed of 25,000 rounds per minute for 25 minutes, which procedure was repeated for at least another time. The [PC] of upper supernatant was fluorometrically determined using precalibrated TD700-fluorometer (Turner Designs, Inc., San Jose, CA, USA). For determination of [chl-*a*], water samples for chlorophyll-*a* extraction were concentrated onto 0.45 μm pore size acetate filters (Whatman) and later extracted in 90% acetone for about 24 hours. For water samples collected between 2005 and 2008, [chl-*a*] was fluorometrically determined by a TD700-fluorometer (Li et al., 2010) and for

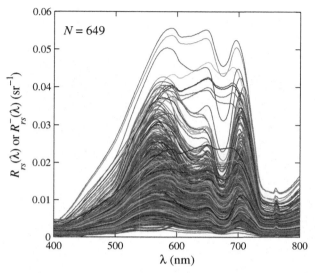

FIGURE 8.2 Remote sensing reflectance collected in three Indiana central reservoirs between 2005 and 2010. In 2005 and 2006, the above surface remote sensing reflectance $R_{rs}(\lambda)$ was measured using ASD spectrometer. In 2007, 2008, and 2010, the below surface remote sensing reflectance $R_{rs}^-(\lambda)$ was measured using Ocean Optics USB4000.

water samples collected in 2010, [chl-*a*] was determined using approach described in Ritchie (2008). Finally, for determination of [TSM], 100−300 ml of water was filtered onto preashed (530°C for 1.5 hours), prefiltered (200 ml Milli-Q water), and preweighed GF/F filters (Whatman) according to the turbidity of samples. The filters with particles were then dried at 105°C for 1.5 hours and weighed with an electronic balance. The TSM weight was calculated as the difference between weight of the filter with particles on and weight of the blank filter. The [TSM] was computed as the ratio of the TSM weight to the volume of filtered water. The statistics of [PC], [chl-*a*], [PC]:[chl-*a*], and [TSM] is shown in Table 8.2.

8.4.2 Evaluation of the Estimation Accuracy

The estimation accuracy for all selected [PC] models is evaluated by relative error (RE) and mean of relative error (MRE) defined in Eq. (8.18) and (8.19), respectively.

$$\mathrm{RE} = 100\% \times \frac{|X_i' - X_i|}{X_i}, \tag{8.18}$$

$$\mathrm{MRE} = 100\% \times \sum_{i=1}^{N} \frac{|X_i' - X_i|}{X_i} \Big/ N, \tag{8.19}$$

TABLE 8.2 [a] Statistics of Chlorophyll-*a* ([chl-*a*]), Phycocyanin ([PC]), and Total Suspended Matter ([TSM]) Concentrations as well as [PC]:[chl-*a*]

Date	Site	[chl-*a*] (mg m^{-3})			[PC] (mg m^{-3})			[PC]:[chl-*a*]			[TSM] (g m^{-3})		
		Min	Max	Avg.	Min	Max	Avg.	Min	Max	Avg.	Min	Max	Avg.
Sep	MR	18.02	168.58	55.81	2.19	135.06	43.37	0.08	1.68	0.68	3.86	31.13	11.03
2005	GR	34.71	118.90	71.45	25.20	185.06	94.57	0.35	2.33	1.34	6.30	22.00	15.80
May–Oct 2006	ECR	2.93	107.46	33.64	0.73	234.28	58.47	0.05	5.19	1.38	3.14	38.24	11.69
	MR	21.30	128.70	52.30	3.33	370.95	88.71	0.08	4.82	1.69	5.83	34.23	15.21
	GR	23.10	182.55	70.09	2.55	210.20	82.82	0.09	3.85	1.18	6.43	33.25	17.40
Aug–Sep 2007	ECR	16.88	255.00	124.09	30.91	114.05	74.88	0.33	1.83	0.72	10.45	38.14	19.46
	MR	26.88	214.20	101.23	41.24	136.29	81.66	0.41	1.53	1.03	8.39	123.79	30.31
	GR	14.45	193.18	144.04	0.91	149.95	92.59	0.06	0.88	0.58	14.56	38.05	23.55
Jul–Nov 2008	ECR	16.00	285.80	53.40	5.67	131.87	36.94	0.12	3.19	0.85	3.24	17.08	9.29
	MR	27.20	215.40	70.57	48.41	220.43	138.84	0.90	3.56	2.19	3.09	33.77	10.98
	GR	30.20	72.60	51.14	18.73	204.75	132.80	0.47	3.42	2.50	2.34	19.04	7.98
Apr–Oct 2010	ECR	21.62	128.04	54.51	8.58	72.24	32.59	0.17	2.58	0.74	5.83	36.17	13.22
	MR	1.85	129.39	60.21	1.39	146.10	66.44	0.19	1.90	0.97	6.67	30.33	17.16
	GR	8.31	62.12	34.47	6.57	140.76	66.94	0.14	3.45	1.70	5.17	81.17	12.67
All data		1.85	285.80	59.40	0.73	370.95	74.43	0.05	5.19	1.27	2.34	123.79	14.54

MR, Morse Reservoir, GR, Geist Reservoir, and ECR, Eagle Creek Reservoir.
[a]*Cite from Li et al. (2015).*

where X'_i and X_i are the estimated and measured values for sample i respectively, and N is the total sample number. The RE and MRE are selected because they reflect the estimation accuracy of each individual sample in the dataset, which will be needed for analyzing the factors affecting the performance of the selected models.

8.4.3 Evaluation of a Band Ratio Algorithm

As there are many empirical and semi-empirical algorithms available (see Sections 8.3.1 and 8.3.2), they all require recalibration when applied to different dataset other than the one where the models were originally developed. For example, Song et al. (2013) reported that both the spectral bands and statistical relationship between [PC] and band ratio or three-band index were different for data collected in different years even for the same study sites. For the purpose of illustrating such feature of empirical and semi-empirical models, a band ratio algorithm is evaluated using the field dataset because band ratio algorithms are the simplest and one of most representative semi-empirical algorithms that estimates [PC] directly from the ratio of $R_{rs}(\lambda)$ or $R_{rs}^-(\lambda)$ at two different wavelengths. Because it is not meant to determine which algorithm is best among all empirical and semi-empirical algorithms, the earliest band ratio algorithm developed by Schalles and Yacobi (2000) (Eq. 8.3 with $\lambda_1 = 650$ and $\lambda_2 = 625$) is examined using the Indiana field dataset.

The model using this band ratio for lakes in Nebraska, USA cannot achieve an acceptable accuracy when being directly applied to the three central Indiana reservoirs. Fig. 8.3(A) shows the comparison between estimated and measured [PC] using the band ratio model in Schalles and

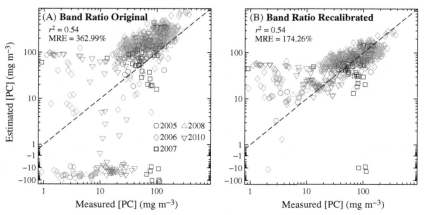

FIGURE 8.3 Comparison between measured and estimated [PC] using band ratio $[R_{rs}^-(650)/R_{rs}^-(625)]$ algorithm. (A) Before calibration to Indiana dataset; and (B) after calibration to the Indiana dataset. The dashed line is 1:1 line in this figure and the following figures.

Yacobi (2000). Although a moderate correlation ($r^2 = 0.54$) is observed, it significantly overestimates the measured [PC] with MRE = 362.99%. In particular, the model performs poorly when [PC] is low as RE can be up to 10^4% and many estimated [PC] become negative. When the model was recalibrated using the Indiana field dataset, the performance of the model using the same band ratio was significantly improved with much lower MRE (=174.26%) and fewer negative estimations of [PC] (Fig. 8.3(B)). The comparison of Figs. 8.3(A) and 8.3(B), before and after site-specific calibration respectively, well indicates the dataset-specific nature of semi-empirical models and the requirements of recalibration when being applied to a new dataset. The still low r^2 and high MRE even after recalibration suggest that different spectral bands should also be determined using iterative searching, which technique can be easily found in studies of semi-empirical algorithms in Section 8.3.2 and thus is not shown in this chapter.

8.4.4 Evaluation of Semi-analytical Models

Among the very few available semi-analytical models for [PC], the model by Simis et al. (2005, 2007) (Eq. 8.14) currently is most widely used and the IIMIW model by Li et al. (2015) (Table 8.1) was suggested to be able to improve the estimation accuracy of low [PC]. Therefore, the two models are evaluated.

The performance of these two semi-analytical models is shown in Fig. 8.4. Unlike empirical and semi-empirical models, no recalibration was done for these semi-analytical models. Fig. 8.4(A) shows the comparison between estimated [PC] by the Simis' model and measured [PC]. Even though no recalibration was applied, the Simis' model already performs better than the calibrated band ratio model shown in Fig. 8.3(B), with higher

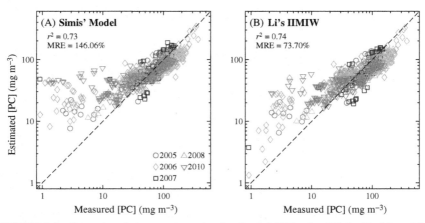

FIGURE 8.4 Comparison between measured and estimated [PC] using (A) the Simis' model (Simis et al., 2007) and (B) the Li's IIMIW model (Li et al., 2015).

$r^2 = 0.73$ and lower MRE = 146.06%. In spite of improved accuracy for low [PC] (e.g., ≤ 10 mg m^{-3}), significant overestimation is still observed for the Simis' model. Meanwhile, the Li's IIMIW model generally performs better than the Simis' model with similar estimation accuracy for relatively high [PC] and better accuracy for low [PC] (Fig. 8.4(B)). The improved accuracy for low [PC] is well reflected by the much lower MRE = 73.70% although r^2 (=0.74) stays similar. It is noted that the $a^*_{pc}(620)$ used in original IIMIW model (Li et al., 2015) was changed to 0.006 m^2 (mg PC)$^{-1}$, very close to the Simis' $a^*_{pc}(620)$, for accommodating the temporal variations over 6 years of field campaigns in study sites of another continent.

8.4.5 Evaluation of Two Baseline Algorithms Using AOP and IOP

To illustrate the advantages to determine [PC] from remotely estimated IOPs other than directly from $R_{rs}(\lambda)$ and $R_{rs}^-(\lambda)$, the performances of Dekker's and a new IOP-based baseline model are compared in Fig. 8.5. The baseline algorithm by Dekker (1993) (Eq. 8.8) is a pioneering semi-empirical algorithm. Because $R_{rs}(\lambda)$ or $R_{rs}^-(\lambda)$ is an AOP, Dekker's baseline algorithm is also called AOP-baseline algorithm in this chapter. This baseline algorithm also requires recalibration when being applying to different study sites due to masking effects of other water constituents (Dekker, 1993). Its performance after site-specific calibration for the Indiana dataset is shown in Fig. 8.5(A). Among the evaluated algorithms in this chapter, this baseline algorithm directly using $R_{rs}(\lambda)$ or $R_{rs}^-(\lambda)$ has the worst performance even after site-specific calibration, with $r^2 = 0.21$ and MRE = 228.86%, although

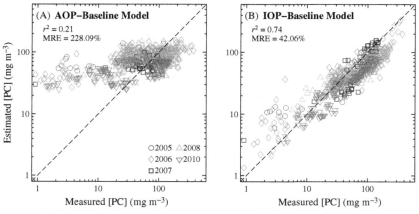

FIGURE 8.5 Comparison between measured and estimated [PC] using (A) calibrated baseline algorithm using $R_{rs}^-(\lambda)$ according to Dekker (1993) (Eq. 8.8) and (B) baseline algorithm using a_{phy} derived from the IIMIW by Li et al. (2015) (Eq. 8.20).

no negative estimations are found. This algorithm again shows very poor performance on estimating low [PC].

The discussion in Sections 8.4.4 implies that semi-analytical models often have a better performance and do not require extensive site-specific calibration. It is because semi-analytical models utilize more physical basis (Ogashawara, 2015) to firstly derive absorption coefficient from $R_{rs}(\lambda)$ or $R_{rs}^-(\lambda)$ and then estimating [PC] from retrieved absorption coefficient, an IOP. Using the same fact that PC has an absorption peak around 620 nm, based on which the Dekker's baseline algorithm was established, Li et al. (2012) attempted to develop a semi-analytical baseline algorithm for [PC] using two three-band indices. However, a few assumptions that do not favor the accurate estimation of [PC] were made, for example, the assumptions of flat spectral $a_{CDM}(\lambda)$ and $b_b(\lambda)$. The issues due to these assumptions were minimized in the Li's IIMIW model and a new IOP-baseline model can be developed using the similar physical basis shown in Eq. 8.7:

$$[PC] = a_{pc}(620)/0.0043, \qquad (8.20a)$$

where

$$a_{pc}(620) = a_x(620) - a_x(600) - \frac{620 - 600}{648 - 600}[a_x(648) - a_x(600)]. \qquad (8.20b)$$

The $a_x(\lambda)$ in Eq. 8.20b can be either $a_{nw}(\lambda)$ or $a_{phy}(\lambda)$ derived from $R_{rs}(\lambda)$ or $R_{rs}^-(\lambda)$ using the Li's IIMIW model shown in Table 8.1. A few negative values were found for [PC] ≤ 10 mg m^{-3} when using a_{nw} (not shown). The occurrences of negative values of estimated [PC] might be due to the effects of $a_{CDM}(\lambda)$ because $a_{nw}(\lambda) = a_{phy}(\lambda) + a_{CDM}(\lambda)$. Other than that, IOP-baseline models using both $a_{nw}(\lambda)$ and $a_{phy}(\lambda)$ achieve similar performance. Fig. 8.5(B) shows the comparison between estimated [PC] using $a_{phy}(\lambda)$-baseline model and measured [PC]. Much better performance is achieved by the IOP-baseline model compared to AOP-baseline model. In fact, the IOP-baseline algorithm performs the best among the five evaluated models in this chapter with lowest MRE (=42.06%). In particular, it has the best estimation accuracy for samples with [PC] ≤ 10 mg m^{-3} although overestimation for low [PC] is still observed. Therefore, this new IOP-baseline model is appended to Table 8.1 as extended steps of IIMIW model for [PC] estimation.

8.4.6 Discussion of Factors Influencing the Remote Estimation of Phycocyanin

Several factors may influence the modeling accuracy of [PC] retrieval using remotely sensed data. The first factor that may affect the remote estimation of [PC] is the vertical distribution of cyanobacteria within the water column.

Many cyanobacteria have the capability to migrate in the water column (Paerl and Huisman, 2009) and the stratification of cyanobacterial biomass could result in different remote sensing reflectance (Kutser et al., 2008). As a result, it may affect the retrieval of [PC] from remote sensing reflectance. Nevertheless, the [PC] of the Indiana dataset represents the average [PC] within the euphotic layer because water samples were collected in multiple depths from surface to the euphotic depth and mixed together. In this case, the effect of stratification may not be significant for this dataset. However, it is worth further effort to study how vertical distributions of cyanobacterial biomass may influence the estimation accuracy of [PC].

It was suggested that $a_{CDM}(\lambda)$ can affect the estimation accuracy of [PC] (Li et al., 2012, 2015; Mishra et al., 2013). For example, the Simis' model (Simis et al., 2005, 2007) ignores the effects of $a_{CDM}(\lambda)$ by assuming $a_{CDM}(620) \approx 0$. Li et al. (2015) found that such assumption may lead to over-estimation of low [PC] by 1600%. As estimation of low [PC] is important for early warning of cyanobacteria (Hunter et al., 2009), it is of significance to minimize the effects of $a_{CDM}(\lambda)$ to improve the accuracy in estimation of low [PC]. Efforts were made by Mishra et al. (2013) and Li et al. (2015) to develop more sophisticated semi-analytical models to explicitly estimate $a_{CDM}(\lambda)$ and subsequently exclude it from determination of [PC]. Because $a_{CDM}(\lambda)$ was not measured for majority of water samples collected in the Indiana sites, it is not analyzed here how improvements were made by excluding $a_{CDM}(\lambda)$ from estimation of [PC], which otherwise can be found in Li et al. (2015) for only data collected in 2010.

The choice of phycocyanin-specific absorption coefficient at 620 nm, $a_{pc}^*(620)$, can also affect the remote estimation of [PC] (Li et al., 2015; Mishra et al., 2013; Simis et al., 2007). Generally $a_{pc}^*(620)$ varies insignificantly within different species of cyanobacteria (Simis and Kauko, 2012), but it varies with season, cell morphology, and pigment compositions due to changes in nutrient and light conditions (Tandeau de Marsac, 1977). There are not many measurements of $a_{pc}^*(620)$ available in literature and they varies from $0.0032 \text{ m}^2 \text{ (mg PC)}^{-1}$ to $0.0095 \text{ m}^2 \text{ (mg PC)}^{-1}$ (e.g., Dekker, 1993; Jupp et al., 1994; Mishra et al., 2013; Simis et al., 2005, 2007; Simis and Kauko, 2012). The variation in these measured $a_{pc}^*(620)$ is partly due to different methodologies they used to extract PC, determine [PC], and measure $a_{pc}(\lambda)$. Therefore, it is commonly found that $a_{pc}^*(620)$ are different for different models. For example, Simis et al. (2007) lowered $a_{pc}^*(620)$ to $0.007 \text{ m}^2 \text{ (mg PC)}^{-1}$ from $0.0095 \text{ m}^2 \text{ (mg PC)}^{-1}$ in Simis et al. (2005); Mishra et al. (2013) found the mean of $a_{pc}^*(620)$ was $0.0048 \text{ m}^2 \text{ (mg PC)}^{-1}$; and it is changed to $0.006 \text{ m}^2 \text{ (mg PC)}^{-1}$ in the Li's IIMIW model to accommodate the temporal changes over 6 years of field campaign. Yacobi et al. (2015) suggested that $a_{pc}^*(625)$, which is slightly higher than $a_{pc}^*(620)$, stays nearly stable around $0.007 \text{ m}^2 \text{ (mg PC)}^{-1}$ when $[PC] > 10 \text{ mg m}^{-3}$, but it is highly varied when $[PC] \leq 10 \text{ mg m}^{-3}$. They claimed that it is reasonable to estimate [PC] from remote sensing reflectance spectra using a constant $a_{pc}^*(620)$ in mesotrophic and eutrophic inland waters, but it is difficult to estimate low [PC], if not impossible, due to the high variation in $a_{pc}^*(620)$. It is encouraged to investigate more about the spatial and temporal variations in $a_{pc}^*(620)$ using a large dataset obtained by a uniform methodology [e.g., the approach by Simis and Kauko (2012) to determine *in vivo* $a_{pc}^*(\lambda)$], which may in turn improve the accuracy of remotely estimated [PC], in particular low [PC].

The most commonly reported influencing factor is the dominance of cyanobacteria over the entire phytoplankton biomass in the water column, which is represented by [PC]:[chl-*a*] (e.g., Hunter et al., 2010; Li et al., 2015; Mishra et al., 2013; Simis et al., 2007). Chl-*a* and accessory pigments (e.g., chl-*b* and chlorophyll-*c*) all contribute to the phytoplankton absorption at 620 nm (Simis et al., 2007), so [PC]:[chl-*a*] can be an indicator reflecting how much influence on $a_{pc}(620)$ may be caused by chl-*a* and other accessory pigments. Hunter et al. (2010) and Li et al. (2015) observed that errors in estimated PC increase significantly for water samples with [PC]: [chl-*a*] ≤ 0.5 because [PC]:[chl-*a*] ≤ 0.5 indicates cyanobacteria are not the dominant phytoplankton species in the water body. Their observations are supported by Fig. 8.6 in which relative error (RE) of an individual [PC] is analyzed against [PC]:[chl-*a*]. Almost all water samples with overestimation of [PC] by more than 100% have [PC]:[chl-*a*] ≤ 0.5 for all four models. [PC]:[chl-*a*] ≤ 0.5 often suggests low [PC] or the early stage of cyanobacterial blooms, which is critical for issuing early warning of possible cyanobacterial blooms to protect animal and human health (Hunter et al., 2009). Although the Li's IIMIW and IOP-baseline models already improve the estimation accuracy of low [PC] for water samples with [PC]:[chl-*a*] ≤ 0.5 (Fig. 8.6), further attempts are needed to additionally improve the estimation of low [PC] because the estimation error for low [PC] is still high.

8.5 MAPPING PC USING AIRBORNE IMAGES

In 2005, airborne images of Geist Reservoir (GR) and Morse Reservoir (MR) were acquired using the Airborne Imaging Spectrometer for Applications (AISA)-Eagle (Spectral Imaging Ltd. Oulu, Finland) sensor on board a Piper Saratoga airplane owned by the Center for Advanced Land Management Information Technologies (CALMIT), University of Nebraska at Lincoln (UNL). The images were acquired in the spectral region of approximately 392−982 nm with a bandwidth of 9−10 nm, that is, 62 bands in total. The images had 1 m spatial resolution and 1000 m wide swath. The acquired AISA images were then geo-rectified, normalized, mosaicked, and calibrated using ENVI 4.2 software. The detail about acquisition and processing of the AISA images can be found in Li et al. (2010).

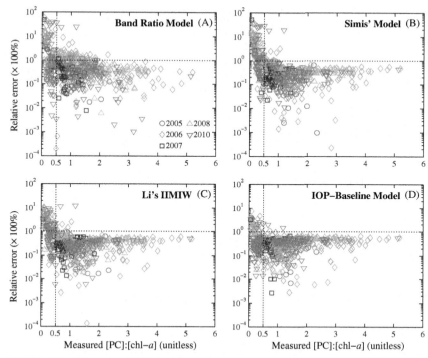

FIGURE 8.6   Influence of [PC]:[chl-*a*] on the accuracy of remotely estimated [PC] for (A) Calibrated band ratio model; (B) the Simis' model; (C) the Li's IIMIW model; and (D) the new baseline model based on a_{phy}.

The processed AISA images were then used to map [PC] in GR and MR. Because AISA-Eagle sensor does not have all the exact required bands for the Simis' and Li's IIMIW models, both of which were developed according to the bands available on MERIS and Ocean Land Colour Instrument satellite sensors, the IOP-baseline algorithm based on $a_{nw}(\lambda)$ was used to generate the maps of [PC] for the two reservoirs. The two maps are shown in Fig. 8.7. The spatial pattern transits smoothly and the ranges of predicted [PC] for GR and MR cover those reported in Table 8.2 for 2005. [PC] in GR were higher at the upright part of the reservoir due to high nutrients inputs and at the lower part of the reservoir due to calm water conditions being suitable for growth of cyanobacteria. In the middle portion of GR, [PC] is relatively low, revealing relatively low nutrients supply and more dynamic water conditions. [PC] in MR generally decrease from upright to lower part of the reservoir. It might be mainly due to the availability of nutrient (e.g., phosphorus) along different parts of the reservoir because the spatial pattern of [PC] correlates with the pattern of phosphorus reported in Song et al. (2012).

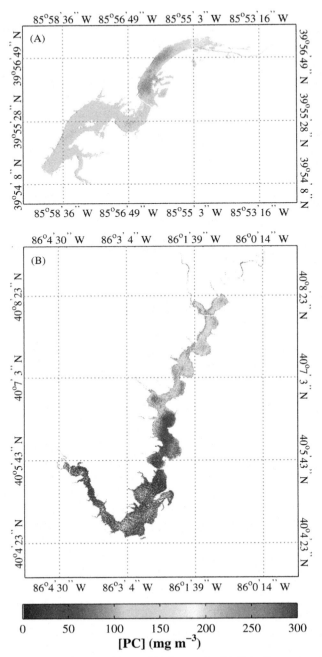

FIGURE 8.7 Maps of [PC] derived from AISA images for (A) Geist Reservoir and (B) Morse Reservoir.

8.6 SUMMARY AND FUTURE WORK

Cyanobacterial blooms are harmful blooms that jeopardize the safety of our daily water supply and recreational activities in lakes and reservoirs due to the production of surface scum, musty odor, and particularly various toxins. Various approaches have been developed to monitor the cyanobacterial abundance in the inland water bodies, including the innovative techniques such as remote sensing that are effective on monitoring the cyanobacterial growth in lakes and reservoirs. The focus of remote estimation of cyanobacteria has been placed on retrieving [PC] from remote sensing measurements.

In this chapter, the physical basis for remote sensing algorithms for [PC] were described in Section 8.2, and based on how the algorithms were developed, these algorithms were classified into three categories: empirical, semi-empirical, and semi-analytical. Their advantages and limitations were discussed and demonstrated by applying several selected algorithms to a large field dataset.

The largest group of algorithms is semi-empirical due to easy development and implementation. However, this type of algorithms generally performs well for the dataset where they were originally developed and a recalibration is often required to achieve satisfactory estimation accuracy when being applying to different site or dataset. Such requirement of statistical regression between constructed radiometric index and [PC] limits the wide use of this type of algorithms.

There are currently three existing semi-analytical algorithms for [PC]. As semi-analytical algorithms utilize most physical basis to firstly derive $a_{pc}(\lambda)$ and then derive [PC] from $a_{pc}(\lambda)$, this type of algorithms have shown best site and dataset transferability. For example, the two selected semi-analytical algorithms by Simis et al. (2005, 2007) and Li et al. (2015) performed well on the field dataset even without a site-specific calibration. Two maps of [PC] were generated from AISA airborne images using a semi-analytical algorithm described in Section 8.4.5, and the two maps well depict the spatial pattern of [PC] in the entire reservoir.

Many potential factors that influence the performance of these reviewed algorithms were discussed in Section 8.4.6. These factors include the vertical distribution of cyanobacterial cells, absorption of colored detritus matter, $a_{pc}^{*}(\lambda)$, and interference of chl-a and accessory pigments (e.g., [PC]:[chl-a] as a proxy). The most pronounced influencing factors may be $a_{pc}^{*}(\lambda)$ and [PC]:[chl-a], especially for improving the estimation accuracy of the low [PC]. An accurate estimation of low [PC] is of importance for issuing early warning of cyanobacterial blooms to the public. However, all examined algorithms in Section 8.4 showed large estimation errors for [PC] ≤ 10 mg m^{-3} although IOP-baseline algorithm achieved substantially improved accuracy within this range of low [PC] [Fig. 8.5(B)]. When [PC] is low, $a_{pc}^{*}(\lambda)$ varies much (Yacobi et al., 2015) and [PC]:[chl-a] < 0.5 (Hunter et al., 2010; Li et al., 2015), which imposes difficulties in accurately estimating [PC].

In summary, a significant progress has been achieved on remote sensing of [PC], that is, bio-optical modeling of PC. In particular, the emerging of semi-analytical algorithms of [PC] greatly promoted the possibility of using remote sensing approaches to detect PC and thus surveillance cyanobacterial abundance in inland water bodies. However, more efforts are still needed to further improve the accuracy of remotely estimated [PC]. Specifically, investigations are required to more comprehensively understand the variations of $a_{pc}^*(\lambda)$ in different waters with different limnological conditions as currently available data of $a_{pc}^*(\lambda)$ are relatively limited and there was no ubiquitous approach to determining $a_{pc}^*(\lambda)$. Attempts are also needed to further explore the feasibility to accurately estimate [PC] when [PC]:[chl-a] is less than 0.5, i.e., when cyanobacteria are not the dominant phytoplankton within the water column.

ACKNOWLEDGEMENTS

The authors would like to thank all of the graduate students and faculty in the Department of Earth Sciences, IUPUI, and the staff of CEES who participated in the sampling excursion and who assisted in the laboratory analyses. In particular, the principle investigator of the project, Prof. Lin Li, is highly appreciated for his permission to use the data collected under his supervision. We also thank the Veolia Water Indianapolis, LLC White River Laboratories for all of their assistance with sample analyses; the University of Nebraska CALMIT laboratory for acquiring the AISA imagery; Indy Parks and Eagle Creek Park for dock space; and Kent Duckwell for launch access to Geist Reservoir. The data were collected under financial supports of Indiana Department of Natural Resources LARE Grant and Veolia Water Company, LLC before 2009 and NASA Energy and Water Cycle Study program (Grant No. NNX09AU87G) in 2010. Thanks also extend to Natural Science Foundation of China (Grant No. 41471290) and "One Hundred Talents" Program from Chinese Academy of Sciences granted to Dr. Kaishan Song.

REFERENCES

Aurin, D.A., Dierssen, H.M., 2012. Advantages and limitations of ocean color remote sensing in CDOM-dominated, mineral-rich coastal and estuarine waters. Remote Sens. Environ. 125, 181–197.

Azevedo, S.M., Carmichael, W.W., Jochimsen, E.M., Rinehart, K.L., Lau, S., Shaw, G.R., et al., 2002. Human intoxication by microcystins during renal dialysis treatment in Caruaru—Brazil. Toxicology 181, 441–446.

Baker, J.A., Entsch, B., Neilan, B.A., McKay, D.B., 2002. Monitoring changing toxigenicity of a cyanobacterial bloom by molecular methods. Appl. Environ. Microbiol. 68 (12), 6070–6076.

Buiteveld, H., Hakvoort, J.H.M., Donze, M., 1994. The optical properties of pure water. SPIE Proc. Ocean Optics XII 2258, 174–183.

Carmichael, W.W., Azevedo, S.M., An, J.S., Molica, R.J., Jochimsen, E.M., Lau, S., et al., 2001. Human fatalities from cyanobacteria: chemical and biological evidence for cyanotoxins. Environ. Health Persp. 109 (7), 663–668.

Chorus, I., Bartram, J., 1999. Toxic cyanobacteria in water: a guide to their public health consequences, monitoring, and management. E&FN Spon., London and New York.

Codd, G.A., Chorus, I., Burch, M.D., 1999. Design of monitoring programmes. In: Chorus, I., Bartram, J. (Eds.), Toxic cyanobacteria in water: a guide to their public health consequences, monitoring, and management. E&FN Spon, London and New York, p. 416.

Codd, G.A., Morrison, L.F., Metcalf, J.S., 2005. Cyanobacterial toxins: risk management for health protection. Toxicol. Appl. Pharmacol. 203, 264–272.

Dash, P., Walker, N.D., Mishra, D.R., Hu, C., Pinckney, J.L., D'Sa, E.J., 2011. Estimation of cyanobacterial pigments in a freshwater lake using OCM satellite data. Remote Sens. Environ. 115 (12), 3409–3423.

Dekker, A.G., 1993. Detection of optical water quality parameters for eutrophic waters by high resolution remote sensing. Ph.D. dissertation. Vrije University, Amsterdam, Netherlands.

Falconer, I.R., Humpage, A.R., 1996. Tumour promotion by cyanobacterial toxins. Phycologia 35, 74–79.

Falconer, I.R., Beresford, A.M., Runnegar, M.T.C., 1983. Evidence of liver damage by toxin from a bloom of the blue-green alga Microcystis aeruginosa. Med. J. Aust. 1, 511–514.

Gitelson, A.A., Gritz, Y., Merzlyak, M.N., 2003. Relationships between leaf chlorophyll content and spectral reflectance and algorithms for non-destructive chlorophyll assessment in higher plant leaves. J. Plant Physiol. 160 (3), 271–282.

Gitelson, A.A., Schalles, J.F., Hladik, C.M., 2007. Remote chlorophyll-a retrieval in turbid, productive estuaries: chesapeake Bay case study. Remote Sens. Environ. 109, 464–472.

Gons, H., 1999. Optical teledetection of chlorophyll a in turbid inland waters. Environ. Sci. Technol. 33, 1127–1132.

Gons, H.J., Auer, M.T., Effler, S.W., 2008. MERIS satellite chlorophyll mapping of oligotrophic and eutrophic waters in the Laurentian Great Lakes. Remote Sens. Environ. 112, 4098–4106.

Gordon, H.R., Brown, O.B., Evans, R.H., Brown, J.W., Smith, R.C., Baker, K.S., et al., 1988. A semianalytic radiance model of ocean color. J. Geophys Res. 93, 10909–10924. Available from: http://dx.doi.org/10.1029/JD093iD09p10909.

Guanter, L., Ruiz-Verdu, A., Odermatt, D., Giardino, C., Simis, S., Estelles, V., et al., 2010. Atmospheric correction of ENVISAT/MERIS data over inland waters: validation for European lakes. Remote Sens. Environ. 114, 467–480.

Huang, C.C., Shi, K., Yang, H., Li, Y.M., Zhu, A.X., Sun, D.Y., et al., 2015. Satellite observation of hourly dynamic characteristics of algae with Geostationary Ocean Color Imager (GOCI) data in Lake Taihu. Remote Sens. Environ. 159, 278–287.

Hunter, P.D., Tyler, A.N., Présing, M., Kovács, A.W., Preston, T., 2008a. Spectral discrimination of phytoplankton colour groups: the effect of suspended particulate matter and sensor spectral resolution. Remote Sens. Environ. 112 (4), 1527–1544.

Hunter, P.D., Tyler, A.N., Willby, N.J., Gilvear, D.J., 2008b. The spatial dynamics of vertical migration by Microcystis aeruginosa in a eutrophic shallow lake: a case study using high spatial resolution time-series airborne remote sensing. Limnol. Oceanogr. 53, 2391–2406.

Hunter, P.D., Tyler, A.N., Gilvear, D.J., Willby, N.J., 2009. Using remote sensing to aid the assessment of human health risks from blooms of potentially toxic cyanobacteria. Environ. Sci. Technol. 43, 2627–2633.

Hunter, P.D., Tyler, A.N., Carvalho, L., Codd, G.A., Maberly, S.C., 2010. Hyperspectral remote sensing of cyanobacterial pigments as indicators for cell populations and toxins in eutrophic lakes. Remote Sens. Environ. 114, 2705–2718.

Humpage, A.R., Hardy, S.J., Moore, E.J., Froscio, S.M., Falconer, I.R., 2000. Microcystins (cyanobacterial toxins) in drinking water enhance the growth of aberrant crypt foci in the mouse colon. J. Toxicol. Environ. Health A 61, 155–165.

Jochimsen, E.M., Carmichael, W.W., An, J.S., Cardo, D.M., Cookson, S.T., Holmes, C.E.M., et al., 1998. Liver failure and death after exposure to microcystins at a hemodialysis center in Brazil. N. Engl. J. Med. 338, 873–878.

Jupp, D.L.B., Kirk, J.T.O., Harris, G.P., 1994. Detection, identification and mapping of cyanobacteria—using remote sensing to measure the optical quality of turbid inland waters. Aust. J. Mar. Freshw. Res. 45, 801–828.

Kutser, T., 2004. Quantitative detection of chlorophyll in cyanobacterial blooms by the satellite remote sensing. Limnol. Oceanogr. 49, 2179–2189.

Kutser, T., Metsamaa, L., Dekker, A.G., 2008. Influence of the vertical distribution of cyanobacteria in the water column on the remote sensing signal. Estuar. Coast. Shelf Sci. 78, 649–654.

Kutser, T., Vahtmae, E., Paavel, B., Kauer, T., 2013. Removing glint effects from field radiometry data measured in optically complex coastal and inland waters. Remote Sens. Environ. 133, 85–89.

Lawton, L.A., Beattie, K.A., Hawser, S.P., Campbell, D.L., Codd, G.A., 1994. Evaluation of assay methods for the determination of cyanobacterial hepatotoxicity. In: Codd, G.A., Jeffries, T.M., Keevil, C.W., Potter, E. (Eds.), Detection methods for cyanobacterial toxins. The Royal Society of Chemistry, Cambridge, United Kingdom, pp. 111–116.

Le, C.F., Li, Y.M., Zha, Y., Sun, D.Y., Yin, B., 2009. Validation of a quasi-analytical algorithm for highly turbid eutrophic water of meiliang bay in taihu lake, China. IEEE Trans. Geosci. Remote Sens. 47, 2492–2500.

Le, C.F., Li, Y.M., Zha, Y., Wang, Q., Zhang, H., Yin, B., 2011. Remote sensing of phycocyanin pigment in highly turbid inland waters in Lake Taihu, China. Int. J. Remote Sens. 32, 8253–8269.

Lee, Z.P., Carder, K.L., Arnone, R.A., 2002. Deriving inherent optical properties from water color: a multiband quasi-analytical algorithm for optically deep waters. Appl. Optics 41, 5755–5772.

Li, L., Sengpiel, R.E., Pascual, D.L., Tedesco, L.P., Wilson, J.S., Soyeux, E., 2010. Using hyperspectral remote sensing to estimate chlorophyll-a and phycocyanin in a mesotrophic reservoir. Int. J. Remote Sens. 31, 4147–4162.

Li, L., Li, L., Shi, K., Li, Z., Song, K., 2012. A semi-analytical algorithm for remote estimation of phycocyanin in inland waters. Sci. Total Environ. 435, 141–150.

Li, L., Li, L., Song, K., Li, Y., Tedesco, L.P., Shi, K., et al., 2013. An inversion model for deriving inherent optical properties of inland waters: establishment, validation and application. Remote Sens. Environ. 135, 150–166.

Li, L., Li, L., Song, K., 2015. Remote sensing of freshwater cyanobacteria: an extended IOP Inversion Model of Inland Waters (IIMIW) for partitioning absorption coefficient and estimating phycocyanin. Remote Sens. Environ. 157, 9–23.

Lunetta, R.S., Schaeffer, B.A., Stumpf, R.P., Keith, D., Jacobs, S.A., Murphy, M.S., 2015. Evaluation of cyanobacteria cell count detection derived from MERIS imagery across the eastern USA. Remote Sens. Environ. 157, 24–34.

Lyu, H., Wang, Q., Wu, C., Zhu, L., Yin, B., Li, Y.M., et al., 2013. Retrieval of phycocyanin concentration from remote-sensing reflectance using a semi-analytic model in eutrophic lakes. Ecol. Inform. 18, 178–187.

Metsamaa, L., Kutser, T., Strombeck, N., 2006. Recognising cyanobacterial blooms based on their optical signature: a modelling study. Boreal Environ. Res. 11, 493–506.

Mishra, S., 2012. Remote Sens. of Harmful Algal Bloom. Ph.D. Thesis. Mississippi State University, MS, USA.

Mishra, S., Mishra, D.R., 2014. A novel remote sensing algorithm to quantify phycocyanin in cyanobacterial algal blooms. Environ. Res. Lett. 9 (11), 114003(9pp). http://dx.doi.org/10.1088/1748-9326/9/11/114003.

Mishra, S., Mishra, D.R., Schluchter, W.M., 2009. A novel algorithm for predicting phycocyanin concentrations in cyanobacteria: a proximal hyperspectral remote sensing approach. Remote Sens. 1, 758–775.

Mishra, S., Mishra, D.R., Lee, Z.P., Tucker, C.S., 2013. Quantifying cyanobacterial phycocyanin concentration in turbid productive waters: a quasi-analytical approch. Remote Sens. Environ. 133, 141–151.

Mishra, S., Mishra, D.R., Lee, Z.P., 2014. Bio-optical inversion in highly turbid and cyanobacteria-dominated waters. IEEE Trans. Geosci. Remote Sens. 52 (1), 375–388.

Mobley, C.D., 1994. Light and water: radiative transfer in natural waters. Academic press, San Diego, USA.

Mobley, C.D., 1999. Estimation of the remote-sensing reflectance from above-surface measurements. Appl. Optics 38, 7442–7455.

Mueller, J.L., Fargion, G.S., 2002. Ocean optics protocols for satellite ocean color sensor validation. Revision 3 (2), 171–179.

Ogashawara, I., 2015. Terminology and classification of bio-optical algorithms. Remote Sens. Lett. 6 (8), 613–617.

Ogashawara, I., Mishra, D.R., Mishra, S., Curtarelli, M.P., Stech, J.L., 2013. A performance review of reflectance based algorithms for predicting phycocyanin concentrations in inland waters. Remote Sens. 5 (10), 4774–4798.

Ogashawara, I., Alcântara, E.H.D., Stech, J.L., Tundisi, J.G., 2014. Cyanobacteria detection in Guarapiranga Reservoir (São Paulo State, Brazil) using Landsat TM and ETM+ images. Revista Ambiente & Água 9 (2), 224–238.

Paerl, H.W., Huisman, J., 2008. Climate---blooms like it hot. Science 320, 57–58.

Paerl, H.W., Huisman, J., 2009. Climate change: a catalyst for global expansion of harmful cyanobacterial blooms. Environ. Microbiol. Rep. 1, 27–37.

Paerl, H.W., Otten, T.G., 2013. Harmful cyanobacterial blooms: causes, consequences, and controls. Microb. Ecol. 65 (4), 995–1010.

Paerl, H.W., Paul, W.J., 2012. Climate change: links to global expansion of harmful cyanobacteria. Water Res. 46, 1349–1363.

Pizarro, S.A., Sauer, K., 2001. Spectroscopic study of the light-harvesting protein C-Phycocyanin associated with colorless linker peptides. Photochem. Photobiol. 73 (5), 556–563, 2001.

Qi, L., Hu, C., Duan, H., Cannizzaro, J., Ma, R., 2014. A novel MERIS algorithm to derive cyanobacterial phycocyanin pigment concentrations in a eutrophic lake: theoretical basis and practical considerations. Remote Sens. Environ. 154, 298–317.

Randolph, K., Wilson, J., Tedesco, L., Li, L., Pascual, D.L., Soyeux, E., 2008. Hyperspectral remote sensing of cyanobacteria in turbid productive water using optically active pigments, chlorophyll a and phycocyanin. Remote Sens. Environ. 112, 4009–4019.

Reynolds, C.S., 1987. Cyanobacterial water blooms. Adv. Bot. Res. 13, 67–143.

Ritchie, R.J., 2008. Universal chlorophyll equations for estimating chlorophylls a, b, c, and d and total chlorophylls in natural assemblages of photosynthetic organisms using acetone, methanol, or ethanol solvents. Photosynthetica 46, 115−126.

Ruiz-Verdu, A., Simis, S.G.H., de Hoyos, C., Gons, H.J., Pena-Martinez, R., 2008. An evaluation of algorithms for the remote sensing of cyanobacterial biomass. Remote Sens. Environ. 112, 3996−4008.

Sarada, R., Pillai, M.G., Ravishankar, G.A., 1999. Phycocyanin from Spirulina sp: influence of processing of biomass on phycocyanin yield, analysis of efficacy of extraction methods and stability studies on phycocyanin. Process Biochem. 34, 795−801.

Schalles, J.F., Yacobi, Y.Z., 2000. Remote detection and seasonal patterns of phycocyanin, carotenoid and chlorophyll pigments in eutrophic waters. Archiv für Hydrobiologie Special Issues Advances in Limnology 55, 153−168.

Simis, S.G.H., Kauko, H.M., 2012. In vivo mass-specific absorption spectra of phycobilipigments through selective bleaching. Limnol. Oceanogr.: Meth. 10, 214−226.

Simis, S.G.H., Peters, S.W.M., Gons, H.J., 2005. Remote sensing of the cyanobacterial pigment phycocyanin in turbid inland water. Limnol. Oceanogr. 50, 237−245.

Simis, S.G.H., Ruiz-Verdu, A., Dominguez-Gomez, J.A., Pena-Martinez, R., Peters, S.W.M., Gons, H.J., 2007. Influence of phytoplankton pigment composition on remote sensing of cyanobacterial biomass. Remote Sens. Environ. 106, 414−427.

Singh, N.K., Sonani, R.R., Rastogi, R.P., Madamwar, D., 2015. The phycobilisomes: an early requisite for efficient photosynthesis in cyanobacteria. EXCLI J. 14, 268−289.

Sivonen, K., 1996. Cyanobacterial toxins and toxin production. Phycologia 35, 12−24.

Smith, V.H., 1986. Light and nutrient effects on the relative biomass of blue-green algae in lake phytoplankton. Can. J. Fish. Aquat. Sci. 43 (1), 148−153.

Song, K., Li, L., Li, S., Tedesco, L., Hall, B., Li, L., 2012. Hyperspectral remote sensing of total phosphorus (TP) in three central Indiana water supply reservoirs. Water Air Soil Poll. 223 (4), 1481−1502.

Song, K., Li, L., Li, Z., Tedesco, L., Hall, B., Shi, K., 2013. Remote detection of cyanobacteria through phycocyanin for water supply source using three-band model. Ecol. Inform. 15, 22−33.

Song, K., Li, L., Tedesco, L.P., Li, S., Hall, B.E., Du, J., 2014. Remote quantification of phycocyanin in potable water sources through an adaptive model. ISPRS J. Photogramm. Remote Sens. 95, 68−80.

Sun, D., Li, Y., Wang, Q., Le, C., Lv, H., Huang, C., et al., 2012. A novel support vector regression model to estimate the phycocyanin concentration in turbid inland waters from hyperspectral reflectance. Hydrobiologia 680, 199−217.

Sun, D., Hu, C., Qiu, Z., Shi, K., 2015. Estimating phycocyanin pigment concentration in productive inland waters using Landsat measurements: a case study in Lake Dianchi. Opt. Express 23 (3), 3055−3074.

Tandeau de Marsac, N., 1977. Occurrence and nature of chromatic adaptation in cyanobacteria. J. Bacteriol. 130, 82−91.

Vincent, R.K., Qin, X.M., McKay, R.M.L., Miner, J., Czajkowski, K., Savino, J., et al., 2004. Phycocyanin detection from LANDSAT TM data for mapping cyanobacterial blooms in Lake Erie. Remote Sens. Environ. 89, 381−392.

Woźniak, M., Bradtke, K.M., Darecki, M., Krężel, A., 2016. Empirical model for phycocyanin concentration estimation as an indicator of cyanobacterial bloom in the optically complex coastal waters of the Baltic Sea. Remote Sens. 8 (3), 212−234. Available from: http://dx. doi.org/10.3390/rs8030212.

Yacobi, Y.Z., Köhler, J., Leunert, F., Gitelson, A., 2015. Phycocyanin-specific absorption coefficient: eliminating the effect of chlorophylls absorption. Limnol. Oceanogr.: Meth. 13 (4), 157−168.

Chapter 9

Bio-optical Modeling and Remote Sensing of Aquatic Macrophytes

Tim J. Malthus
CSIRO Oceans and Atmosphere, Brisbane, QLD, Australia

9.1 INTRODUCTION

Aquatic macrophytes, defined here as emergent, floating-leaved or sub-merged macroscopic plant species with distinct roots and shoots, play a highly important role in freshwater ecosystem structure and functioning and the services these ecosystems provide (Carpenter and Lodge, 1986). Macrophytes create important habitats for fish and zooplankton and their presence is often associated with clear water conditions where their influence on shading, chemical inhibition, and zooplankton populations can serve to limit the growth of phytoplankton populations (Mitchell, 1989). Their presence also reduces water velocity and increases sedimentation, which in turn reduces turbidity (Ginn, 2011). Increased plant biomass can affect an entire lake by altering biological communities, water and sediment chemistry, and nutrient/energy cycling (Malthus et al., 1990; Ginn, 2011).

As "integrators" of the general trophic conditions present in a water body aquatic macrophytes can serve as valuable indicators of environmental and trophic status (e.g., Sondergaard et al., 2010; Poikane et al., 2014) and as a result a number of monitoring frameworks established around the world include macrophyte species, cover, and condition as fundamental components for the assessment of ecosystem ecological assessment (e.g., Ruiz et al., 2011). Examples of such frameworks include the European Water Framework Directive (EU-WFD), the US Clean Water Act (US-CWA), and the National Water Quality Management Strategy Guidelines for Australia and New Zealand (NWQMS, 2000).

Macrophytes are also regarded as potentially useful indicators of water and/or sediment quality. Eutrophication, the present of excess nutrients in an aquatic system, can lead to the disappearance of macrophytes (especially

Bio-optical Modeling and Remote Sensing of Inland Waters.
DOI: http://dx.doi.org/10.1016/B978-0-12-804644-9.00009-4

263

submersed) as light penetration is reduced due to excessive growth of phyto-plankton (Mitchell, 1989). Eutrophication can also stimulate macrophytic growth to the point where they themselves become the nuisance organism, present in excessive amounts [nuisance blooms of macrophytes may also be the result of invasive macrophytic species which frequently displace endemic species leading to changes in ecological assemblage (Carignan and Kalff, 1980; Madsen et al., 1991)]. Both of these responses to eutrophication may significantly affect environmental, recreational, and commercial amenity of the waterbody (through reduced water transparency, clogged waterways for fishing, swimming, boating and navigation, and draw off for urban supply or irrigation (e.g., Sondergaard et al., 2010).

Macrophyte indicators (frequently expressed in terms of cover, coloniza-tion depth, species abundance, and plant volume/biomass), typically show general relationships to indicators of eutrophic status, commonly expressed as Total Phosphorus or chlorophyll (Chl-a) concentrations, the latter used as an expression of phytoplankton biomass (Brucet et al., 2013). However, the link to eutrophic state and nutrient abundance is not always straightforward; most relationships show a high degree of variability. This is not surprising when macrophytes are often responding to other environmental variables unrelated to nutrient loading; macrophyte cover and biomass may also show a high degree of correlation such as lake morphometric indicators such as lake area, slope, mean depth (a proxy for light levels), water movement, and exposure to wind and wave action (Duarte and Kalff, 1986; Madsen et al., 2001; Sondergaard et al., 2010; Ginn, 2011; Middelboe and Markager, 1997).

Despite their mixed status as an indicator of trophic status, macrophytes are clearly an important component of aquatic ecosystems and their presence in anything but nuisance levels is taken and an indicator of "good ecological status." The monitoring of macrophytes and their variations in cover is thus an important process against which to assess the progress of lake manage-ment plans (Ginn, 2011), but the causes leading to their variation may not be so easily diagnosed (NWQMS, 2000).

Traditional approaches for surveying the distribution and biomass of aquatic macrophytes have relied on quadrat and transect-based methods that are very similar to those used in terrestrial vegetation surveys albeit assisted by subaqua methods. Such methods are difficult logistically, time consuming and costly, and other more time saving field survey efforts have included the use of grabs, grappling hooks, and underwater viewing devices. Field-based surveys are also often subject to high observer error rates (Willby et al., 2009) and may fail to capture the full heterogeneity of macrophyte commu-nities (Hunter et al., 2010). If the distribution of plants over large areas are to be accurately mapped in often inaccessible lakes and rivers, remote sens-ing could offer a time and cost-saving alternative, potentially being capable of rapidly providing the required synoptic information over large areas.

However, such methods still need to be integrated into monitoring frameworks such as the EU-WFD and the US-CWA (Reyjol et al., 2014).

There is a paucity of bio-optical modeling studies for freshwater macrophytes in the literature. Instead, the aims of this chapter are to review the current status of remote sensing of aquatic macrophytes, including bio-optical modeling. Reviews of a range of remote sensing techniques for monitoring higher vegetation in aquatic systems have been published elsewhere (Adam et al., 2009; Ashraf et al., 2010; Klemas, 2013; Silva et al., 2008); the purpose of this chapter is thus to review recent progress in the field and to highlight priorities for future research including bio-optical modeling.

9.2 SPECTRAL CHARACTERISTICS OF AQUATIC MACROPHYTES

Aquatic macrophytes may be separated into three main groups based upon principal growth habits. Some species remain principally *submersed* throughout their life cycles. *Floating-leaved* plants such as the water lilies have leaves floating on the water surface and *emergent* macrophytes penetrate through the water surface (Cook, 1996; Malthus and George, 1997). Plants typical of each growth habit are shown in Fig. 9.1. Hunter et al. (2010) introduced a fourth *"partially emergent"* class recognizing the marked differences in morphology between tall monocotyledonous emergent species and other species with comparatively little emergent growth. Growth habit is important as it has a particular influence on the spectral signatures observed over aquatic macrophytes and of the influence of the aquatic medium in which they sit. Typically, growth habit also has a spatial component with emergent species found closest to the shore, merging further out into a littoral zone dominated by floating-leaved or mixed communities, then to deeper waters in which submersed species are only found.

Making measurements of the spectral reflectances of macrophyte species in situ requires the use of a spectroradiometer and a rigorous attention to methods, particularly for the measurement of submersed species where both above-surface and subsurface factors affect the reflectance obtained. Robust methods for the determination of spectral reflectance above macrophyte canopies can be found in Pinnel (2006) and Wolf et al. (2013). Accurate measurements are difficult to acquire; care is needed to account for boat or diver stability, not to disturb sediments with paddles and anchors, the influence of the water column, wave action, overhead sky conditions, and accurate measurement of distance from the surface to the species being measured (Wolf et al., 2013). Measurement should not only concentrate on the determination of the spectral reflectance characteristics of the plants of interest but also of the attenuation and other optical

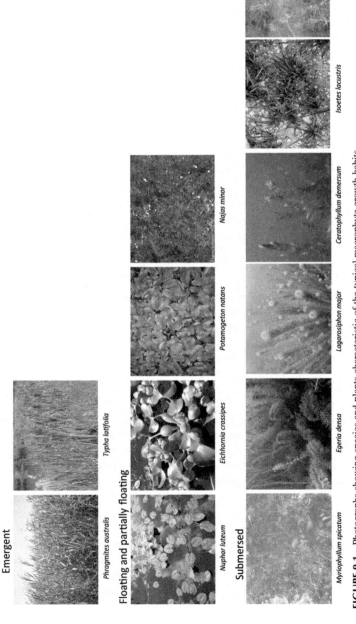

Emergent

Phragmites australis   Typha latifolia

Floating and partially floating

Nuphar luteum   Eichhornia crassipes   Potamogeton natans   Najas minor

Submersed

Myriophyllum spicatum   Egeria densa   Lagarosiphon major   Ceratophyllum demersum   Isoetes lacustris   Chara vulgaris

FIGURE 9.1 Photographs showing species and plants characteristic of the typical macrophyte growth habits.

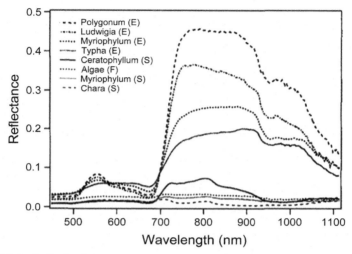

FIGURE 9.2 Stylized reflectances measured above differing aquatic macrophytes species with different growth habits. *Taken from Penuelas et al. (1993).*

characteristics of the water column surrounding the vegetation, including its optical water quality parameters (Pinnel, 2006).

Most spectral measurements of submersed freshwater macrophytes have been made in the 400 to 900 nm spectral range beyond which strong attenuation of light in near infrared (NIR) wavelengths limits reflectance from submersed species. However, useful signals beyond this range may be obtained from emergent vegetation (Hestir et al., 2015).

Typical reflectances from species of different growth habits are displayed in Fig. 9.2. Spectra of aquatic macrophytes show reflectance features typical of their terrestrial counterparts, with low reflectance in the blue and red regions of the visible spectrum due to absorption by photosynthetic pigments, higher reflectances in the green part of the spectrum and the transition via the "red-edge" to relatively high reflectances in the NIR largely as a result of a high degree of scattering of light within individual leaves. Along with plant pigments, differences in leaf morphology (including internal structure) and water content will also influence reflectance properties (Silva et al., 2008). Inter-specific differences in (a) individual plant structure (expressed in differences in leaf distribution, leaf density and orientation) and (b) in canopy structure (plant biomass and the density of cover) will also influence overall spectral reflectance properties (Everitt et al., 2011; Silva, et al., 2008). Absorption by the surrounding water will also influence reflectances, particularly in the NIR. Differences are also evident as a result of changes in phenology of different species (e.g., Fernandes et al., 2013; Wolf et al., 2013).

The above factors act differently depending on the growth habit of specific species to significantly influence the spectral signatures observed. These differences are treated separately below:

- **Emergent and partially emergent species** most closely resemble and indeed overlap with the reflectance characteristics of terrestrial vegetation with prominent green reflectance peaks due to internal leaf scattering, low blue and red reflectance due to absorption by plant photosynthetic pigments, and prominent NIR reflectance the result of strong internal leaf scattering and amount of vegetation present. Near to mid-infrared bands are attenuated by the occurrence of underlying water and wet soil; plants with lower cover and density in their canopies will have more exposed water in the background signal that would contribute to lower NIR reflectance (Everitt et al., 2011). The overlap of reflectance with terrestrial components can affect the ability to discriminate emergent aquatic macrophytes from surrounding terrestrial vegetation (Silva et al., 2008).

  Differences in the spectral reflectance features between emergent species can be subtle but with the collection of sufficiently high spectral resolution reflectances, features exist upon which to be able to statistically separate, and thus discriminate different emergent species (Schmidt and Skidmore, 2003).

- Depending on the density of leaves on the water surface and the proximity of underlying plant structures to the surface, **floating-leaved species** typically exhibit intermediate levels of reflectance in the NIR and short wave infrared (SWIR) wavelengths (Fig. 9.2), between fully submersed and emergent growth habits. Their spectra may be similar to those of sparse emergent and **partially emergent** species and may be confused with the reflectances observed over dense cyanobacterial blooms (Fig. 9.3). However, in comparison with aquatic macrophytes, cyanobacterial blooms show deeper water absorption troughs around 970 nm and lower reflectance at SWIR bands, suggesting that the effects of water absorption on cyanobacterial blooms are stronger than those on aquatic macrophytes (Oyama et al., 2015).

- Most measurements of **submersed species** have concentrated on reflectance measurements made above the water surface where the overlying water column can significantly influence the reflectances measured. While relevant to determining the extent and depths to which submersed species can be observed, above-surface measurements give little indication of the true differences in reflectance properties of submersed species per se. There are very few studies which have examined specific spectral signatures in situ (Wolf et al., 2013), largely as a result of the challenges in making such measurements accurately without disturbance of the plants themselves, their epiphytic growths, and of the surrounding sediments.

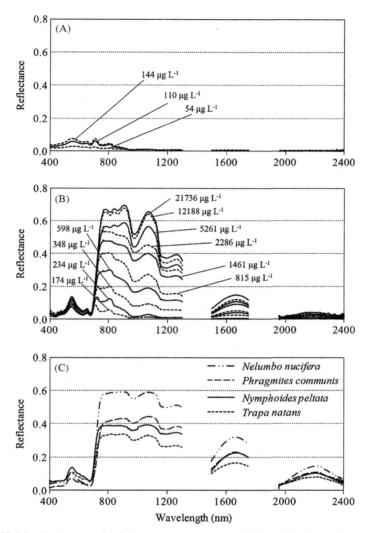

FIGURE 9.3 In situ spectral reflectances measured above (A) open lake water, (B) cyanobacterial blooms, and (C) floating-leaved and emergent macrophytes. Figures accompanying the curves in (A) and (B) indicate varying chl-*a* concentrations obtained at the time of measurement. *Taken from Oyama et al. (2015).*

As illustrated in Fig. 9.4 above-surface reflectances from submersed macrophytes are typically reduced by the strong absorption of light by the water column where with increasing depth, the reflectance may increasingly resemble that of the surrounding water and bottom sediments (Han and Rundquist, 2003; Hunter et al., 2010). In the visible part of the spectrum, green reflectance is prominent as the green region provides greater

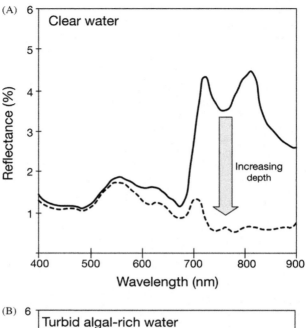

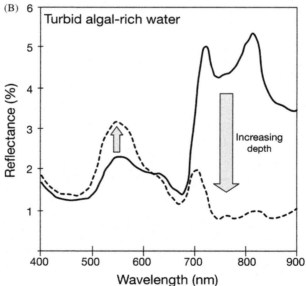

FIGURE 9.4 Influence of increasing water depth (10–110 cm) on above-surface spectral reflectances of macrophytes in contrasting clear water and turbid, algal-rich water. *Redrawn from the experimental results of Han and Rundquist (2003) using* Ceratophyllum demersum.

light penetration in waters and reflectance in the NIR is low due to water absorption (Silva et al., 2008). Thus, along with depth, the optical properties of the water column, in terms of color due to dissolved organic matter and phytoplankton and the presence of turbidity from varying concentrations of suspended sediments or phytoplankton, determine the extent to which the presence of submersed plants may be detected. The presence of epiphytic algal communities growing on the macrophytes will further influence the reflectances measured (Hestir et al., 2008).

Using an experimental tank to examine the change in surface spectral reflectance with changing depth of a submersed macrophyte species in both clear and algal rich water, (Han and Rundquist, 2003) demonstrated the influence of increased water absorption at longer NIR wavelengths. This is summarized in Fig. 9.4 where under clear water conditions reflectance decreases across all wavelengths with depth in proportion to absorption by water itself, light at longer wavelengths (700 nm and above) is strongly attenuated. In turbid, algal rich waters, there is an increase in reflectance with increasing water depth due to scattering up to ~ 600 nm, beyond this reflectance decreases due to water absorption. Han and Rundquist (2003) concluded that submersed aquatic vegetation is harder to remotely sense when concentrations of algae increase, where the signal becomes "concentrated" around the green reflectance peak at 550 nm and the signals are confounded by the increasing volume reflectance from surrounding algal-rich water and decreasing reflectance from the submerged macrophyte.

The main challenge of remote sensing of submerged aquatic plants is thus to isolate plant signal from the overall water column interference and from optical variability of the substrate.

The physiological status of vegetation also contributes to variation in plant spectral signatures (Silva et al., 2008), which can be the result of different phenological growth stages (e.g., canopy sprouting, maturing, flowering, and decay) or of physiological stress (Everitt et al., 2011). Wolf et al. (2013) measured in situ remote sensing reflectances of submersed aquatic macrophytes in German waterbodies over a growing season and found a dependence of the spectral shapes on plant phenological states (Fig. 9.5). For the species shown, when macrophytes were developing early in the growing season (May/June) reflectances were brighter as more background sediment was exposed; the canopy reached full cover across August and September but in October the canopy collapsed and contained fewer pigments contributing to a flattening of the reflectance curve at this time. Wolf et al. (2013) found that phenological differences such as those observed could improve the discrimination of the species themselves as well as from background sediment reflectances. Exploiting such phenological "spectral fingerprints" through multitemporal measurement over a

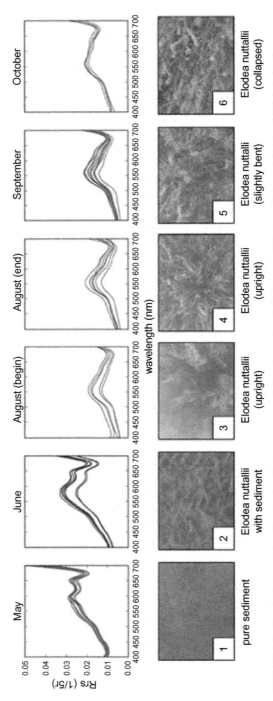

FIGURE 9.5 Seasonal time series of spectral reflectances measured above canopies of *Elodea nuttallii* between May and October 2011 in Lake Tegernsee, Germany, with corresponding photographs of the targets measured. *After Wolf et al. (2013).*

growing season thus improves the ability to discriminate species and potentially would reduce problems associated with misclassification, but this still relies on spectral features being particularly distinct to aid discrimination (Fyfe, 2003; Villa et al., 2013).

Exploiting the richness of data provided, a number of methods have been used to explore the information in in situ high spectral resolution reflectances for their ability to discriminate or classify species and growth habits. Analytical methods have included: separability assessments (Fernandes et al., 2013; Villa et al., 2015), discriminant analysis (Everitt et al., 2011; Malthus and George, 1997; Peñuelas et al., 1993), spectral derivatives (Becker et al., 2005; Peñuelas et al., 1993), principal component analysis (Wolf et al., 2013), multiple comparison range tests (Everitt et al., 2011), analysis of variance (Fyfe, 2003), and nonparametric methods (Schmidt and Skidmore, 2003). Most techniques have been used as methods of data reduction to determine wavelengths or wavelength regions which have most impact on discriminating species (e.g., Fyfe, 2003; Schmidt and Skidmore, 2003; Wolf et al., 2013) or growth forms (Malthus and George, 1997; Peñuelas et al., 1993) and of macrophytes from surrounding sediment or surrounding water column (Wolf et al., 2013). All the methods employed are effectively methods to quantify the distances between different vegetation types in hyperspectral space (Schmidt and Skidmore, 2003).

The various methods used have successfully demonstrated that, on the basis of high spectral resolution reflectance data, discrimination between species is possible to some degree, but that in some cases the differences can be subtle; the extent to which some differences observed in situ would translate to discriminatory power from sensing platforms at altitude and over the spatial resolution of pixels is still being tested (next section). The results further demonstrate the high degree of intraspecific variability in spectral reflectances (e.g., Wolf et al., 2013), which ultimately suggests there is a limit to discrimination on the basis of spectral signature alone (Hestir et al., 2008). Discrimination may be improved by the use of multitemporal reflectance data but results can be specific to the time of measurement; when attempting to differentiate 9 different species over three measurement dates, Everitt et al. (2011) found only one waveband (405 nm) identified consistently identified over the three dates.

Furthermore, results for the optimum bands to differentiate species appear to differ depending on specific study sites, the analysis performed, whether tests evaluate the discriminating power of each band individually (e.g., multiple range comparisons), or whether the analysis is designed to select a subset of bands that best discriminate species (e.g., stepwise discriminant analysis). For emergent and floating-leaved species, the greatest range of variation and hence discriminating power appears to lie in the NIR and red-edge regions of the spectrum (Adam et al., 2009; Malthus and George, 1997; Schmidt and Skidmore, 2003; Thito et al., 2015). These signals are largely linked to the amount of green biomass of the plants on or above the water surface and their influence on masking the absorption properties of the underlying water column. In the visible region, differences in pigment content for emergent species can be further differentiated using continuum removal (Schmidt and Skidmore, 2003).

Underwater plants were clearly distinguished by their lower absolute reflectances, especially in the NIR, while wavelengths in other parts of visible range can be used to differentiate species (Fyfe, 2003) where narrow hyperspectral channels can be used to exploit subtle differences between species on the basis of leaf optical properties reflecting equally subtle differences in pigment content (Adam et al., 2009; Peñuelas et al., 1993).

In conclusion, the results from a number of studies of the in situ reflectance of aquatic macrophytes demonstrates the high degree of spectral similarity and hence, the high degree of correlation, between aquatic macrophyte species. However, there are differences which allow the differentiation between submerged, floating, and emergent growth habits and species which can be used to form the basis of discrimination using reflectance parameters, but the ability to differentiate species within these groups is also highly dependent on plant biomass and cover, water depth, bottom type, the type and abundance of aquatic species, and surrounding water quality. Few studies exist from which to draw definitive conclusions on changes in reflectance in species over time and hence of the spectral variation occurring within single species and the influence of surrounding environmental factors upon this. The key drivers of reflectance properties for aquatic vegetation are clearly the same as for those of terrestrial vegetation (e.g., chlorophyll and other pigments, biomass, leaf water content and density), but that the ability to accurately differentiate these characteristics is significantly affected by the depth and turbidity of the overlying water column, the presence of epiphytes and the underlying substrate reflectance (Kemp et al., 2004, Adam et al., 2009; Hunter et al., 2010). The complicating influences of such physical and environmental variability on the growth and spectral expression led Hestir et al. (2008) to describe the normal situation of above-surface reflectances from aquatic macrophytes as one of "mixed degraded spectra" in which intraspecies variation that can result in overlapping spectral features between co-occurring species and which ultimately leads to mixed pixels with spectrally similar constituents. This ultimately challenges the degree to which species will be mapped using remote sensing platforms.

9.3 APPLICATION OF REMOTE SENSING SYSTEMS

The previous section has highlighted the challenges in interpreting the optical signals emanating from aquatic macrophytes and the effect of surrounding environmental influences in determining whether aquatic macrophytes can be differentiated per se and if individual species can be differentiated on the basis of their spectral reflectance. That challenge is further exacerbated when such measurements are upscaled to the spatial measurement acquired by airborne and satellite systems where a number of other factors can determine success of the approach. Atmospheric influences (Gao, 2016), prevailing weather conditions at the time of acquisition, variable sun angle and its interaction with sensor view angles and surface water effects (e.g. sunglint and wave action) can have a critical impact on the degree to which remote observation of such systems can be applied (Hestir et al., 2008). As the previous section highlighted, the contribution of submersed aquatic vegetation to the water-leaving reflectance decreases as the depth and turbidity of the water column increases; the proportion of signal received by a remote sensing platform attributable to the presence of aquatic vegetation may be small. Over the scale of a pixel, significant biological heterogeneity may also impact the detection.

Ideally, images acquired for the purposes of detecting the presence and properties of aquatic macrophytes should be radiometrically calibrated and orthorectified, atmospheric corrected and water column corrected. However, the literature demonstrates a widespread variation in the extent to which these corrections have be applied, which limits the extent to which general comparisons can be made between studies and appropriate and definitive conclusions made.

A number of studies have used **analogue or digital aerial photography** for large-scale mapping of macrophytes in lakes (Malthus et al., 1990; Marshall and Lee, 1994; Orth and Moore, 1983; Windham and Lathrop, 1999). Such methods may be of high resolution which may facilitate the identification of individual species, but earlier studies relied on laborious and potentially subjective, visual interpretation to achieve this (e.g., Husson et al., 2014; Malthus et al., 1990; Marshall and Lee, 1994). The recent widespread availability of low cost but high-quality digital photographic systems along with low cost drones and unmanned aerial vehicles (UAVs) to deploy them (either above or below the water surface) has seen a resurgence in the use of such systems in the environmental monitoring domain (Anker et al., 2014; Flynn and Chapra, 2014; Husson et al., 2014). Computer-aided image interpretation can be applied to digital photographs to speed up the time of analysis and reduce the subjectivity but digital and analogue photographic methods are often limited by the lack of spectral information necessary to effect accurate classification (Flynn and Chapra, 2014; Hunter et al., 2010; Silva et al., 2008).

Acoustic methods, in the form of either **multibeam** or single beam **echosounders**, deployed in situ, have been shown to provide an accurate solution to mapping both the presence and amount of macrophyte species (Rotta et al., 2012, 2016; Valley et al., 2005). For mapping in turbid waters, such systems have been shown to be more accurate than airborne optical Multi-detector Electro-optical Imaging Sensor (MEIS) data for detecting macrophyte growth forms (Vis et al, 2003) and effective in differentiating percentage cover and canopy height (Rotta et al., 2012). However, such systems cannot generally be used in shallow waters ($< \sim 70$ cm, Vis et al., 2003); depending on the nature of the lake system and the distribution of species in shallow waters, the method must then be deployed in conjunction with other remote sensing methods to provide full system coverage. Acquisition is also potentially more time consuming than digital optical methods, where transect lines need stitching to achieve full coverage otherwise interpolation methods can be used to infer macrophyte cover between transect runs.

As part of a wider study, the application of **satellite** systems to aquatic macrophyte mapping was recently and briefly reviewed by Guerschman et al. (2016); Table 9.1 provides an overview of the abilities of current and future optical satellite sensors to differentiate the growth habits of different macrophytes. Different satellite systems trade off temporal frequency of coverage versus spatial, spectral and radiometric resolution; this influences their usefulness for macrophyte mapping. Spatial resolution ultimately determines the size of the water body and macrophyte patch that may be differentiated. Spectral resolution (number, width, and placement of spectral bands) ultimately determines the degree to which macrophyte species may be discernible from the surrounding water body and from each other and radiometric resolution determines the lowest level of radiance or reflectance that the sensor can reliably detect per spectral band (but in general, more modern sensors have a higher radiometric sensitivity than older sensors).

A number of satellite systems have been tested for their ability to detect macrophyte species and their characteristics across a range of water bodies worldwide (Table 9.1). Coarse spatial resolution systems (e.g., MODIS and Sentinel 3) lack sufficient spatial resolution to produce an accurate extent map. Coarse spectral resolution systems (e.g., Landsat and SPOT) have shown the ability to map macrophytes where water quality allows across a range of water bodies and macrophyte growth habits and species (Sakuno and Kunii, 2013; Shuchman et al., 2013) but such systems may be too coarse both spatially and spectrally to distinguish the fine ecological divisions and gradients between vegetation units in inland systems required in operational monitoring. Macrophyte patches need to be large and sufficiently dense for successful identification. Heblinski et al. (2010) state that the minimum dimension of the patches should be at least twice the length of a pixel diagonal. Landsat 8, with its increased radiometric resolution and additional visible band should provide improved ability in inland water environments (e.g., Lymburner et al., 2016). Sentinel 2, offers improvements in spectral, spatial, and temporal resolution over medium resolution sensors that may improve the accuracy with which macrophytes and their properties may be mapped to allow the operational monitoring of macrophyte dynamics over time (Hedley et al., 2012).

Higher spatial resolution satellites generally provide increased levels of accuracy compared to moderate resolution systems (e.g., Shuchman et al., 2013)

TABLE 9.1 Existing and Near Future Satellite Sensors Systems Together with a Qualitative Assessment of their Applications to Macrophyte Mapping and Monitoring

Sensor	Spatial Resolution (Pixel Size)	Spectral Bands (Water-Relevant Range, 400–1000 nm)	Revisit Frequency	Raw Data Cost (US$ per km²)	Launch Date	End Date	Emergent	Floating Leaved	Submersed	Indicative Studies, where Available
Ocean-coastal sensors										
MERIS	1.2 km and 0.3 km	15	2 days	Free	Mar 02	May 12	❸	❹	❹	Shuchman et al. (2013)
MODIS-Aqua & Terra	250 m–1 km	2–9[a]	Daily	Free	Dec 99		❹	❸	❸	Cheruiyot et al. (2014); Silva et al. (2010); Shuchman et al. (2013)
OCM-2	300 m	15	2–3 days	Free	Sep 09		❸	❹	❹	Liu et al. (2015)
Suomi-VIIRS	740 m	7	Twice daily	Free	Oct 11		❹	❹	❹	
Sentinel-3	300 m	21	Daily (with two satellites)	Free	Mar 16		❸	❹	❹	

(Continued)

TABLE 9.1 (Continued)

Sensor	Spatial Resolution (Pixel Size)	Spectral Bands (Water-Relevant Range, 400–1000 nm)	Revisit Frequency	Raw Data Cost (US$ per km²)	Launch Date	End Date	Emergent	Floating Leaved	Submersed	Indicative Studies, where Available
Mid-spatial resolution										
LANDSAT 1–7	30 m	4	16 days	Free	Jul 72		2	3	3	Nelson et al. (2006)
										Oyama et al. (2015)
										Shuchman et al. (2013)
LANDSAT 8	30 m	5	16 days	Free	Sep 13		2	2	2	Villa et al. (2015)
SPOT 1–5	10–20 m	3	26 days	1.1	Feb 86	Mar 15	2	3	3	Schmidt and Witte (2010)
SPOT 6–7	6 m	3	3 days	1.5	Sep 12		2	3	3	
SENTINEL-2	10–60 m bands	10	10 days per sensor; 5 days with two S2s	Free	Jun 15		2	2	2	
ASTER	15–90 m	4	16 days	Free	Dec 99		2	2	3	Sakuno and Kunii (2013)
										Giardino et al. (2010)
ALOS AVNIR2	10 m	4	14 days	Free	May 14		2	2	3	Jiang et al. (2012); Villa et al. (2015)

High-spatial resolution

	Resolution		Revisit		Launch				References
RapidEye	6.5 m	5	Daily	1.5	Aug 08	❷	❷	❷	
IKONOS, QuickBird, GeoEye etc.	2–4 m	3–4	Programmable 60 days to 2–3 days	5 to 15	10, 1999 and onward	❷	❷	❷	Villa et al. (2013); Heblinski et al. (2010); Dogan et al. (2009); Yuan and Zhang (2008)
WORLDVIEW-2	2 m spectral to 0.5 m B&W	8	Programmable 60 days to 1 day	30	Oct 09	❶	❶	❶	Villa et al. (2013); Shuchman et al. (2013); Robertson et al. (2015); Whiteside and Bartolo (2015)
WORLDVIEW-3	1.24 m spectral to 0.50 m B&W	8	Programmable 60 days to 1 day	30	2014	❶	❶	❶	

Hyperspectral

	Resolution		Revisit		Launch				References
EnMap	30 m	90	Programmable (once per 4 days)	Free(?)	2017	❷	❷	❷	

(Continued)

TABLE 9.1 (Continued)

Sensor	Spatial Resolution (Pixel Size)	Spectral Bands (Water-Relevant Range, 400–1000 nm)	Revisit Frequency	Raw Data Cost (US$ per km²)	Launch Date	End Date	Emergent	Floating Leaved	Submersed	Indicative Studies, where Available
PRISMA	20 m spectral to 2.5 m B&W	60	25 days	Free?	2017		2	2	2	
Airborne systems			(Bandwidth)							
CASI	Variable[a]	15–288	1.9 nm	Mission specific	1993		1	2	3	Hunter et al. (2010)
Aviris	Variable[a]	~110	10 nm	Mission specific	1987		1	2	3	Underwood et al. (2006)
Hymap	Variable[a]	64	15 nm	Mission specific	1996		1	2	3	Hestir et al. (2008)
APEX	Variable[a]	114	8 nm	Mission specific	2015		1	2	3	Villa et al. (2014) Bolpagni et al. (2014)
Synthetic aperture radar										
ERS 1	30 m	C Band, VV polarization	35 days	Free	1991 (ERS-1)	2000 (ERS-1)	3	4	4	
ERS 2	30 m	C Band, VV polarization	35 days	Free	1995 (ERS-2)	2011 (ERS-2)	3	4	4	

Sensor	Resolution[a]	Band/polarization	Revisit	Commercial or Free	Launch	Suitability			References
ALOS PALSAR	10–100 m	L Band, HH, VV, HV, VH??	46 days	Commercial or R&D	2011	❶	❹	❹	Evans et al. (2010)
Radarsat 2	30 m	C Band, fully polarimetric	24 days	Commercial	2007	❷	❹	❹	Costa (2004); Silva et al. (2010); Robertson et al. (2015)
ALOS 2	3–100 m	L band, HH, VV, HV, VH	14 days	Commercial	May-14	❶	❹	❹	
Cosmo-Skymed	5–100 m	X Band, HH, VV, HV, VH	16 days	Commercial or R&D	??	❸	❹	❹	
TerraSAR-X/Tandem-X	25 cm–40 m	X Band	3 days	Commercial or R&D	2007 (2010)	❸	❹	❹	
Sentinel 1	5–20 m	C Band, HH, VV, HV, VH	12 days	Free??	Mar 14	❷	❹	❹	

Selected studies, which have investigated the applications of the sensors, are also given. A range of commonly studied airborne sensors is also given. ❶: highly suitable, ❷: suitable, ❸: potential, and ❹: not suitable.

[a]Dependent on altitude.

Adapted and modified from Guerschman et al. (2016).

and the potential to differentiate submersed aquatic vegetation structure and growth type from sediment in the littoral zone. Significant interest has been shown in the additional spectral bands offered by the WorldView 2 and 3 sensors where 8 spectral bands are available across the visible and NIR regions. These include additional blue, yellow, and red-edge bands, which may have benefit in improving macrophyte classification over other high spatial resolution sensors that employ only four spectral bands (Whiteside and Bartolo, 2015). Carle et al. (2014) found WorldView-2 imagery capable of differentiating macrophyte species to level comparable to that achieved using hyperspectral imagery.

Airborne optical systems (e.g., CASI, MIVIS, HyMAP, AVIRIS, and sensors deployed on UAVs) provide a level of agility (such as rapid mobilization to take advantage of favorable acquisition conditions) over satellite systems but other influences such as changing sun angles and wind speeds may contribute to variable effects over multiple flight lines; careful flight planning is thus required to ensure these effects do not confound region-wide analysis (Hestir et al., 2008). Bresciani et al. (2009) provide an example of the use of such data in highlighting trends and their relation to eutrophication and lake management and Hestir et al. (2008) an example of use in system-wide assessment. Airborne sensors generally offer considerably higher spectral resolution that satellite systems which in turn has generally lead to improved differentiation of macrophyte characteristics compared to satellite systems (Adam et al., 2009; Giardino et al., 2015). Becker et al. (2005, 2007) derived optimal band combinations from hyperspectral scanner data and optimal spatial resolutions for wetland vegetation classification. They recommended an optimal spatial resolution of 2 m and a minimum of 7 m, with strategically located bands in the visible and near infrared wavelength regions.

Synthetic aperture radar (SAR) data also have merit in providing monitoring of aquatic macrophytes and wetlands independent of weather systems (Table 9.1, e.g., Silva et al., 2010). Dielectric signal differences arise due to the presence of water as surface water and within vegetation (Silva et al., 2008), thus allowing for detection between dry and flooded vegetation; such systems thus allow for the mapping of flooding extent as well as of the extent of emergent vegetation (Costa, 2004; Evans et al., 2010). The lack of penetration of microwaves into water prevents detection of submersed macrophyte species. SAR wavelength bands include X (3 cm wavelength), C (5.6 cm), S (10 cm), L (23 cm), and P (75 cm) bands and increasingly, SAR systems may detect defined polarizations in same (HH or VV) or cross-polarization modes (e.g., HV or VH). Depending on wavelength, the intensity of the radar backscatter is directly related to the roughness and, combined with volumetric scattering, wavelength, and polarization, provides specific vegetation responses and hence information on canopy characteristics (Evans et al., 2010; Kasischke and Bruhwiler, 2003; Klemas, 2013; Martinez and Le Toan, 2007; Robertson et al., 2015).

9.4 DISCRIMINATION AND CLASSIFICATION

A range of different analytical techniques has been explored to produce cover maps for aquatic macrophytes in inland systems. These include traditional classification techniques such as unsupervised methods (Albright and Ode, 2011; Dogan et al., 2009) and maximum likelihood classification methods that, although often precise, are effectively limited to application on single scenes (Giardino et al., 2010; Silva et al., 2008). Although accuracy assessments are often qualitative in such studies, validation accuracies have been shown to range ranging between 70 and 96% (Dörnhöfer and Oppelt, 2016; Silva et al., 2008).

Mapping approaches include empirical classifications (e.g., Hunter et al., 2010; Oppelt et al., 2012) and semi-analytical, spectral unmixing algorithms (e.g., Dierssen et al., 2003, Dörnhöfer and Oppelt, 2016; Giardino et al., 2015, McKinna et al., 2015). Emergent vegetation is often assessed through classification approaches and band ratios (e.g., Bresciani et al., 2011; Villa et al., 2013). Other methods investigated include linear spectral mixture analysis (Giardino et al., 2015), spectral angle mapping (Flynn and Chapra, 2014), principal components analysis (Zhang, 1998), random forests (Corcoran et al., 2013), and fuzzy methods (Heblinski et al., 2010). Comparisons have generally been made to percentage coverages of single species or to discrete growth habitat classes. The use of multispectral sensors with high spatial resolution or airborne hyperspectral data has been shown to improve classification accuracy, but few of these studies have investigated depth-invariant indices (e.g., Brooks et al., 2015; Roessler et al., 2013; Shuchman et al., 2013). Physics-based classification methods were originally proposed by Heege et al. (2004).

In high spatial resolution data in particular, macrophyte stands may exhibit variations in spectral characteristics which present challenges to traditional spectral classifiers that rely on a defined spectral uniqueness. This has drawn others to investigate **object-based image analysis** (OBIA) which attempts to initially group adjacent pixels into spectrally homogeneous regions, or objects, representing meaningful ground entities with intrinsic spatial, textural, and contextual attributes and where within-object variance is less than between-object variance (Blaschke, 2010; Fernandes et al., 2013). During the subsequent classification step, these objects can be further aggregated and labeled to overcome the problem of pixel heterogeneity. Along with spectral information, additional attributes in the remotely sensed data, such as geometric, topographic, and textural attributes, can be utilized to improve overall object classification (Whiteside and Bartolo, 2015). For emergent species, the peculiar spectral and geometric traits, at leaf, canopy and stand level, can make OBIA a very suitable technique for management purposes (Fernandes et al., 2013). Furthermore, the use of combined multidate optical and SAR methods in OBIA has been shown to improve macrophyte discrimination and classification than each method separately deployed (Robertson et al., 2015; Silva et al., 2010). However, while OBIA can allow for the identification of discrete density classes such as "low," "medium," or "high," it is less useful for more appropriate quantitative canopy-scale descriptors such as LAI (Hedley et al., 2016). Furthermore, in common with other spectral classification methods, the thresholds that determine the classifications are not readily known or interpreted for their physical basis (Hedley et al., 2016).

The limitations of traditional statistical-based image classification techniques, such as minimum distance or maximum likelihood methods, in their inability to effectively use the full information content, particularly in high dimensional remotely sensed data, and the strict assumptions they make about the distributions of such data, has led others to explore nonparametric machine learning methods as alternatives. Such methods include support vector machines, artificial neural networks, random forests, and classification trees. The value of such approaches is that they can accommodate raw reflectance data and associated data such as vegetation indices (Vis). However, counter to expectations, support vector machine classification has not apparently provided significantly different results from maximum likelihood classification when applied to aquatic macrophyte mapping across all growth habits (e.g., Carle et al., 2014; Hunter et al., 2010) and may be limited in the ability to cope with hyperspectral data and potentially noisy data (Hestir et al., 2012). Neural networks remain to be tested for the application to aquatic macrophyte mapping but are limited by the network model chosen and the quality of the training data supplied. Random forests are computationally fast and can output classification confidence (Corcoran et al., 2013).

Rule based methods (sometimes called **decision trees**), based on the dichotomous splitting of data using thresholds of the most relevant variables in the data, allow for the incorporation of a variety of data sources (e.g., remotely sensed data from different sources, multitemporal image data, bathymetric data, and categorical data) and multiple classification methods in a hierarchical binary structure to improve overall macrophyte classification accuracy (e.g., Hestir et al., 2008; Whiteside and Bartolo, 2015). A combination of approaches may be used at different levels (e.g., continuum removal, spectral unmixing, and Spectral Angle Mapping) and different phenological states may also be accommodated in the decision tree, thus better accounting for the variability inherent in dynamic ecosystems and with minimal training requirements. Such approaches have the ability to account for variability in large datasets (Hestir et al., 2008) and have been shown to overcome the confusion that may exist in purely spectral based approaches (e.g., 78% accuracy achieved versus 64% using random forests classification; (Whiteside and Bartolo, 2015). Decision trees based on manually set thresholds have merit especially where there are classes that are difficult to map based on statistical methods alone.

Classification methods lose their power when applied in a retrospective context to historical datasets. Historical data also often lack the ground reference data with which to train or to validate the classification results. The imagery used could be acquired by a different sensor, and/or will most likely be obtained at a different time of year, under different sun angles, atmospheric conditions, and ground conditions. This makes extension of classification developed on one image at one point in time effectively equivalent to an unsupervised classification process (Hestir et al., 2012). The problem is further exacerbated in application to aquatic macrophyte monitoring given the variations in reflectance properties due to growth habit and amount [so-called "intrinsic" factors (Luo et al., 2014] and the additional influences on water column depth and turbidity ("extrinsic" factors independent of macrophyte condition), as discussed in the previous section.

Hestir et al. (2012) tested an ensemble classification tree in its ability to classify historic hyperspectral airborne data when trained on a single year's data and found it could classify to an accuracy comparable to other classification methods. Success factors included the large variability encompassed by the training set, which likely encompassed sufficient intrinsic and extrinsic variation evident in the historical data, the time interval between the training data set and the image to be classified, and the careful choice of training data. Other studies have investigated the improvement brought about using methods to modify the thresholds applied over multiyear image datasets (Jiang et al., 2012; Luo et al., 2014; Zhao et al., 2012).

In summary, whilst classification of emergent species may be straightforward, extrinsic factors dominate the ability to distinguish and identify submersed macrophytes with any accuracy (e.g., (Hestir et al., 2012). For all growth habits, differences between species can be subtle which will also limit classification accuracy. Careful attention to timing of data acquisition is required to maximize spectral differences and hence improve classification; inclusion of phenological state may also help.

9.5 DETERMINATION OF MACROPHYTE BIOPHYSICAL PROPERTIES

A number of studies have investigated the application of image datasets to mapping the quantitative distribution of biophysical properties of macrophytic systems (e.g., plant abundance/density, cover, biomass, leaf area index, and productivity). Zhang (1998), using the first and second principal components of a transformed Landsat TM image, estimated the biomass of submerged stands in Honghu Lake, China. Nelson et al. (2006) investigated relationships between littoral percent plant cover in Michigan lakes and Landsat 5 TM data using binomial and multinomial logistic regression. Above water plant categories were easier to predict than below water elements; the inclusion of additional water column quality information did not improve model predictions. Changes in reflectance of submersed species with varying coverage are mainly observed in the green and NIR spectral regions (Yuan and Zhang, 2008).

In general, predictions have been restricted to broad classes of cover or biomass (e.g., sparse, moderate, or dense). For example, Shuchman et al. (2013) were only able to use Landsat data to provide a qualitative estimate of "dense" and "less dense" categories in optically shallow waters; in this study estimating biomass directly from a given Landsat image was not feasible due to the often very small differences in radiance values as a function of biomass, the limited radiometric fidelity of the Landsat sensor, and imprecise atmospheric correction. Similar results have been obtained for high spatial resolution data; Heblinski et al. (2010) were able to map six classes differing in density and growth type using Quickbird data with inclusion of methods to account for varying water depth. However, vegetation patches needed to be large (at least twice the length of a pixel diagonal) and adequately dense for successful identification. The low spectral resolution of the QuickBird sensor limited the effectiveness of the classification algorithms.

9.5.1 Use of Indices

In the terrestrial environment, VIs based on combinations of spectral bands, have provided considerable power as indicators of vegetation amount, growth, and vigor, being shown to be quantitatively related to a number of canopy biophysical characteristics such as leaf area index, green biomass, cover, and light absorption (Huete et al., 2002). In aquatic vegetation studies, VIs have been found to be useful, particularly as indicators of the biomass and vigor of emergent species (Bresciani et al., 2009; Tian et al., 2010; Villa et al., 2013) and of their pigment composition and photosynthetic performance (Peñuelas et al., 1993).

Along with the factors which influence the performance of VIs noted for terrestrial vegetation (e.g., canopy characteristics and structure, community species composition, spatial scale, background composition, and exposure), other factors also considerably influence the radiometric response of aquatic vegetation including: background water which varies in both depth and turbidity and variable bottom influences (Silva et al., 2008; Wolf et al., 2013). Indices also need to distinguish macrophytes from surrounding water especially when that water may contain high cyanobacterial concentrations, which may have similar spectral characteristics, particularly in the NIR (Oyama et al., 2015).

In studies of aquatic vegetation, common indices applied include the normalized difference vegetation index (NDVI, e.g., Oyama et al., 2015; Peñuelas et al., 1993) the enhanced vegetation index (EVI, e.g., Villa et al., 2013), the floating algal index (FAI, e.g., Hu, 2009; Oyama et al., 2015), the normalized difference water index (NDWI, e.g., Andrew and Ustin, 2008; Hestir et al., 2008), the normalized difference aquatic vegetation index (NDAVI), the water adjusted vegetation index (WAVI, Villa et al., 2015; Villa et al., 2014), and the NIR-green angle index (NGAI, Tian et al., 2010). The formulations of these indices are cited in Table 9.2. Few indices have been derived specifically for aquatic macrophytes; exceptions include WAVI, NDAVI, and NGAI.

VIs have effectively described the functionality, vigor, and greenness of common reed beds, proving useful for highlighting changes in canopy characteristics over growing seasons and over years (Villa et al., 2013, 2015). Many of these indices incorporate NIR and SWIR wavelengths, which show sensitivity to biomass and density and the presence of above-surface or floating-leaved canopy components (Table 9.2). Combinations of these wavelengths are also least sensitive to atmospheric effects compared to shorter wavelengths in the visible spectrum. NDVI may also saturate at high canopy biomass, which may limit its use to estimate biomass in very dense canopies such as water hyacinth (Robles et al., 2015). EVI proved more effective than NDVI for emergent reed bed vegetation (Villa et al., 2013). Indices combined from hyperspectral data appeared to show greater utility in differentiation of surface canopy components compared to multispectral data (Tian et al., 2010). NGAI had apparent advantages in estimating density over the NDVI (increase in explained variance by 18 percentage levels, Tian et al., 2010). Index data have also proved effective as additional layers in classification schema, where their inclusion has been shown to improve classification performance (Davranche et al., 2010; Hestir et al., 2008; Villa et al., 2015). However, it remains to be seen if such indices are capable of mapping the biophysical characteristics of *submersed* macrophyte species.

Villa et al. (2014) tested a range of VIs on aerial and satellite data acquired with varying spatial (2−90 m) and spectral resolutions over northern Italian lakes dominated by emergent and floating-leaved species. In most cases, the WAVI performed best, demonstrating its usefulness as an effective tool to map aquatic vegetation and to separate it from terrestrial and open water components. The integration of multisource (multiplatform and multifrequency) and multitemporal remotely sensed data has been shown in a number of studies to provide greater utility in minimizing confusion over and above the use of single date optical imagery traditionally used for wetland classification (Corcoran et al., 2013; Giardino et al., 2010; Liu et al., 2015; Silva et al., 2008). A general increment in the ability to separate macrophyte classes is observed when spectral reflectances are combined in the form of VIs, especially when multitemporal observations of changes in indices are available over a growing season; single observations at the peak of the growing season perform poorly in comparison (Villa et al., 2015). VIs also have utility in detecting changes across longer time scales (Giardino et al., 2010).

TABLE 9.2 Selected Indices Applied to Aquatic Macrophyte Reflectances in a Number of Studies

Index	Abbreviation	Formulation	Function	Citation
Normalized difference vegetation index	NDVI	$NDVI = \dfrac{(R_{Red} - R_{NIR})}{(R_{Red} + R_{NIR})}$	Vegetation vigor and separation, sensitive to chlorophyll	Tucker (1979)
Enhanced vegetation index	EVI	$EVI = 2.5\dfrac{(R_{NIR} - R_{Red})}{(R_{NIR} + 6R_{Red} - 7.5R_{Blue} + 1)}$	Vigor of vegetation, compensating for atmospheric disturbances, more responsive to canopy structure	Huete et al. (1997)
Normalized difference water index	NDWI	$NDWI = (R_{NIR} - R_{SWIR})/(R_{NIR} + R_{SWIR})$	Sensitive to plant water content; utility in separation of emergent, floating-leaf and submerged growth habits; less sensitive to atmospheric scattering effects	Gao (1996)
Normalized difference aquatic vegetation index	NDAVI	$FAI = R_{NIR} - [R_{Red} + (R_{SWIR} - R_{Red})] \times \dfrac{(\lambda_{NIR} - \lambda_{Red})}{(\lambda_{SWIR} - \lambda_{Red})}$	Specific for aquatic vegetation; incorporates visible blue to exclude distortions due to water; useful for emergent vegetation	Villa et al. (2013)
Floating algae index	FAI	$NDVI = \dfrac{(R_{Red} - R_{NIR})}{(R_{Red} + R_{NIR})}$	For detection of cyanobacterial scums	Hu (2009)
Near infrared-green angle index	NGAI	$NGAI = Slope^*A_{NIR}$, where $Slope = R_{NIR} - R_{Green} + 0.03$, and $A_{NIR} = Cos^{-1}\left[\dfrac{a^2 + b^2 + c^2}{2ab}\right]$ where a, b and c are Euclidian distances between vertices at three bands (NIR1 at 820 nm, NIR2 at 977 nm) and green (554 nm)	Angle index, useful for discriminating soil and vegetation and tracking soil moisture	Tian et al. (2010)
Water adjusted vegetation index	WAVI	$WAVI = (1 + L)\left[\dfrac{R_{NIR} - R_{Blue}}{R_{NIR} + R_{Blue} + L}\right]$ where L values range 0 to 1	As for NDAVI but includes a correction for background effects; utility for emergent vegetation	Villa et al. (2014)

9.6   BIO-OPTICAL MODELING OF AQUATIC MACROPHYTES

As outlined in the previous sections, aquatic macrophytes are not as easily detectable as terrestrial vegetation (Silva et al., 2008). In situ studies have made considerable advances in understanding of the spectral reflectances above macrophyte canopies, particularly for emergent and floating-leaved species. Given the challenges in collecting accurate in situ measurements over macrophytes, bio-optical modeling represents a means by which further understanding of the physical interaction between electromagnetic radiation and macrophytes and the influence of the environment on this can be studied.

For submersed plant species, the volume reflectance from a submerged plant canopy of macrophytes can be divided into three components: light that is reflected from the water column, light that is reflected from the plant material, and light that is reflected from the underlying substrate (Ackleson and Klemas, 1987). A bio-optical model attempts to simulate the cumulative effect of photo transport through and in contact with each of these components including the transition across the water-air interface.

Assuming a homogeneous medium, solar radiation, and the attenuation of light through a plant canopy can be modeled using Beer's law for transmission of light through a turbid medium. A summary of the bio-optical models found in the literature that have been developed for aquatic macrophytes as well as those for seagrasses in coastal systems is presented in Table 9.3. This table reflects the low number of modeling studies that have been undertaken specifically focusing on macrophytes in both the freshwater and marine environments. These models differ mainly in the sophistication with which macrophyte are represented, ranging from treatment as a simple single layer albedo to a full structural representation.

In the earliest example of bio-optical modeling Ackleson and Klemas (1987) adapted a single-scattering volume reflectance model (Philpot, 1981) to represent the interaction between the three main components of the signal from submerged vegetation (water, bottom, and plants). This simple approach was used to investigate the influences of changes in canopy cover and water depth on volume reflectance from a water column containing submersed aquatic vegetation. Using the assumption that substrate reflectance was very different from reflectance of the submersed vegetation; they were able to show the transition from signals dominated by canopy density in shallow waters to one where water column volume reflectance ultimately dominates in optically deep water. The greater the difference between substrate and vegetation reflectance and optically deep water, the greater the effect depth will have upon volume reflectance.

Semi-analytical methods typically represent macrophytes by a single end-member reflectance spectrum (essentially a single layer without identifying leaves and stems, etc.) but which can be varied additively with a substrate

TABLE 9.3 Summary Table of the Bio-Optical Models Developed for Freshwater Aquatic Macrophytes or Seagrass Communities

Model Type	How Macrophytes Included	Species	Research Purpose	Reference
Single scattering volume reflectance	Single albedo measurement	Zostera (seagrass)	Interactions of percent canopy cover and water depth	Ackleson and Klemas (1987)
Semi-analytical	Endmember reflectance spectrum	Potamogeton anguillanus and Cladophora sp.	Influence of water column depth on NDVI	Sakuno and Kunii (2013)
Semi-analytical	Endmember reflectance spectrum	Unspecified seagrass	Determine substrate detectability and separability over a range of depths and water qualities	Botha et al. (2013)
Semi-analytical	Endmember reflectance spectrum	Seagrass, coral and other substrates	Simulation of Sentinel 2 satellite detection	Hedley et al. (2012)
Coupled leaf-canopy turbid medium model	Leaf and canopy 1D model with background water reflectance	Unspecified emergent vegetation and floating-leaved	Test new vegetation indices	Villa et al. (2014)
Coupled leaf-canopy turbid medium model	Leaf and canopy 1D model with water column bio-optical model	Posidonia (seagrass)	Influence of depth, density, and turbidity on subsurface reflectance	Plummer et al. (1997)
Coupled leaf-canopy turbid medium model	Leaf and canopy 1D model with water column bio-optical model	Scirpus spp.	Canopy density—Retrieval of biophysical or biochemical parameters	Zhou et al. (2015)

Model	Description	Species	Purpose	Reference
Two-flow model of plane irradiance distribution	Distribution of leaf area and biomass with height based on morphometric analysis—Canopy plate model	*Thalassia and Zostera* (seagrasses)	impacts of variations in canopy architecture on irradiance distribution and canopy photosynthesis	Zimmerman (2003)
Monte Carlo	Distribution of leaf area and biomass with height based on morphometric analysis—Canopy plate model	*Posidonia* (seagrass)	Biomass impacts and depth on surface reflectance	Malthus et al. (1997)
Monte Carlo	Canopy plate model	*Scirpus* spp.	Retrieval of biophysical or biochemical parameters	Zhou et al. (2015)
3D radiative transfer model	Canopy structures modeled with physical dynamic model	*Thalassia* (seagrass)	Light absorption, LAI, depth, and BRDF	Hedley and Enríquez (2010)
				Hedley et al. (2016)

Note that models used for retrieval purposes are not included in this table.

spectrum to allow for simulations in canopy "density." These offer a means of reasonably simple set up and fast computation. If well constructed, such models have been shown to be accurate in comparison to other more complicated computationally complicated models (e.g., Zhou et al., 2015; Hedley et al., 2012). The models have been shown to adequately allow for investigation of spectral band placement and varying water column depth and turbidity on reflectance measured above the canopy and on VIs. For example, Sakuno and Kunii (2013) used a radiative transfer model to investigate the influence of water column depth on NDVI in the presence of submersed species but only in shallow water depths. Although detection sensitivity changed with species, the models showed that NDVI was only sensitive down to ~ 50 cm depth; beyond this depth the water column itself significantly reduced NIR reflectance.

Approaches using *turbid medium models* have utilized common methods employed in modeling terrestrial vegetation canopies, typically built around the PROSPECT and SAIL family of coupled leaf and canopy models (Jacquemoud et al., 2009). Aquatic vegetation are included as one or more horizontal layers of vegetation without individualizing leaves or stems but in which LAI and distribution of leaf angles can be accounted for. In essence, such models are one dimensional in space (e.g., Plummer et al., 1997; Villa et al., 2014). By way of example, Villa et al. (2014) used a coupled leaf-canopy radiative transfer turbid medium model to simulate top of canopy reflectances for emergent species under varying structural and optical properties of the canopy and leaves. The model also included soil/water background influences and was able to accommodate arbitrary sun-target-sensor geometries typically encountered in Italian lakes. A variety of different scenarios was simulated separately for sparse and dense vegetation canopies and their effects on different VIs investigated. While the work predominantly focused on the separation of terrestrial from aquatic vegetation, they were able to show the advantages of the specifically derived aquatic VIs NDAVI and WAVI to improve discrimination of sparse to medium density and medium to dense emergent and floating-leaved cover, respectively. Notably, they also showed that the most influential parameter contributing to VIs values for sparse vegetation was LAI and average leaf slope, while for dense vegetation it was pigment concentration and average leaf slope.

Whilst simplistic in formulation the two mechanistic modeling methods above are appropriate for modeling a horizontally homogeneous canopy; they do not take spatial or vertical structure explicitly into account. However, while not useful for spatial prediction, they have value for the rapid investigation of the effects of canopy density (as LAI or cover) and overlying water column depth and turbidity on reflectance signals (Zhou et al., 2015) and for modeling temporal changes in reflectance such as phenological changes via the manipulation of both cover and leaf biochemistry. Turbid medium models such as the SAIL family have been successfully applied to study emergent and floating-leaved species with little modification where the background water column can be represented by the background reflectance (Villa et al., 2014, 2015). However, their formulation does not account for attenuation by water between leaves in the submersed macrophyte case.

To date, those few studies which have undertaken modeling in freshwater macrophytic environments have used relatively simplified approximations of canopy structure and have not included the directional interaction of light with the canopy or the variation in optical properties among macrophyte leaves (Hedley and Enríquez, 2010). Explicit, *three-dimensional models* consisting of crown architectural elements (e.g., Hedley and Enríquez, 2010; Malthus et al., 1997, Zhou et al., 2015, Hedley et al., 2016) attempt to represent the structure as canopy components (as surfaces or shells) which can be varied in density and position. Such approaches can also include leaf optical properties, which requires empirical estimation. To solve the radiative transfer Monte Carlo methods can be applied which exploit the concept of light as discrete photons and use ray tracing to achieve a solution. However, while easily understood in terms of radiative processes such methods are somewhat computationally inefficient. Alternatively, nonstatistical, deterministic methods are markedly more computationally efficient but come at the expense of mathematical complexity (Mobley, 1994).

Whilst the modeling has been used to good effect to further understanding of the influences of macrophyte cover, biomass and water column depth and turbidity on reflectances observed over macrophyte stands, evaluation of outputs is typically not thorough, particularly with respect to accuracy (e.g., Zimmerman, 2003). Sensitivity studies to determine the interactions between the key parameters driving the models are required, especially in freshwater studies. Few in depth comparisons of the performances of the different modeling approaches have also been performed (an exception being Zhou et al., 2015).

The above analysis discussed the use of models in the forward sense, for predicting water-leaving remote sensing reflectance. Accurate reflectance models can be further extended into algorithms using *inversion* of reflectance measured at altitude for the purposes of retrieving one or more biophysical properties (e.g., Lee et al., 1998, Dierssen et al., 2003). The properties retrieved are those which achieve the best match between the measured and modeled reflectances. Examples of such inversion methods applied to shallow water environments include optimization techniques (Giardino et al., 2012; Heege et al., 2004) and matrix inversion (Brando et al., 2009). Other methods include the use of look up tables (Hedley et al., 2009; Mobley et al., 2005), neural networks (Nagamani et al., 2012), and least squares statistical methods (McKinna et al., 2015, Hedley et al., 2016). Such models have the benefit of simultaneous retrieval of a number of parameters including depth, concentration of optically active water quality components in the water column, bottom reflectance and potentially macrophyte biophysical properties (Dekker et al., 2011). The models generally show greater agreement in shallower clear water than deeper or more turbid water but can be used to "flag" regions of low accuracy on the basis of low confidence in retrievals for such pixels (Brando et al., 2009). Other applications of such bio-optical models include their use to generate synthetic data for unmixing or classification purposes and for bathymetric retrieval (Brando et al., 2009).

In summary, there is to date a paucity of bio-optical modeling studies for freshwater macrophytes in the literature and the field would benefit from additional investigations being undertaken. Modeling allows the investigation of external influences across a greater range of canopy and background conditions and macrophyte types than otherwise might be possible to measure in the field (Dekker et al., 2005). Such approaches allow for the systematic exploration of sensitivities of canopy reflectance to variations in the model input values (Villa et al., 2014). Further modeling studies are required to enable:

- The ability to better understand the limits of depth detection of submersed species in the presence of varying water turbidity. These studies are also required to understand the extent to which macrophyte communities are light limited by constituents in the water column—this would assist water managers to set water quality targets necessary to achieve water column light penetration goals to ensure the continued viability of submersed macrophyte beds (Biber et al., 2007).
- The means to extend in situ measurement studies to investigate the limits to spectral separation between species, growth habits and separation of emergent species from surrounding riparian terrestrial vegetation (Botha et al., 2013; Hedley and Enríquez, 2010; Hedley et al., 2012).
- The ability to understand the influence of species/growth habit, phenology, water depth, and quality on high resolution spectral reflectance (Malthus et al., 1997) and in so doing understand the relation between macrophyte biophysical properties and spectral reflectance and potential Vis (Villa et al., 2014).
- Greater understanding of the factors that may lead to classification error, for example, differences between less dense canopies in deep water versus dense canopies in shallow water.
- Investigation of the influence of acquisition characteristics (e.g., sun-sensor geometry, surface effects) on the ability to retrieve macrophyte properties.

9.7 DISCUSSION AND PRIORITIES FOR FURTHER RESEARCH

Remote sensing of aquatic macrophtyes has seen considerable progress in the last decade and has proved helpful in overcoming the limitations of traditional survey methods to assess whole lake distributions of macrophyte biomass and cover. As macrophyte habitats are influenced by variations in local factors such as substrate composition, wave action, light penetration, and nutrient availability the method is useful to assess the patchy distribution of macrophytes across entire lakes. With recent satellite data coming on stream (Landsat 8, Sentinel 2 series), the method may also provide a promising tool for assessing macrophyte distributions across wider geographic regions. However, the technique is not yet sufficiently developed to form the basis of a worldwide assessment of macrophyte status and trends.

Aquatic vegetation exhibit distinct spectral signatures across a range of differing growth habits and we have moved some way to understanding the influence of canopy density and cover on signals recorded over macrophyte communities, and of their interactions with varying depth and water column properties. Progress has also been made on classification methods and the development of VIs, which allow for the discrimination of macrophytes from other influences as well as relationships with plant vigor; these have been most successful for emergent and floating-leaved species. There remains a paucity of studies and results for submersed vegetation with relatively inconclusive results on species discrimination and only the assessment of broad categories of vegetation vigor. There are even fewer studies, which have used the power of bio-optical modeling to further understanding of the reflectances possible to be recorded above macrophytes and the means by which macrophyte growth habit and vigor can be assessed.

In short, we are still a long way off an understanding that allows us to use remote sensing technology adequately for macrophyte management purposes and to better understand macrophyte dynamics and their role in aquatic ecosystem function such as their role in the carbon cycle. This section lays out a number of research priorities for fruitful further research to further the understanding that is required.

9.7.1 In Situ Measurement of Spectral Signatures

Compared to terrestrial vegetation, the reflectance of aquatic plants themselves is poorly understood and what exists is patchy. Further, high quality systematic studies of the in situ spectral signatures above macrophytes of all growth habits and species is required, similar to those obtained by Wolf et al. 2013), along with the collection of their biophysical properties. These measurements should be made at high spectral resolution such that spectral signatures and their similarities and differences can be understood. Measurements need to be accumulated into spectral libraries for different species, obtained at different times of the growing season and of their substrates. Spectral libraries will help overcome the current lack of pure endmember spectra for many macrophyte species. Concurrent in situ reflectance datasets are also required for calibration and validation of satellite data.

Reflectances measured over submersed species need to be made both with and without the effects of the overlying water column. Surface measurements are required to validate model outputs and relate to observations made from remote sensing platforms; measurements made just above and within canopies are required to provide parameterization for models and to validate model outputs at this level, before signals are modified by the overlying water column. Measurements are also needed of the optical characteristics of the overlying water column and, where possible, of leaf level optical properties.

The development of spectral libraries and associated descriptive phenological libraries are underway in some studies (Wolf et al., 2013) and plant life histories need to be accounted for to capture key phenological attributes such as flowering and senescence (Hestir et al., 2015). Only when we undertake such characterization will we be able to adequately identify sources of variability observed in macrophytes and the overlying water column and be able to understand the inter- and intra-specific differences in spectral reflectance and relate these to factors such as species structure, depth of observation, growth habit, and overlying water quality (Schmidt and Skidmore, 2003).

9.7.2 Signature Analysis

Collections of spectral libraries can be then used to conduct analyses to understand what are the appropriate spectral bands—how many and where located—that may be required to separate species and to discriminate macrophyte species from surrounding extraneous factors (e.g., surrounding water column and substrate) and to determine the constraints where environmental variation (depth and water composition) and instrument noise ultimately limits detectability. Such analyses are required to determine the narrow wavelength regions, which may form the basis of spectral classifications, and the degree to which these can be applied in the presence of spectral variations within and between species. Only when this is undertaken will we begin to understand the high rate of misclassification observed for submersed and floating-leaved species.

Hyperspectral data has the potential for more accurate classification where spectrally similar vegetation species might exist and also to unmix a greater number of endmembers. In the context of macrophyte monitoring and variability, the complexity of the problem necessitates hyperspectral studies from the outset which may thus enable retrievals across a wider range of conditions and environments (Hestir et al., 2015). Detection of submersed macrophytes is very much limited to the visible spectral region but expands into the infrared for emergent and floating-leaved growth habits. Methods are required to reduce the spectral dimensionality and spectral noise in the hyperspectral data (Botha et al., 2013; Hestir et al., 2015; Yang et al., 2012).

Further work is required into the range of methods that can be applied to reduce this dimensionality including spectral separability and distance measures, spectral angle mapping, clustering methods, derivative analysis, mixture analysis, classification/decision trees, and other machine learning methods (Botha et al., 2013; Fernandes et al., 2013; Kotta et al., 2014; Reichstetter et al., 2015) and to assist in the elimination of false positives such as the misclassification of algal blooms and scums. It is highly likely that minimum waveband sets will vary dependent on species, growth habit, and phenological stage and of course the datasets analyzed must adequately encompass the existing spectral variability present in such dynamic targets and sufficiently encompasses the variation in background factors.

9.7.3 Bio-optical Modeling

Modeling assists us to extend the understanding in theory of the measurements and relationships made in situ to help overcome the challenges in obtaining reflectance over such systems in the field. Accurate models for a number of species are required such that we can model reflectances with confidence and begin to better understand spectral signatures, the nature of their changes with water depth, water column turbidity and growing season, and to better understand the limits of detection for different species (e.g., Hedley et al., 2016). Surface effects and viewing geometries also need to be studied (e.g., Mobley, 1999). Optical modeling is perhaps the most effective approach to derive this understanding and in particular determine if quantitative information can be derived (Dekker et al., 2001).

Radiative transfer studies are needed but the potential complexity of the radiative transfer system required, involving three-dimensional canopy structures in combination with an overlying water column should not be overlooked (e.g., Hedley et al., 2016). Such studies are required to understand the influences on hyperspectral reflectance and to convolve such data for satellite sensor spectral resolutions to understand their ability to resolve key features.

9.7.4 Relationships with Biophysical Properties

The extent to which biophysical parameters of macrophytes can be assessed using remote sensing is poorly understood and undeveloped. Some work has been undertaken for emergent species but the extent of methods and limits for detection for floating-leaved and submersed species remains to be addressed. The degree to which anything other than broad categories of submersed macrophyte biomass may be assessed needs to be determined and the influence of limiting factors determined. Bio-optical modeling can also help significantly in this direction (Hedley et al., 2016).

9.7.5 Inversion Algorithms

Progress has been made with the use of specifically derived VIs to discriminate macrophytes and to relate to biophysical variables such as vegetation amount, growth, and vigor. However, the application of inversion algorithms largely has been overlooked and warrants further investigation for shallow water inland systems and for macrophyte discrimination. Candidate models exist in the methods undertaken by (Brando et al., 2009; Gege, 2014; Giardino et al., 2012; Hedley et al., 2016; Villa et al., 2014) but will need adaptation to allow for the inclusion of macrophytes with different growth habits. By allowing for simultaneous retrieval of key relevant parameters including depth, water column constituents and bottom reflectance this family of algorithms is better placed to take into account variations across lakes to assist in the development of algorithms with regional application. Retrievals will be further improved with the availability of accurate bathymetric data for lakes; inclusion of depth information has been shown to ultimately increase mapping accuracies (Ackleson and Klemas, 1987; Plummer et al., 1997). The approaches also allow for the incorporation of retrieval uncertainty to be reported for each satellite pixel.

9.7.6 Assessment of Remote Sensing Platforms

Along with the challenges of surface effects and water column depth and turbidity, aquatic macrophyte monitoring using remote sensing platforms also includes issues associated with spatial, spectral, and radiometric resolutions. In assessment of the application of sensors for inland water monitoring the needs for macrophyte mapping are often not included (Mouw et al., 2015). However, if the goals of the recommendations above are met, progress will have been achieved into what might be the ideal characteristics in terms of spectral, spatial, radiometric, and temporal resolutions required of sensors for mapping the distributions and biophysical properties of aquatic vegetation in inland waters. The WAVI and NDAVI indices, specifically adapted for aquatic vegetation, may be straightforward to apply across the bands available on a broad range of multispectral systems and could be adapted for application at both high resolution and medium resolution scales (Villa et al., 2015).

Airborne campaigns may not meet the objective of operational cost-effective spatial and temporal coverage. However, UAVs hold some promise as a tool for localized survey application (Flynn and Chapra, 2014; Visser et al., 2013). Further studies investigating the application of more recent satellite sensors (Landsat 8, WorldView3, and Sentinel 2) to macrophyte mapping are warranted, particularly into the separability of bottom covers using the limited spectral data available in these sensors. Landsat 8 has improved radiometric resolution over earlier sensors in the Landsat series and this is proving useful for inland water quality studies (Lymburner et al., 2016). WorldView 3 has eight band spectral data which may allow for improved spectral discrimination of macrophyte growth habits and species. The higher resolution multispectral data delivered by sensors typical of this platform can in some cases provide improved classifications than when compared to hyperspectral imagery with moderate spatial resolution through reduced spectral mixing (Carter et al., 2009).

Although the arguments for the adoption of hyperspectral data are strong, the benefits of such data may not necessarily be substantial enough to warrant the additional investment. Sentinel 2 shows considerable promise over other medium resolution sensors through improved spectral band placement and narrow bandwidths which may prove useful to cope with spectrally complex benthic reflectances, along with improvements in spatial resolution and instrument noise. However, combining Sentinel 2 spectral data acquired at multiple spatial resolutions (10−60 m) will need to be carefully handled (Hedley, 2012).

The synergy between optical and acoustic sensors and the information offered by each also warrants further investigation (Malthus and Karpouzli, 2009) as does the use of texture and spatial context information provided in higher spatial resolution data (Fernandes et al., 2013; Visser et al., 2013).

9.7.7 Regional Assessment/Global Monitoring

We are far from a capability to be able to study the patterns of macrophyte distributions across space and their variations with time. This capability is required to understand the spread of invasive macrophytic species, the patterns of change in macrophyte distributions across landscapes and the resilience of macrophytic communities to external forces and environmental degradation such as eutrophication. Remote sensing is the ideal tool with which to undertake such large scale inventories and to assess change (e.g., Rebelo et al., 2009) but progress here will be hindered as long as the algorithms and tools to undertake this remain undeveloped; the lack of in situ data for calibration and validation also hinders development of large scale mapping capability and retrospective analysis (Hedley et al, 2016). The results of the work done so far, while valuable, remain incompatible and inconsistent. Regional methods will only be developed if we can arrive at robust, standardized methods to assess morphoecological gradients, structural complexity and functional status of macrophyte dominated habitats (Villa et al., 2015).

With the current generation of medium resolution multispectral sensors better suited to measurement over aquatic targets (Landsat 8, and Sentinels 2A and 2B) we are approaching a stage where the implementation of more regional scale mapping of macrophyte distributions and phenologies could be realized. The high spatial heterogeneity of macrophytes and their presence in relatively small water bodies and wetland areas, particularly in temperate zones, means the method will not be supported by operational low resolution remote sensing platforms (of 300–1000 m pixel resolution (Villa et al., 2015). However, so far much of the studies have focused on targets of limited extent, mostly consisting of a single study area. Exceptions include the work of Nelson et al. (2006) who used Landsat data to map four macrophyte groups in 13 small low turbidity lakes (but this study highlights much to improve on if this is to become a reliable tool) and Villa et al. (2015) which attempted comparisons across 4 European and 1 Chinese lake.

9.7.8 Role in Management

When the greater understanding and tools highlighted in the recommendations above are in place we will be in a place where remote sensing tools may make a sound contributions to the management of macrophytes in inland water systems, which is surely the ultimate the goal of this research. Along with costs and programmatic support, data continuity and product accuracy were perceived as key impediments to the adoption of remote sensing for water quality management in the US Environmental Protection Agency (Schaeffer et al., 2013). The same issues, notably confidence in the methodology, will apply to adoption of the tools for macrophyte monitoring. The need for wider ecosystem-level integrated management and monitoring programs to better understand ecosystem dynamics and regime shifts, the impacts of management regimes, the distribution of invasive species and their impacts on biodiversity and water quality, will provide the drivers for greater adoption of remote sensing methods (Hedley et al., 2012; Santos et al., 2009).

REFERENCES

Ackleson, S., Klemas, V., 1987. Remote sensing of submerged aquatic vegetation in lower Chesapeake Bay—a comparison of Landsat MSS to TM imagery. Rem. Sens. Environ., 22, 235–248.

Adam, E., Mutanga, O., Rugege, D., 2009. Multispectral and hyperspectral remote sensing for identification and mapping of wetland vegetation: a review. Wetl. Ecol. Manag., 18, 281–296.

Albright, T.P., Ode, D.J., 2011. Monitoring the dynamics of an invasive emergent macrophyte community using operational remote sensing data. Hydrobiol., 661, 469–474.

Andrew, M., Ustin, S., 2008. The role of environmental context in mapping invasive plants with hyperspectral image data. Rem. Sens. Environ., 112, 4301–4317.

Anker, Y., Hershkovitz, Y., Ben Dor, E., Gasith, A., 2014. Applicaiton of aerial digital photography for macrophyte cover and composition in small rural streams. River Res. Appl., 30, 925–937.

Ashraf, S., Brabyn, L., Hicks, B.J., Collier, K., 2010. Satellite remote sensing for mapping vegetation in New Zealand freshwater environments: a review. NZ Geographer 66, 33–43.

Becker, B., Lusch, D., Qi, J., 2005. Identifying optimal spectral bands from in situ measurements of Great Lakes coastal wetlands using second-derivative analysis. Rem. Sens. Environ., 97, 238−248.

Becker, B., Lusch, D., Qi, J., 2007. A classification-based assessment of the optimal spectral and spatial resolutions for Great Lakes coastal wetland imagery. Rem. Sens. Environ., 108, 111−120.

Biber, P.D., Gallegos, C.L., Kenworthy, W.J., 2007. Calibration of a bio-optical model in the North River, North Carolina (Albemarle−Pamlico Sound): a tool to evaluate water quality impacts on seagrasses. Estuar. Coast., 31, 177−191.

Blaschke, T., 2010. Object based image analysis for remote sensing. ISPRS J. Photogramm. Rem. Sens., 65, 2−16.

Bolpagni, R., Bresciani, M., Laini, A., Pinardi, M., Matta, E., Ampe, E., et al., 2014. Remote sensing of phytoplankton-macrophyte coexistence in shallow hypereutrophic fluvial lakes. Hydrobiol., 737, 67−76.

Botha, E.J., Brando, V.E., Anstee, J.M., Dekker, A.G., Sagar, S., 2013. Increased spectral resolution enhances coral detection under varying water conditions. Rem. Sens. Environ., 131, 247−261.

Brando, V.E., Anstee, J.M., Wettle, M., Dekker, A.G., Phinn, S.R., Roelfsema, C., 2009. A physics based retrieval and quality assessment of bathymetry from suboptimal hyperspectral data. Rem. Sens. Environ., 113, 755−770.

Bresciani, M., Giardino, C., Longhi, D., Pinardi, M., Bartoli, M., Vascellari, M., 2009. Imaging spectrometry of productive inland waters. Application to the lakes of Mantua. Ital. J. Rem. Sens., 41, 147−156.

Bresciani, M., Sotgia, C., Fila, G.L., Musanti, M., Bolpagni, R., 2011. Assessing common reed bed health and management strategies in Lake Garda (Italy) by means of Leaf Area Index measurements. Ital. J. Rem. Sens., 43, 9−22.

Brooks, C., Grimm, A., Shuchman, R., Sayers, M., Jessee, N., 2015. A satellite-based multi-temporal assessment of the extent of nuisance Cladophora and related submerged aquatic vegetation for the Laurentian Great Lakes. Rem. Sens. Environ., 157, 58−71.

Brucet, S., Poikane, S., Lyche-Solheim, A., Birk, S., 2013. Biological assessment of European lakes: ecological rationale and human impacts. Freshwater Biol. 58, 1106−1115.

Carignan, R., Kalff, J., 1980. Phosphorus sources for aquatic weeds-water or sediments. Science 207, 987−989.

Carle, M.V., Wang, L., Sasser, C.E., 2014. Mapping freshwater marsh species distributions using WorldView-2 high-resolution multispectral satellite imagery. Int. J. Rem. Sens., 35, 4698−4716.

Carpenter, S., Lodge, D., 1986. Submerged macrophytes-carbon metabolism, growth regulation and role in macrophyte-dominated ecosystems cover image effects of submersed macrophytes on ecosystem processes. Aquat. Bot., 26, 341−370.

Carter, G., Lucas, K., Blossom, G., Lassitter, C., Holiday, D., Mooneyhan, D., et al., 2009. Remote sensing and mapping of Tamarisk along the Colorado River, USA: a comparative use of summer-acquired Hyperion, Thematic Mapper and Quickbird data. Rem. Sens, 1, 318−329.

Cheruiyot, E.K., Mito, C., Menenti, M., Gorte, B., Koenders, R., Akdim, N., 2014. Evaluating MERIS-based aquatic vegetation mapping in Lake Victoria. Rem. Sens., 6, 7762−7782.

Cook, C., 1996. Aquatic Plant Book. SPB Academic Publishing, Amsterdam.

Corcoran, J., Knight, J., Gallant, A., 2013. Influence of multi-source and multi-temporal remotely sensed and ancillary data on the accuracy of random forest classification of wetlands in Northern Minnesota. Rem. Sens 5, 3212−3238.

Costa, M., 2004. Use of SAR satellites for mapping zonation of vegetation communities in the Amazon floodplain. Int. J. Rem. Sens., 25, 1817−1835.

Davranche, A., Lefebvre, G., Poulin, B., 2010. Wetland monitoring using classification trees and SPOT-5 seasonal time series. Rem. Sens. Environ., 114, 552−562.

Dekker, A., Brando, V., Anstee, J., Pinnel, N., Kutser, T., Hoogenboom, E., et al., 2001. Imaging spectrometry: basic principles and prospective applications. In: VanDerMeer, F., DeJong, S. (Eds.), Imaging Spectrometry: Basic Principles and Prospective Applications. Springer, Berlin, pp. 307−359.

Dekker, A., Phinn, S., Anstee, J., Bissett, P., Brando, V.E., Casey, B., et al., 2011. Intercomparison of shallow water bathymetry, hydro-optics, and benthos mapping techniques in Australian and Caribbean coastal environments. Limnol. Oceanogr-Meth., 9, 396−425.

Dekker, A.G., Brando, V.E., Anstee, J.M., 2005. Retrospective seagrass change detection in a shallow coastal tidal Australian lake. Rem. Sens. Environ., 97, 415−433.

Dierssen, H.M., Zimmerman, R.C., Leathers, R.A., Downes, T.V., Davis, C.O., 2003. Ocean color remote sensing of seagrass and bathymetry in the Bahamas Banks by high-resolution airborne imagery. Limnol. Oceanogr 48, 444−455.

Dogan, O.K., Akyurek, Z., Beklioglu, M., 2009. Identification and mapping of submerged plants in a shallow lake using Quickbird satellite data. J. Environ. Manage., 90, 2138−2143.

Dörnhöfer, K., Oppelt, N., 2016. Remote sensing for lake research and monitoring—recent advances. Ecol. Indic., 64, 105−122.

Duarte, C., Kalff, J., 1986. Littoral slope as a predictor of the maximum biomass of submerged macrophyte communities. Limnol. Oceanogr., 31, 1072−1080.

Evans, T., Costa, M., Telmer, K., Silva, T., 2010. Using ALOS/PALSAR and RADARSAT-2 to map land cover and seasonal inundation in the Brazilian Pantanal. IEEE J. Sel. Top. Appl., 3, 560−575.

Everitt, J.H., Yang, C., Summy, K.R., Owens, C.S., Glomski, L.M., Smart, R.M., 2011. Using in situ hyperspectral reflectance data to distinguish nine aquatic plant species. Geocarto Int., 26, 459−473.

Fernandes, M.R., Aguiar, F.C., Silva, J.M.N., Ferreira, M.T., Pereira, J.M.C., 2013. Spectral discrimination of giant reed (Arundo donax L.): a seasonal study in riparian areas. ISPRS J. Photogramm. Rem. Sens., 80, 80−90.

Flynn, K.F., Chapra, S.C., 2014. Remote sensing of submerged aquatic vegetation in a shallow non-turbid river using an unmanned aerial vehicle. Rem. Sens., 6, 12815−12836.

Fyfe, S., 2003. Spatial and temporal variation in spectral reflectance are seagrass species spectrally distinct? Limnol. Oceanogr., 48, 464−479.

Gao, B., 1996. NDWI—a normalized difference water index for remote sensing of vegetation liquid water from space. Rem. Sens. Environ., 58, 257−266.

Gao, B.-C., 2016. Atmospheric Correction for Inland Waters. Chapter 3 of this book.

Gege, P., 2014. WASI-2D: a software tool for regionally optimized analysis of imaging spectrometer data from deep and shallow waters. Comput. Geosci., 62, 208−215.

Giardino, C., Bresciani, M., Villa, P., Martinelli, A., 2010. Application of remote sensing in water resource management: the case study of Lake Trasimeno, Italy. Wat. Res. Manag., 24, 3885−3899.

Giardino, C., Candiani, G., Bresciani, M., Lee, Z., Gagliano, S., Pepe, M., 2012. BOMBER: a tool for estimating water quality and bottom properties from remote sensing images. Comput. Geosci., 45, 313−318.

Giardino, C., Bresciani, M., Valentini, E., Gasperini, L., Bolpagni, R., Brando, V.E., 2015. Airborne hyperspectral data to assess suspended particulate matter and aquatic vegetation in a shallow and turbid lake. Rem. Sens. Environ., 157, 48–57.

Ginn, B.K., 2011. Distribution and limnological drivers of submerged aquatic plant communities in Lake Simcoe (Ontario, Canada): utility of macrophytes as bioindicators of lake trophic status. J. Great Lakes Res., 37, 83–89.

Guerschman, J., Donohue, R., Van Niel, T., Renzullo, L., Dekker, A., Malthus, T., et al., 2016. Earth observations for monitoring water resources. In: Garcia, L., Rodríguez, D., Wijnen, M., Pakulski, I. (Eds.), Earth Observation for Water Resources Management: Current Use and Future Opportunities for the Water Sector. The World Bank Group, Washington, pp. 78–143.

Han, L., Rundquist, D.C., 2003. The spectral responses of Ceratophyllum demersum at varying depths in an experimental tank. Int. J. Rem. Sens., 24, 859–864.

Heblinski, J., Schmieder, K., Heege, T., Agyemang, T.K., Sayadyan, H., Vardanyan, L., 2010. High-resolution satellite remote sensing of littoral vegetation of Lake Sevan (Armenia) as a basis for monitoring and assessment. Hydrobiol., 661, 97–111.

Hedley, J., Enríquez, S., 2010. Optical properties of canopies of the tropical seagrass Thalassia testudinum estimated by a three-dimensional radiative transfer model. Limnol. Oceanogr., 55, 1537–1550.

Hedley, J., Roelfsema, C., Phinn, S., 2009. Efficient radiative transfer model inversion for remote sensing applications. Rem. Sens. Environ., 113, 2527–2532.

Hedley, J., Roelfsema, C., Koetz, B., Phinn, S., 2012. Capability of the Sentinel 2 mission for tropical coral reef mapping and coral bleaching detection. Rem. Sens. Environ., 120, 145–155.

Hedley, J., Russell, B., Randolph, K., Dierssen, H., 2016. A physics-based method for the remote sensing of seagrasses. Rem. Sens. Environ., 174, 134–147.

Heege, T., Bogner, A., Pinnel, N., 2004. Mapping of submerged aquatic vegetation with a physically based process chain. In: Kramer, E. (Ed.), Proceedings SPIE 5233 Remote Sensing of the Ocean and Sea Ice, September 8–12, 2003. SPIE, The International Society for Optical Engineering, Barcelona, Spain.

Hestir, E.L., Khanna, S., Andrew, M.E., Santos, M.J., Viers, J.H., Greenberg, J.A., et al., 2008. Identification of invasive vegetation using hyperspectral remote sensing in the California Delta ecosystem. Rem. Sens. Environ., 112, 4034–4047.

Hestir, E.L., Greenberg, J.A., Ustin, S.L., 2012. Classification trees for aquatic vegetation community prediction from imaging spectroscopy. IEEE J. Sel. Top. Appl., 5, 1572–1584.

Hestir, E.L., Brando, V.E., Bresciani, M., Giardino, C., Matta, E., Villa, P., et al., 2015. Measuring freshwater aquatic ecosystems: the need for a hyperspectral global mapping satellite mission. Rem. Sens. Environ., 167, 181–195.

Hu, C., 2009. A novel ocean color index to detect floating algae in the global oceans. Rem. Sens. Environ., 113, 2118–2129.

Huete, A., Liu, H., Batchily, K., van Leeuwen, W., 1997. A comparison of vegetation indices over a global set of TM images for EOS-MODIS. Rem. Sens. Environ., 59, 440–451.

Huete, A., Didan, K., Miura, T., Rodriguez, E., Gao, X., Ferreira, L., 2002. Overview of the radiometric and biophysical performance of the MODIS vegetation indices. Rem. Sens. Environ., 83, 195–213.

Hunter, P., Gilvear, D., Tyler, A., Willby, N., Kelly, A., 2010. Mapping macrophytic vegetation in shallow lakes using the Compact Airborne Spectrographic Imager (CASI). Aquat. Conserv., 20, 717–727.

Husson, E., Hagner, O., Ecke, F., 2014. Unmanned aircraft systems help to map aquatic vegetation. Appl. Veg. Sci., 17, 567−577.

Jacquemoud, S., Verhoef, W., Baret, F., Bacour, C., Zarco-Tejada, P.J., Asner, G.P., et al., 2009. PROSPECT+SAIL models: a review of use for vegetation characterization. Rem. Sens. Environ., 113, S56−S66.

Jiang, H., Zhao, D., Cai, Y., An, S., 2012. A method for application of classification tree models to map aquatic vegetation using remotely sensed images from different sensors and dates. Sensors 12, 12437−12454.

Kasischke, E., Bruhwiler, L., 2003. Emissions of carbon dioxide, carbon monoxide, and methane from boreal forest fires in 1998. J. Geophys. Res., 107, FFR 2-1-FFR 2-14.

Kemp, W.M., Batiuk, R., Bartleson, R., Bergstrom, P., Carter, V., Gallegos, C.L., et al., 2004. Habitat requirements for submerged aquatic vegetation in Chesapeake Bay: water quality, light regime, and physical-chemical factors. Estuaries 27, 363−377.

Klemas, V., 2013. Remote sensing of emergent and submerged wetlands: an overview. Int. J. Rem. Sens., 34, 6286−6320.

Kotta, J., Kutser, T., Teeveer, K., Vahtmae, E., Parnoja, M., 2014. Predicting species cover of marine macrophyte and invertebrate species combining hyperspectral remote sensing, machine learning and regression techniques. PLoS One, 8, e63946.

Lee, Z., Carder, K.L., Mobley, C.D., Steward, R.G., Patch, J.S., 1998. Hyperspectral remote sensing for shallow waters. I. A semi-analytical model. Appl. Opt., 37, 6329−6338.

Liu, X., Zhang, Y., Shi, K., Zhou, Y., Tang, X., Zhu, G., et al., 2015. Mapping aquatic vegetation in a large, shallow eutrophic lake: a frequency-based approach using multiple years of MODIS data. Rem. Sens., 7, 10295−10320.

Luo, J., Ma, R., Duan, H., Hu, W., Zhu, J., Huang, W., et al., 2014. A new method for modifying thresholds in the classification of tree models for mapping aquatic vegetation in Taihu Lake with satellite images. Rem. Sens., 6, 7442−7462.

Lymburner, L., Botha, E., Hestir, E., Anstee, J., Sagar, S., Dekker, A., et al., 2016. Landsat 8: providing continuity and increased precision for measuring multi-decadal time series of total suspended matter. Rem. Sens. Environ., 185, 108−118.

Madsen, J.D., Sutherland, J.W., Bloomfield, J.A., Eichler, L.W., Boylen, C.W., 1991. The decline of native vegetation under dense Eurasian watermilfoil canopies. J. Aquat. Plant Manag., 29, 94−99.

Madsen, J.D., Chambers, P.A., James, W.A., Koch, E.W., Westlake, D.F., 2001. The interaction between water movement, sediment dynamics and submersed macrophytes. Hydrobiol., 444, 71−84.

Malthus, T., George, D., 1997. Airborne remote sensing of macrophytes in Cefni Reservoir, Anglesey, UK. Aquat. Bot., 58, 317−332.

Malthus, T., Karpouzli, E., 2009. On the benefits of using both dual frequency side scan sonar and optical signatures for the discrimination of coral reef benthic communities. In: Silva, S.R. (Ed.), Advances in Sonar Technology. I-Tech Education and Publishing, Vienna, pp. 165−190.

Malthus, T., Best, E., Dekker, A., 1990. An assessment of the importance of emergent and floating-leaved macrophytes to trophic status in the Loosdrecht Lakes (the Netherlands). Hydrobiol., 191, 257−263.

Malthus, T., Ciraolo, G., La Loggia, G., Clark, C., Plummer, S., Calvo, S., et al., 1997. Can biophysical properties of submerged macrophytes be determined by remote sensing? Fourth International Conference on Remote Sensing for Marine and Coastal Environments: Technology and Applications. ERIM, Orlando, Florida, pp. 562−571.

Marshall, T., Lee, P., 1994. Mapping aquatic macrophytes through digital image-analysis of aerial photographs—an assessment. J. Aquat. Plant Manag., 32, 61–66.

Martinez, J., Le Toan, T., 2007. Mapping of flood dynamics and spatial distribution of vegetation in the Amazon floodplain using multitemporal SAR data. Rem. Sens. Environ., 108, 209–223.

McKinna, L.I.W., Fearns, P.R.C., Weeks, S.J., Werdell, P.J., Reichstetter, M., Franz, B.A., et al., 2015. A semi-analytical ocean color inversion algorithm with explicit water column depth and substrate reflectance parameterization. J. Geophys. Res. Oceans 120, 1741–1770.

Middelboe, A., Markager, S., 1997. Depth limits and minimum light requirements of freshwater macrophytes. Freshwat. Biol., 37, 553–568.

Mitchell, S., 1989. Primary production in a shallow eutrophic lake dominated alternately by phytoplankton and by submerged macrophytes. Aquat. Bot., 33, 101–110.

Mobley, C.D., 1994. Light and Water: Radiative Transfer in Natural Waters. Academic Press.

Mobley, C.D., 1999. Estimation of the remote-sensing reflectance from above-surface measurements. Appl. Opt., 38, 7442–7455.

Mobley, C., Sundman, L., Davis, C., Bowles, J., Downes, T., Leathers, R., et al., 2005. Interpretation of hyperspectral remote-sensing imagery by spectrum matching and look-up tables. Appl. Opt., 44, 3576–3592.

Mouw, C.B., Greb, S., Aurin, D., DiGiacomo, P.M., Lee, Z., Twardowski, M., et al., 2015. Aquatic color radiometry remote sensing of coastal and inland waters: challenges and recommendations for future satellite missions. Rem. Sens. Environ., 160, 15–30.

Nagamani, P., Chauhan, P., Sanwlani, N., Ali, M., 2012. Artificial neural network (ann) based inversion of benthic substrate bottom type and bathymetry in optically shallow waters—initial model results. J. Ind. Soc. Rem. Sens., 40, 137–143.

Nelson, S.A.C., Cheruvelil, K.S., Soranno, P.A., 2006. Satellite remote sensing of freshwater macrophytes and the influence of water clarity. Aquat. Bot., 85, 289–298.

NWQMS, 2000. Australian and New Zealand guidelines for fresh and marine water quality, The guidelines Australian and New Zealand Environment and Conservation Council, Volume 1. Agriculture and Resource Management Council of Australia and New Zealand, p. 314.

Oppelt, N., Schulze, F., Bartsch, I., Doernhoefer, K., Eisenhardt, I., 2012. Hyperspectral classification approaches for intertidal macroalgae habitat mapping: a case study in Heligoland. Opt. Eng., 51, 111703.

Orth, R., Moore, K., 1983. Submersed vascular plants—techniques for analyzing their distribution and abundance. Mar. Tech. Soc. J., 17, 38–52.

Oyama, Y., Matsushita, B., Fukushima, T., 2015. Distinguishing surface cyanobacterial blooms and aquatic macrophytes using Landsat/TM and ETM+ shortwave infrared bands. Rem. Sens. Environ., 157, 35–47.

Peñuelas, J., Gamon, J., Griffin, K., Field, C., 1993. Assessing community type, plant biomass, pigment composition, and photosynthetic efficiency of aquatic vegetation from spectral reflectance. Rem. Sens. Environ., 46, 110–118.

Philpot, W., 1981. A Radiative Transfer Model for Remote Sensing of Vertically Inhomogeneous Waters. PhD Thesis. University of Delaware, Newark, DE, p. 40.

Pinnel, N., 2006. A Method for Mapping Submerged Macrophytes in Lakes Using Hyperspectral Remote Sensing. Unpubl. thesis. Department fur Okologie und Okosystemmanagement, Technical University of Munich, p. 136.

Plummer, S., Malthus, T., Clark, C., 1997. Adaptation of a canopy reflectance model for subaqueous vegetation: definition and sensitivity analysis. Fourth International Conference on Remote Sensing for Marine and Coastal Environments: Technology and Applications. ERIM, Orlando, Florida, pp. 149–158.

Poikane, S., Portielje, R., van den Berg, M., Phillips, G., Brucet, S., Carvalho, L., et al., 2014. Defining ecologically relevant water quality targets for lakes in Europe. J. Appl. Ecol., 51, 592−602.

Rebelo, L.M., Finlayson, C.M., Nagabhatla, N., 2009. Remote sensing and GIS for wetland inventory, mapping and change analysis. J. Environ. Manage 90, 2144−2153.

Reichstetter, M., Fearns, P., Weeks, S., McKinna, L., Roelfsema, C., Furnas, M., 2015. Bottom reflectance in ocean color satellite remote sensing for coral reef environments. Rem. Sens., 7, 16756−16777.

Reyjol, Y., Argillier, C., Bonne, W., Borja, A., Buijse, A., Cardoso, A., et al., 2014. Assessing the eco- logical status in the context of the European Water Framework Directive: where do we go now? Sci. Tot. Env., 497-498, 332−344.

Robertson, L.D., King, D.J., Davies, C., 2015. Object-based image analysis of optical and radar variables for wetland evaluation. Int. J. Rem. Sens., 36, 5811−5841.

Robles, W., Madsen, J.D., Wersal, R.M., 2015. Estimating the biomass of waterhyacinth Eichhornia crassipes) using the normalized difference vegetation index derived from simulated Landsat 5 TM. Inv. Plant Sci. Manag., 8, 203−211.

Roessler, S., Wolf, P., Schneider, T., Melzer, A., 2013. Multispectral remote sensing of invasive aquatic plants using RapidEye. In: Krisp, J., Meng, L., Pail, R., Stilla, U. (Eds.), Earth Observation of Global Changes (EOGC). Springer, Berlin, Heidelberg, pp. 109−123.

Rotta, L.H.S., Mishra, D.R., Alcântara, E.H., Imai, N.N., 2016. Analyzing the status of submerged aquatic vegetation using novel optical parameters. Int. J. Rem. Sens., 37, 3786−33810.

Rotta, L., Imai, N., Batista, L., Boschi, L., Galo, M., Velini, E., 2012. Hydro-acoustic remote sensing in submerged aquatic macrophyte mapping. Planta Daninha 30, 229−239.

Ruiz, C., Martinez, G., Toro, M., Camacho, A., 2011. A Review: macrophytes in the assessment of Spanish lakes ecological status under the Water Framework Directive (WFD). Ambientalia.

Sakuno, Y., Kunii, H., 2013. Estimation of growth area of aquatic macrophytes expanding spontaneously in Lake Shinji using ASTER Data. Int. J. Geosci., 04, 1−5.

Santos, M., Khanna, S., Hestir, E., Andrew, M., Rajapakse, S., Greenberg, J., et al., 2009. Use of hyperspectral remote sensing to evaluate efficacy of aquatic plant management. Inv. Plant Sci. Manag., 2, 216−229.

Schaeffer, B., Schaeffer, K., Keith, D., Lunetta, R., Conmy, R., Gould, R., 2013. Barriers to adopting satellite remote sensing for water quality management. Int. J. Rem. Sens., 34, 7534−7544.

Schmidt, K.S., Skidmore, A.K., 2003. Spectral discrimination of vegetation types in a coastal wetland. Rem. Sens. Environ., 85, 92−108.

Schmidt, M., Witte, C., 2010. Monitoring aquatic weeds in a river system using SPOT 5 satellite imagery. J. Appl. Rem. Sens., 4, 043528.

Shuchman, R.A., Sayers, M.J., Brooks, C.N., 2013. Mapping and monitoring the extent of submerged aquatic vegetation in the Laurentian Great Lakes with multi-scale satellite remote sensing. J. Great Lakes Res., 39, 78−89.

Silva, T.S., Costa, M.P., Melack, J.M., Novo, E.M., 2008. Remote sensing of aquatic vegetation: theory and applications. Environ. Monit. Assess., 140, 131−145.

Silva, T.S.F., Costa, M.P.F., Melack, J.M., 2010. Spatial and temporal variability of macrophyte cover and productivity in the eastern Amazon floodplain: a remote sensing approach. Rem. Sens. Environ., 114, 1998−2010.

Sondergaard, M., Johansson, L.S., Lauridsen, T.L., Jorgensen, T.B., Liboriussen, L., Jeppesen, E., 2010. Submerged macrophytes as indicators of the ecological quality of lakes. Freshwat. Biol., 55, 893−908.

Thito, K., Wolski, P., Murray-Hudson, M., 2015. Spectral reflectance of floodplain vegetation communities of the Okavango Delta. Wetl. Ecol. Manag., 23, 637−648.

Tian, Y.Q., Yu, Q., Zimmerman, M.J., Flint, S., Waldron, M.C., 2010. Differentiating aquatic plant communities in a eutrophic river using hyperspectral and multispectral remote sensing. Freshwat. Biol., 55, 1658−1673.

Tucker, C., 1979. Red and photographic infrared linear combinations for monitoring vegetation. Rem. Sens. Environ., 8, 127−150.

Underwood, E., Mulitsch, M., Greenberg, J., Whiting, M., Ustin, S., Kefauver, S., 2006. Mapping invasive aquatic vegetation in the Sacramento-San Joaquin Delta using hyperspectral imagery. Environ. Monit. Assess., 121, 47−64.

Valley, R.D., Drake, M.T., Anderson, C.S., 2005. Evaluation of alternative interpolation techniques for the mapping of remotely-sensed submersed vegetation abundance. Aquat. Bot., 81, 13−25.

Villa, P., Laini, A., Bresciani, M., Bolpagni, R., 2013. A remote sensing approach to monitor the conservation status of lacustrine Phragmites australis beds. Wetl. Ecol. Manag., 21, 399−416.

Villa, P., Mousivand, A., Bresciani, M., 2014. Aquatic vegetation indices assessment through radiative transfer modeling and linear mixture simulation. Int. J. Appl. Earth Obs., 30, 113−127.

Villa, P., Bresciani, M., Bolpagni, R., Pinardi, M., Giardino, C., 2015. A rule-based approach for mapping macrophyte communities using multi-temporal aquatic vegetation indices. Rem. Sens. Environ., 171, 218−233.

Vis, C., Hudon, C., Carignan, R., 2003. An evaluation of approaches used to determine the distribution and biomass of emergent and submerged aquatic macrophytes over large spatial scales. Aquat. Bot., 77, 187−201.

Visser, F., Wallis, C., Sinnott, A.M., 2013. Optical remote sensing of submerged aquatic vegetation: opportunities for shallow clearwater streams. Limnologica, 43, 388−398.

Whiteside, T., Bartolo, R., 2015. Mapping aquatic vegetation in a tropical wetland using high spatial resolution multispectral satellite imagery. Rem. Sens., 7, 11664−11694.

Willby, N., Pitt, J., Phllips, G., 2009. The ecological classification of UK lakes using aquatic macrophytes. Environment Agency Science Report, UK, Report no. SC010080/R2.

Windham, L., Lathrop, R., 1999. Effects of Phragmites australis (common reed) invasion on aboveground biomass and soil properties in brackish tidal marsh of the Mullica River, New Jersey. Estuaries 22, 927−935.

Wolf, P., Rößler, S., Schneider, T., Melzer, A., 2013. Collecting in situ remote sensing reflectances of submersed macrophytes to build up a spectral library for lake monitoring. Eur. J. Rem. Sens., 46, 401−416.

Yang, C., Goolsby, J.A., Everitt, J.H., Du, Q., 2012. Applying six classifiers to airborne hyperspectral imagery for detecting giant reed. Geocarto Int., 27, 413−424.

Yuan, L., Zhang, L.-Q., 2008. Mapping large-scale distribution of submerged aquatic vegetation coverage using remote sensing. Ecol. Inf., 3, 245−251.

Zhang, X., 1998. On the estimation of biomass of submerged vegetation using Landsat thematic mapper (TM) imagery: a case study of the Honghu Lake, PR China. Int. J. Rem. Sens., 19, 11−20.

Zhao, D., Jiang, H., Yang, T., Cai, Y., Xu, D., An, S., 2012. Remote sensing of aquatic vegetation distribution in Taihu Lake using an improved classification tree with modified thresholds. J. Environ. Manage., 95, 98−107.

Zhou, G., Niu, C., Xu, W., Yang, W., Wang, J., Zhao, H., 2015. Canopy modeling of aquatic vegetation - a radiative transfer approach. Rem. Sens. Environ 163, 186−205.

Zimmerman, R., 2003. A biooptical model of irradiance distribution and photosynthesis in seagrass canopies. Limnol. Oceanogr 48, 568−585.

Index

Printed in the United States
By Bookmasters